Methods
in Cell Wall
Cytochemistry

T0200362

Methods
in Cell Wall
Cytochemistry

K.V. Krishnamurthy

Professor
Department of Plant Science
Bharathidasan University
Tiruchirappalli, India

CRC Press
Taylor & Francis Group
Boca Raton London New York

CRC Press is an imprint of the
Taylor & Francis Group, an **informa** business

CRC Press
Taylor & Francis Group
6000 Broken Sound Parkway NW, Suite 300
Boca Raton, FL 33487-2742

First issued in paperback 2019

© 1999 by Taylor & Francis Group, LLC
CRC Press is an imprint of Taylor & Francis Group, an Informa business

No claim to original U.S. Government works

ISBN-13: 978-0-8493-0729-4 (hbk)
ISBN-13: 978-0-367-39986-3 (pbk)

Library of Congress Card Number 99-18125

Library of Congress Cataloging-in-Publication Data

Krishnamurthy, K.V.
      Methods in cell wall cytochemistry / K.V. Krishnamurthy.
            p.   cm.
      Includes bibliographical references and index.
      ISBN 0-8493-0729-5 (alk. paper)
            1. Plant cell walls. 2. Plant cytochemistry — Technique. I. Title.
QK725.K75 1999
571.6′82—dc21
                                                                                    99-18125
                                                                                    CIP

Visit the Taylor & Francis Web site at
http://www.taylorandfrancis.com

and the CRC Press Web site at
http://www.crcpress.com

PREFACE

One of the outstanding features of the plant cell, which distinguishes it from an animal cell, is the presence of an extra protoplasmic cell wall. The characteristics of the cell wall are in a large measure related to the functions imposed on the cell at any particular time. They are also related to the developmental stage of a given cell type. Consequently, many functional deviations that a cell may experience are accompanied by many quantitative and qualitative changes in its cell wall. Therefore, the study of the cell wall is very important for the proper understanding of the functions of the cell itself.

The cell wall scores heavily in matters of practical industrial importance. In the form of timbers, plant cell walls have been exploited as building materials, as pulp for paper, as raw materials for fermented beverages, and as fibers for clothes since prehistoric times. Knowledge about cell wall structure has considerably helped these industries out of their traditional days. Materials extracted from cell walls are used as pastes and glues. Cell walls are clearly also a matter of concern in fruit ripening, quality of vegetables, in the invasion of plants by microbes, and in the utilization of plants as food and feed. Complete knowledge of the cell wall is importantly required in utilizing cell walls more profitably in industry.

The plant cell wall is a structure of great and byzantine complexity in which a bewildering variety of chemical substances are found. The chief components are carbohydrates, which contain many types of monosaccharides or their derivatives linked together as polysaccharides by a variety of different glycosidic bonds. The presence of many other cell-wall components such as proteins, lipids, lignins and phenolics, mineral substances, water, etc. further complicates the cell wall.

A number of techniques have been developed and used to resolve the complexity of the cell wall. These include biochemical procedures, use of bright field, fluorescence, phase contrast, polarization and dark field optics, scanning and transmission electron microscopy, X-ray diffraction techniques, Fourier transform and infrared microspectroscopy, cytochemical methods, autoradiography, and immunological and immunocytochemical procedures. This book essentially deals with cytochemical and immunocytochemical methods, their rationale, and their achievements in adding information about the complexity of the cell wall. Techniques exclusively used for bacterial and other prokaryotic cell walls are beyond the scope of this book.

Cytochemistry combines cytology and analytical biochemistry and thus it allows one to relate chemical activity and function to cytological structure. It also gives more chemical specificity to cytology. Cytochemistry is becoming a rational area of science as its mechanics are increasingly known. It often provides chemical information about the cell, which otherwise may be difficult to obtain even with sophisticated biochemical procedures. Cytochemistry is also advantageous in precisely locating biochemicals within a cell. Hence, this attempt to compile cytochemical techniques as applied to the cell wall has been made.

In planning the content of this book and its arrangement, the author has been influenced by a number of factors. The first and foremost among these are the probable users of this book who will range from fresh graduate students to well-informed researchers. Therefore, efforts have been made to blend basic information with advanced information and techniques. The second factor pertains to available facilities. In thirdworld countries, a researcher is often not equipped with modern gadgets or costly reagents. Hence, simple techniques that could be carried out even in the poorest of laboratories have been provided along with "costlier " techniques.

Many individuals have contributed significantly to the production of this book. A great many of my students have used most of the methods contained herein for a variety of plant materials ranging from algae to angiosperms and have standardized them. I have relied heavily on their standardized procedures, especially with reference to the duration of treatment in different reagents employed in the various techniques. Since the list of students is too long, I am forced to refrain from mentioning the individual names. I am thankful to each one of them. All the credit goes to them and I own the shortcomings. I am particularly grateful to Dr. L.A. Staehelin, University of Colorado, Boulder, for his encouragement, support, and contributory discussion. In fact, a part of the text was completed while the author was a Fulbright Fellow at his laboratory. I am also much indebted to my wife Brindha and son Arvind for all the moral support and their forbearance.

K.V. Krishnamurthy

THE AUTHOR

K.V. Krishnamurthy, Ph.D., is a Professor of plant science in the Bharathidasan University, Tiruchirappalli, India.

He received his B.Sc. degree with a first class from Madras University in 1964 and M.Sc. with a first class and first rank from the same university in 1966. He obtained his Ph.D. degree from Madras University in 1973 with emphasis on Developmental Plant anatomy. Until 1978 he was employed in Tamil Nadu Government Colleges and subsequently entered Bharathidasan University. He has been teaching Botany/Plant Science at the under-graduate and masters levels since 1970. His research areas include plant morphogenesis (both *in vivo* and *in vitro*), reproductive biology, and ecology and environmental botany.

He has been a Fellow of the Linnaean Society, London; Fellow of the Indian Association for Angiosperm Taxonomy; and a Fellow of the National Academy of Sciences, India. He is the Editor of the *Journal of Swamy Botanical Club* and is on the editorial board of *Rheedia* and *Journal of Nature Conservation*.

Dr. Krishnamurthy has over 110 research publications, and 5 books, of which one is on Histochemistry. He was a Fulbright visiting scientist in 1993 at the University of Colorado, Boulder. He has won several prizes and awards for his research contributions.

Contents

Chapter **1**

Cell Wall — Morphology and Chemical Composition

Contents

1.1   Morphology

The cell wall is unique to plants and envelops the protoplast. Its presence or absence serves as a criterion to find out whether a given taxon is a plant or an animal. This characteristic is more reliable than the presence or absence of chlorophyll, as plant cells can change into a heterotrophic way of life by losing chlorophyll. As Frey-Wyssling and Mühlethaler[1] had remarked, the plant cell walls historically played a special role in so far as they gave rise to the origin of the name "cell" itself. When Robert Hooke,[2] in 1667, discovered cells in bottle cork, he observed only cell walls which formed compartmented structures similar to the ones seen in a honeycomb. Had the cell not been discovered through its cell walls in a plant structure, but in an animal such as in a protozoan, it would scarcely have been called *"cell"*: the discipline *"cytology"* would not have been named that way either.

By virtue of its strategic position and structural organization, the cell wall or the *extracellular matrix* is meant to give shape or rigidity as well as protection and support to cells and tissues. It is also involved in the dimensional changes of the cells as these changes inevitably lead to changes in wall structure. The interaction between cell shape, cell growth rate, wall structure, and wall mechanics is influenced in large part by changes in biochemical processes inside the cell and vice versa. The other functions of the cell wall include control of intercellular transport, protection against other organisms, recognition and cell signalling, and storage of food reserves.[3] It is at the heart of many key events in the life of a cell and the organism of which the cell is a part. In other words, "if the cell wall has something to do with the cell, it is equally true that the cell has something to do with the cell wall".[4]

The cell wall is a layered structure formed just outside the plasma membrane. In multicellular plants, the cell walls of adjacent cells are separated, or rather cemented, by an amorphous intercellular layer called the *middle lamella*. Very often the cell walls are classified as *primary* and *secondary*. Such a classification is often beset with problems such as, for example, the exact definitions for primary and secondary walls. The primary wall is defined by some people on the basis of its thinness, relatively undifferentiated nature, and as the one that occurs around the protoplast of the cell at the time of cell enlargement.[5] Walls of collenchymatous cells and of some algae defy this definition and may undergo thickness during the process of cell enlargement. Also, a distinction cannot be made with certainty in some higher plants as there are often transition lamellae. Kerr and Bailey[6] had distinguished primary and secondary walls by the mechanism of wall growth, with primary walls showing *intussusception* and secondary walls *apposition*. Others found that primary walls showed a dispersed texture of microfibrils while secondary walls showed a parallel structure.[1] But later on it was found that these distinctions were also not tenable as both types of structures may be found in both walls. A chemical difference between the two was introduced by researchers working in higher plants. Primary walls lack *lignin, cutin, suberin,* and *wax,* whereas secondary walls contain one or more of these in addition to components of the primary wall. In algae and other thallophytes, such a distinction cannot be made and one can speak of only growing and nongrowing cell walls.[5] However, in view of their establishment in botanical literature, the terms primary and secondary walls are regularly used in describing the cell walls of land plants.

The secondary wall of land plants is laid in certain cell types after the cessation of cell growth. The deposition often involves considerable increase in the thickness of the wall. By nature, the secondary wall does not permit the cell possessing it to enlarge further but it provides considerable strength and mechanical properties to the concerned cell. In cell wall research, it is the secondary wall that has received extensive scientific and technological attention because it is responsible for the special characteristics of raw materials, wood, textile fibers, paper fibers, cellulose, straw, and cork. In the xylary elements of gymnosperms and angiosperms, very often the secondary wall is demarcated into three layers, S_1, S_2, and S_3.[6] S_1 is towards the primary wall and S_3 is towards the cell lumen. Invariably S_2 is the thickest. Under the polarizing microscope, the cell wall of a cell with secondary deposition shows the following characteristics: The middle lamella is *isotropic*, the primary wall is weakly *birefringent*, S_1 and S_3 are very bright, and S_2 is dark. These differences are primarily due to the differences in the orientation of the *microfibrils* which make up the different layers of the cell wall. It is not necessary that all these layers should be developed by a cell. S_3 often may be absent or is replaced (often along with S_2) by a gelatinous layer as in *tension wood* fibers.

The *three-ply* structure of the secondary wall need not be present in all cell types showing secondary wall desposition. In laticifers, the secondary wall shows many submicroscopic lamellae.[7] The same holds good for cotton hairs, where there are as many as 25 lamellae in the secondary walls. In the epidermis of many plants, the secondary wall is deposited in an altogether different fashion with a set of different chemicals. The same can be said of suberised cells of the cork.

1.2 Chemical Composition

The plant cell walls contain several groups of chemical substances. These have been variously classified. Frey-Wyssling and Mühlethaler[1] classified them on the basis of their behavior towards chemicals used in extraction analysis. They recognized the following: (1) *Frame substances*, which are responsible for the mechanical properties of the plant cell wall. These resist hot water and dilute alkali and acid treatments, e.g., *cellulose, chitin, mannans,* etc. (2) *Ground* or *matrix substances*, which are extracted by hot water and dilute alkali or acids or both. These include the *pectins* and *hemicelluloses* (such as *xylans* and *arabinans*). (3) *Incrusting substances*, which are substances left over when frame substances are destroyed. These include *lignins*, *phenols*, and *minerals* such as silica. (4) *Adcrusting substances*, which accumulate on the cell wall surface. These include *callose*, mucilaginous substances of some aquatic taxa, *waxes, suberin,* and *cutin*. A slightly modified classification, recognizing only the *microfibrillar phase* and *matrix phase*, has been followed by many.[8] The former phase is distinguished from the latter by its high degree of crystallinity and its relatively homogenous chemical composition. Cellulose, mannans, and chitins individually make up the microfibrillar phase while all other chemicals constitute the matrix phase. Differential extraction procedures are convenient but not entirely selective in knowing about all chemical substances of the cell wall and the classification of cell wall components is best based on structure rather than on isolation procedures employed.[9] Therefore, cell wall substances are often classified on the basis of their chemical structure into carbohydrates, proteins, phenol-derived compounds, lipid-derived compounds, minerals, etc. This classification has been followed here for the sake of simplicity.

1.2.1 Carbohydrates

Carbohyrates are *polyhydroxyketones* or *polyhydroxyaldehydes* or compounds that can be hydrolyzed to these substances using dilute mineral acids. They all share some chemical properties and some biological functions. The long-recognized biological functions of carbohydrates are in the storage and mobilization of energy and in contributing to the integrity of cells and tissues through their structural and water-binding properties. Some cell wall carbohydrates such as *oligosaccharins* can have cell regulatory functions such as control of gene expression.

Carbohydrates are classified into *monosaccharides* (simple sugars that have aldehyde or ketone groups; e.g., glucose, fructose, etc.), *oligosaccharides* (combined product of 2 to 10 monosaccharides to which they can be hydrolyzed; e.g., sucrose), and *polysaccharides* (high molecular weight compounds with more than 10 units; the majority have 80 to 100 units, while cellulose, on average, has 3000 units). Carbohydrates also form complexes with lipids, proteins, and minerals.

The polysaccharides may be linear or branched. They are often divided into (1) *homopolysaccharides* (or *homoglycans*) which, on hydrolysis, yield only one type of monomer, e.g., cellulose, callose, etc. and (2) *heteropolysaccharides* (or *heteroglycans*), which yield not only monomers containing C, H, and O, but also those which additionally have N and S; e.g., some pectins. Polysaccharides are also classified as *neutral* and *acidic*, the latter either carboxylated or sulphated.

1.2.1.1   *Cellulose*

Cellulose is the most abundant and best-known compound that occurs in plants, forming the main constituent of the cell walls. It is estimated that half of the CO_2 in nature is fixed in the cellulose by the plants. The proportion of cellulose on a weight-to-weight basis varies from 1 to 10% in the primary walls of higher plants and red algae, to 50% in the thick secondary walls of higher plants, and 80% or more in some green algae.[5] Cellulose is mainly responsible for the structural rigidity of the cell walls due to its stability, crystalinity, and microfibrillar nature. Cellulose, on total hydrolysis, yields only β-D glucose residues, which in the intact wall, are linked together linearly by β-(1→4) linkages. The linear chains of monomers (which may achieve chain lengths of up to 15,000 residues, 7.7 μm) form *microfibrils* or bundles of parallel chains which are held together by hydrogen bonds. The cellulose microfibrils are mostly bands of 10 to 25 x 10^{-20} nm wide, often very flat and appear fasciated. X-ray diffraction patterns show that cellulose is composed of repeating units of *cellobiose*, which form a crystal lattice. The unit cells of crystalline cellulose are *monoclinic*, composed of four glucose residues and have the dimensions a = 8.35 Å, b = 10.3 Å, and c = 7.9 Å.[5] The molecular weight of cellulose is estimated to reach up to ca 1,000,000. Cellulose has positive signs of birefringence, little UV absorption, and is *anisotropic* (= bright in plane polarized light).

Cellulose invariably occurs independently without forming complexes with non-carbohydrates. However, in cotton hairs and some other cell types cellulose is known to occur as *glycoproteins*.[10] Cellulose is synthesized by plasma-membrane-bound enzyme complexes.[11]

1.2.1.2   Callose and Other 1, 3 β-Glucans

Callose is a homopolysaccharide that appears to be restricted to plants. It has β-(1→3)-glucan subunits, although evidences are there for the view that callose might be a heterogeneous substance exhibiting source-dependent compositional differences. Callose forms a major component of the walls of pollen tubes, microspore- and megaspore-mother cells, pollen intine, phloem sieve plates,[12,13] endothecium of anthers, many fungal hyphae, cotton fibers, stigmatic surface cells, and in cells that are immediately adjacent to a freshly formed wound.[14,15] In the fungal hyphae, the β-(1→3)-glucan backbone is either substituted at intervals by (1→6)- linked-β-glucosyl residues and the molecules are water soluble, or multiple-branched with varying proportions of (1→3):(1→6)- linkages and the molecules are insoluble.[15] In some yeasts (1→6)-β-glucans alone occur. The β-(1→3), (1→6)-glucan is often known as *mycolaminarin*.[16] Callose may occur either independently or as glycoproteins. Callose is optically isotropic and is soluble in alkali; these two features also distinguish callose from cellulose. A (1→3, 1→4), ß-glucan is formed in the cell walls of many grasses and some monocotyledons.[15] In grasses, this is particularly abundant in the aleurone layer of seeds.

Laminarin, which occurs in the cell walls of brown seaweeds, is apparently a type of callose, as it also has 1,3-linked ß-glucan subunits.[17]

Callose plays an important role during various steps of growth and reproduction as well as in cell and tissue sealing and defense mechanisms against pathogens and other physical and chemical stresses.[18]

1.2.1.3   Chitin

Chitin is a carbohydrate (polyglycan) mainly present in invertebrate animals but also has been recorded in the cell walls of many fungi (up to 25% by dry weight in *Plasmodiophora* species), a few green algae (e.g., *Cladophora* and *Pithophora*), and centric diatoms.[19] Pure chitin is not known to exist. It almost always forms a complex with proteins (in plants) or carbonates (in animals). Chitin is chemically similar to cellulose and the difference between the two lies in the fact that the hydroxyl group on the second carbon atom is replaced by an acetyl amino group in chitin. In other words, chitin is a polymer of 2-acetoamido-2-Deoxy-D-Glucopyranose, each residue being 1→4 linked to form linear chains. The long chains may achieve a molecular weight of the same order of magnitude as that of cellulose. Crystalline chitin forms a unit cell that is orthorhombic, having the dimensions, a = 4.76 Å, b = 10.28 Å, and c = 18.85 Å.[20] Chitin is even more insoluble in water and many organic solvents and more resistant than cellulose.

Chitosan, a deacetylated form of chitin, has been reported in the walls of some fungi such as species of *Phycomyces* and *Mucor* and many members of Mucorales.[21]

1.2.1.4   *Pectins*

Pectins, along with hemicelluloses, were earlier classified under carboxylated polysaccharides. It is often difficult to distinguish pectins from hemicelluloses and a number of investigators treat them as one. The definitions of pectins and hemicelluloses are difficult and are partly based on procedures used to extract and isolate them and partly on the chemical nature of the extraction (see detailed discussion in Selvandran et al.[22] and Wilkie[23]). In this book, pectins are treated as distinct from hemicelluloses.

Pectins are acidic, non-cellulosic polysaccharides present both in cell walls (primary and secondary) and in the cytoplasm. They also occur abundantly in the middle lamella, where they function in the adhesion of adjacent cells. The principle component of pectin is *D-galactopyranose* (*galacturonans*). A type of pectin in which *D-galactopyranose* residues are replaced by *D-galactopyranosyl-uronic acid* residues (*galacturonic acids*) is called *pectic acid*. If uronic acid residues are present as methy esters, the pectin is termed pectinic acid. Pectinic acids are very easily extractable with water and, therefore, possess considerable swelling properties. On the other hand, the pectic acids, which frequently contain calcium or magnesium salts of hexosyluronic acid residues (galacturonic acids), are less soluble and, therefore, require the use of agents such as chelators for their extraction. Pectins, pectic acids, and pectinic acids together constitute the pectic substances.

The most characterized pectic polysaccharides of cell walls are *polygalacturonic acid (PGA*, sometimes referred to as *homogalacturonan*), *Rhmnogalacturonan I* (RG I), and *RG-II*. PGA is a polymer of α–(1→4)-linked D-galactosyluronic acid interrupted periodically with a single α–(1→2)-linked rhamnosyl residue. The length of an uninterrupted galacturonan block appears to be consistent within a single type of wall, but varies between walls of different types. PGA has two major chemical forms: The galactosyluronic acid residues can contain carboxyl group at C-6 (*non-esterified pectin*), or they may be methylesterified at C-6. The degree of esterification varies greatly and within a single wall there are several domains where the degree of pectin esterification is modified.[24] It is generally believed that PGA is synthesized and secreted into the cell wall in a highly methyl esterified form where it is acted upon by pectin methyl esterases, which de-esterifies blocks of consequently galactosyl uronic acid residues. Adjacent blocks of such residues, which are polyionic, can bind inter- or intra-molecularly through Ca^{++} bridging and form stable pectin gels.[25] RG-I backbone consists of multiple (up to 300) repeats of the disaccharide α-D-Gal A-(1→4)-α-Rha- (1→2). The remaining sugars (~1400) are constituents of the sugar side chains of RG-I. About 50% of the backbone rhamnosyl residues are substituted at 0 to 4 with sugar side chains. The distribution of substituted rhamnoasyl residues along the backbone appears to occur with regular periodicity. The side chains are composed of arabinosyl, galactosyl, and fucosyl residues. PGA and RG-I are covalently linked *in muro*. Such linked cell walls are referred to as PGA/RG-I. RG-II is a relatively short polymer of 60 residues. This polymer contains a backbone of PGA and a number of unique sugars, including 2-*O*-methyl-fucose, 2-*O*-methyl-xylose, apiose, aceric acid, and 3-deoxymannoctulosonic acid. Structural regularity in the molecules was established by the presence of two heptasaccharide repeats, which together constitute 60% of the molecule. The reader is advised to refer to standard works for the detailed chemistry of these pectic substances.[8,9]

Signs of birefringence is 0 in pectic substances; they also show little UV absorption. Pectins are isotropic. Several functions have been ascribed to the pectic polysaccharides. Pectins have been believed to be mediators of wall porosity and to help in intercellular transport of materials.[26] They also play an important role in cell wall hydration, adhesion of adjacent cells, and wall plasticity during cell elongation. Pectins are also involved in fruit ripening. They also have a further role in the recognition of reactions between plant cells and some of their bacterial and fungal pathogens.[25] That pectic polysaccharides can even alter the developmental pathway of cells in tissue culture systems has also been shown.[27] Recent evidences show that pectins are not amorphous components of cell walls but they seem to have a fair degree of order in their orientation.[4]

1.2.1.5 Hemicelluloses

The hemicelluloses refer to an ill-defined group of polysaccharides that form the wall materials of plants with the exclusion of cellulose, lignin, and pectins with which they are closely associated. They often encrust cellulose fibrils and in vascular plants are intimately associated with lignin.[28-31] A few hemicelluloses are also present in algae.[32] Hemicelluloses are classified according to the monosaccharide residues present. The known hemicelluloses of cell walls include D-xylans, D-glucans, mannans, xyloglucans, glucomanns, galactomannans, glucuoronomannans, glucuronoxylans, arabinoxylans, arabinogalactans, glucoronoarabinoxylans, etc.

Xyloglucan (XG) is the most studied neutral hemicelluloses of cell walls, making 20% of the dry weight of the primary wall of dicots and 2% of monocots.[33] It consists of a backbone of from 300 to 3000 β-(1→4)-linked D-glucosyl residues of which 65 to 75% are substituted at C-6, with side chains consisting of either a single xylosyl residue or with a trisaccharide of xylose, galactosyl, and fucosyl residues.[34] The side chains are distributed in a highly regular pattern along the backbone such that three consecutive xylosylated glucosyl residues are followed by a single unxylosylated glucosyl residue. The molecular mass of XG varies between 7600 to 330,000. The xylans are a major hemicellulose in monocot cell walls (15 to 20%) but a minor one in dicots (2%). Glucuronoxylans are the principal hemicellulose polymers in primary walls of the grasses but form only small amounts in primary walls of dicots. They have a backbone of β-(1→4)-linked D-xylosyl residues. The backbone xylosyl residues may be substituted at 0 to 3 with α-L-arabinofuranose or at 0 to 2 with α-D-glucosyluronic acid. The D-glucans [β-(1→3) (1→4)-D-glucans] are a hemicellulose uniquely present in cell walls of grasses. Approximately 70% of the linkages are β-(1→4) and 30% are β-(1→3). The molecular weight varies between 20 and 1000 kD. Details on the chemistry of the hemicelluloses are found in Bret and Waldron.[8]

The hemicelluloses, in general , are extractable with warm alkali (0.4 to 4.0 M NaOH or KOH, sometimes added with 0.5 M borate), but not with chelating agents. On hydrolysis, they yield different hexoses (in addition to glucose, mannose, and galactose), pentoses (xylose, arabinose), and uronic acids (galacturonic acid, glucuronic acid). These components appear in widely differing relative amounts and in numerous combinations. Only a few hemicelluloses, like special xylans, glucans, and mannans, are known in fibrillar form and are suited to X-ray diffraction analysis.[5] Many hemicelluloses have zero birefringence and little UV absorption.

Hemicelluloses probably have a structural role. A network of cellulose and xyloglucans is the major contributor to the tensile strength of the cell wall.[35] Hemicelluloses form storage cell wall polysaccharides and occur as thick deposits in the endosperm cell walls of several seeds. They also have a regulatory function, especially in cell elongation. Cell wall loosening and elongation are shown to be due to xyloglucans[34-37] and D-glucans.

Both the hemicelluloses and pectins are synthesised and processed in the Golgi apparatus and packaged into secretory vesicles for transport to the cell wall.[11,38]

1.2.1.6   Alginic Acid

Alginic acid is an important component of brown seaweeds, occurring in both intercellular regions and the cell walls.[17,39] It also occurs at the surface of the algal thallus, often forming thick deposits. It is a mucilaginous, carboxylated polysaccharide[40] that prevents desiccation of the seaweed when exposed to air; it is also involved in ion exchange function. Alginate is a linear polyuronate compound of D-mannopyranosyluronic acid and L-glucopyranosyluronic acid residues, the proportion of the two varying in different taxa. Alginic acid is present as salts of calcium or sodium. Since it produces X-ray diffraction patterns,[41] alginic acid is likely to be crystalline in nature.

1.2.1.7   Sulphated Polysaccharides

This group of polysaccharides is common in algae. *Fucoidins* (including *sargassan, ascophyllan, glucaranoxylofucan*, etc.), *carrageenan, agar*, and green algal sulphated polysaccharides are the well-known sulphated polysaccharides. Fucoidins are a family of polydisperse heteromolecules, containing in addition to fucose varying proportions of galactose, mannose, xylose, uronic acid, and glucuronic acid. These are extremely complex and highly branched macromolecules.

Carrageenan has an alternating sequence of $(1\rightarrow3)$-linked β-D-galactopyranosyl and $(1\rightarrow4)$-linked β-D-galactopyranosyl residues,[42] containing various degrees and sites of sulphation. Agar contains a neutral gelling component *agarose* and an anionic fraction containing sulphate groups. Agarose is believed to consist of chains having alternating $(1\rightarrow3)$ and $(1\rightarrow4)$ linkages. The predominant monosaccharide of carragenans and agar is galactose, although xylose and mannose may also be present.[32] The green alga sulphated polysaccharides fall into three categories: glucuronoxylorhamnans, glucuronoxylorhamnogalactans, and xyloarabinogalactans.[17] Several functions have been suggested for the sulphated polysaccharides: (1) their ability to imbibe water might be considered as a protective pillow against the physical buffeting that seaweeds undergo by wave action, (2) they provide protection against desiccation, and (3) the anionic character of these polysaccharides might serve as a sort of ion exchange material and/or might help to sequester certain ions.[32]

1.2.2 Glycoproteins

Glycoprotein is a term used, often too generally, to apply to any macromolecule, that contains carbohydrate and protein. It has a protein chain consisting of L-α-amino acids; covalently attached to this protein backbone and pendent to it is/are the carbohydrate part(s) of the molecule consisting of oligosaccharide chains. These chains are usually branched and can contain neutral monosaccharides (D-glucose, D-galactose, D-mannose, or L-fucose) or basic monosaccharides (2-amino-2-deoxy-D-mannose or 2-amino-2-deoxy D-galactose).

Most proteins found in the cell walls are glycoproteins, the protein fraction alone constituting about 10% of the mass of the primary cell walls.[43] The nature of their interactions, attachment, development and/or biochemical functions, and their relation to the overall cell wall architecture remains uncertain.[44] Cell wall glycoproteins range from enzymes such as isozymes of peroxidase, phosphatases, etc. to the structural proteins such as *extensins*. Evidence exists to indicate that many major plant cell wall polysaccharides including cellulose contain small amounts of covalently bound proteins and should be considered glycoproteins.

1.2.2.1 *Structural Proteins*

The structural proteins are usually rich in hydroxyproline. These glycoproteins are broadly classified into three groups on the basis of the size of the sugar prosthetic groups.[45]

Lectins.

Lectins are defined as specific sugar-binding proteins of non-immune origin. They are also known as *agglutins, phytoagglutins, phytohaemoagglutins,* or *prolectins.* The first three names were coined because of the ability of the lectins to agglutinate the blood groups. Lectins have been found in almost every plant tissue, although they are widely distributed in seeds, especially members of Leguminosae, Solanaceae, Euphorbiaceae, Lamiaceae, Poaceae, etc. Lectins have also been reported in other groups of plants (Algae to Gymnosperms), bacteria, and also in vertebrate and invertebrate animals.[46-48] Some of the common lectins are *concanavalin A (Con A), Ricin, wheat germ agglutin, peanut agglutin,* etc. Although rich in hydroxyproline, at least some of the lectins have serine, cystine, alanine, and glycine as major amino acids.

Lectins may play an important role in a variety of recognition phenomena, like in legume root-rhizobium, pollen-stigma, and root-mycorrhizal fungal interactions. This is possible through their specific sugar binding properties but without inducing any chemical change in the ligand. They are also implicated in plant defense against microbes.[49] Yet, the biological role of lectins is far from being understood.

Because of the specific sugar binding properties, lectins are advantageously used in the cytochemical localization of carbohydrates.

Arabinogalactan Proteins (AGPs).

These are large-molecular weight polysaccharides attached to proteins with the resultant molecule being 80 to 90% carbohydrate.[43,45,50-52] These glycoproteins resemble the extensins structurally and are rich in hydroxyproline, serine, alanine, glycine, galactose, and arabinose. The protein moiety is acidic. The peptide sequence is alanine-hydroxyproline.[45,51] The polysaccharide portion is similar to arabinogalactan II. The main carbohydrate linkages are Hyp-Gal, Ser-Gal, and Hyp-Ara.

Arabinogalactan proteins are associated with the cell walls (probably as a pectin-binding protein[53]) of all the groups of land plants examined including the Hepaticae (Bryophytes).[54] These proteins are often present extracellularly and are water-soluble. They have been reported in the extacellular fluid of many suspension-cultured tissues, stylar canals, xylem cells, gummy exudates, etc. The exact role of these proteins is not clear. They have been implicated in a nutritive role in pollen tube growth, in assisting pollen adhesion to stigma,[55] in plant defense,[45] in incompatibility reactions expressed in the style, in suppressing local cell proliferation,[56] in tissue differention by deciding the fate of cells derived from meristems,[57] reflecting early developmental cell patterning and cell positions,[24] in flower sex development and differentiation,[58] and in the differentiation of embryogenic calli.[59,60]

Hydroxyproline-Rich Proteins or Extensins.

Extensin is a normally insoluble, basic, hydroxyproline-rich protein attached very tightly to the cell wall polysaccharides of many plants, including algae.[32,46] In *Chlamydomonas* and other members of the green algal group Volvocales, the cell wall does not contain a fibrillar skeletal polysaccharide but instead is composed of crystalline aggregates of extensin-like proteins only.[61,62] Extensin is also referred to as *Hyp-protein* (HRGP) because of the rich (40%) hydroxyproline content (a feature of collagens of the animal extracellular matrix). The remarkable feature of extensin is its insolubility. It cannot be extracted from cell walls with any of the conventional protein solvents such as salt solutions, detergents including 3% SDS in the presence of 1% 2-mercaptoethanol at 100°C, phenol-acetic acid, water (2:1:1), cold aqueous acids or alkalis, chelating agents, or even with anhydrous hydrogen fluoride, a protein solvent *par excellence*.[63] The insolubility is thought to be the result of covalent cross-linking of the extensin to the cell wall polysaccharides, through *isoditryrosine*, a phenolic dimer. The latter is sensitive to $NaClO_2$ and, therefore, extensin can be solubilized from cell walls by a treatment of mildly acidified $NaClO_2$. The extensins are characterised by a repetitive Ser-$(Hyp)_4$ peptide sequence along with (Tyr-Lys-Tyr-Lys) and (Thr-Pro-Val) peptide sequences. They are shown to occur as rod-like structures of *ca* 85 nm long[64] and are stabilized by glycosylation with 1 to 4 arabinosyl residues.[45] About 50% of the weight of extensin is protein and the molecular weight is around 40,000.

In flowering plants, extensins have been reported in the cell walls of many types of cells such as tracheary elements,[65] phloem,[66,67] sclerenchyma,[44] and vascular cambium.[68]

An extensin-1 and extensin-2 have been purified and reported from carrot tissues[69,70] while up to four extensins have been known in tomato.[45]

Extensins are synthesised in ER and are processed in the Golgi before being transported to the cell walls.[71]

The following are the functions attributed to the extensins: (1) Perform a structural role in strengthening cell walls at the completion of cell expansion by forming a cross-linked insoluble matrix. (2) Function as protective and mechanical elements as they often occur abundantly in cell walls of sclerenchyma.[44] (3) Limit wall extension and thereby control cell growth. (4) Protect the cell from wounded surface (see Bacic et al.[72]) and pathogens.[73] Two possible mechanisms are involved in extensin's role in providing resistance to plants against microbes. The first suggests that extensins, by virtue of their tight cross-linking, render the wall indigestible by invading pathogens. The second suggests that extensins, as polycations, agglutinate microbes and thereby prevent their spread.[74] (5) Since extensin production is triggered by ethylene, red light, heat shock, gravity, glutothione, culture conditions, etc., they may likely be the combat chemicals under various stress conditions.[75] (6) They may serve as the xylogenic factor in vessel element formation.

Proline-Rich Proteins (PRPs) .

A number of plants are now known to have cell walls with proline-rich proteins.[46] The occurrence and activity of this family of proteins were found to be developmentally regulated with both organ-specific and stage-specific expressions.[76] A PRP is known in *Nicotiana* styles whose expression is restricted to the transmitting cells.[77]

Glycine-Rich Glycoproteins (GRPs).

These glycoproteins are common in plants like *Petunia* and bean, especially in walls of cells in vascular tissues.[33,46,78]

Threonine-Rich Glycoproteins (TRPs) .

These have been reported in the cell walls of certain monocotyledons.[79]

Thionins.

These are a novel class of highly abundant glycoproteins with antifungal activity present in the cell walls of seeds of plants like barley,[80] maize, wheat, and oat as well as in the leaves of various dicots.[81] They are naturally occurring inducible plant glycoproteins.

Pistil-Specific Extensin-Like Proteins (PELPs).

These are reported in the transmitting tissue of the styles of tobacco.[82] The PELPs differ from previously described extensins by a low tyrosine content, a lower copy number of the Ser-(Pro) 4 motif, and less repetitive nature. Another class of pistil specific glycoproteins that control compatability reactions and that are products of the S-genes for incompatibility have been reported to be restricted to the stigmatic papillae, as in the case of *Brassica*.[83] In *Nicotiana,* the glycoproteins, which are products of S-gene expression during incompatibility reaction, are in fact now known to be ribonucleases.[84] These enzymes, in some way, halt pollen tube growth.

One of the very exciting discoveries in recent years has been the detection of two small proteins, called *Expansins*, which produce cell wall loosening and elongation without detectable enzyme activity.[24,85] These two proteins of *ca* 30 kD have been isolated from cucumber hypocotyls.[86]

1.2.2.2 Enzymes

Various enzymes are located in the primary walls of higher plants. Some, but not all, of these can be solubilized with salts, indicating ionic binding of these to cell wall polymers. The few wall enzymes that have been structurally examined have been found to be glycoproteins.

Peroxidase (E.C. 1. 11. 1. 7).

The peroxidases are widespread in plant tissues and abundant in cell walls.[87,88] Freely soluble as well as apparently ionically or even co-valently bound peroxidases are present. The peroxidases can catalyse phenolic cross-links between macromolecules such as lignin, protein, hemicellulose, and ferulic acid. They can also cross-link proteins by oxidative deamination of lysine.[44] Even for the synthesis of isoditityrosyl residues (a phenolic dimer of extensin), peroxidases are important. Peroxidases restrict wall growth by rigidification. They are also involved in lignification.[46,89] Stylar peroxidases are important in incompatibility reactions, as has been shown in *Petunia*.[90] Peroxidases are also important in stress responses, especially to those involving heat. There are also evidences to indicate their involvement in defense reactions.[44]

At this juncture, it is important to mention that cell wall bound peroxidase often requires H_2O_2 to bring about the polymerization of the C_6-C_3 units of lignin precursors into a high polymer lignin.[91] H_2O_2 is often present in cell walls of cells that are destined to undergo lignification as well as of those that are subjected to wounding and pathogen entry.[92]

Polyamine Oxidase (E.C. 1.4.3.4).

Polyamine oxidase uses polyamine substrates and releases H_2O_2 as one of the resultant products.[93,94] Since H_2O_2 is localized in cell walls, it is natural to expect the enzyme producing it (polyamine oxidase) to also be in the cell wall. In fact, it has been shown to be a cell wall enzyme in oat leaf cell walls.[93] This enzyme is shown to be closely active with peroxidase in the lignosuberization process during fungal entry into a cell.[95]

Phosphatases (E.C. 3.1.3.36).

Phosphatases are present in cell walls, in addition to being cytoplasmic.[46,96,97] The phosphatases originally denoted a large group of enzymes that hydrolyzed a range of phosphate esters at an optimum acid pH and released inorganic phosphates. This is too general a definition, since it encompasses a wide range of enzyme activities with pH optimum between 4.5 and 6.5. In view of this, and because of the frequent employment of non-physiological substrates in cytochemical (and biochemical) studies, a form of nomenclature arose in which the activities were described according to the substrate used, e.g., acid β-glycero phosphatases, acid naphthol AS-BI phosphatase, acid nitrophenyl phosphatase, etc. It is clear that more than one enzyme can work on a given substrate of the type commonly employed with plant tissues.[98-101] This is still not a satisfactory classification, though it aids in comparing published data.

Functions of the phosphatases in the cell wall are not very clear since cellular phosphate esters usually do not leave the cells. They are believed to act as "Reporters" about the conditions within the wall.[44] They are also suggested to have a role in intercellular transport, since plasmodesmata showed significant activity.[102] The possible role of acid phosphatase in microbe-plant interaction has been discussed in Boller.[103]

Protease (E.C. 3.4.21.19).

This enzyme is reported to be present in cell walls.[46,74] Its natural substrate in the cell wall is not known.

Nitrate Reductase (E.C. 1.6.6.1).

This enzyme is reported to be present in the cell wall-plasmalemma region of *Neurospora crassa* cells.[104] The exact significance of the presence of this enzyme in the fungal wall is not known.

Malate Dehydrogenase (E.C. 1.1.1.37).

This enzyme is known to be present in small amounts in cell walls.[105] This provides NADH for the reduction of O_2 to H_2O_2 necessary during lignification. It also helps in the cross-linking of structural proteins.

Prolyl Hydroxylase (E.C. 1.14.11.2).

It is one of the key enzymes involved in the formation of hyp-proteins of cell walls. It is helpful in proline hydroxylation and recognizes the secondary structure of the substrate, the polyproline II-helix.[106] This enzyme has been found in the cell walls of *Chlamydomonas*, bean, carrot etc.[107,108]

Glycosyl Hydrolases and Transferases.

A variety of these enzymes has been reported in the cell walls.[46,97] These include exoglycosidases such as β-glucosidase (E.C. 3.2.1.21) β-xylosidase (E.C. 3.2.1.37), β-galactosidase (E.C. 3.2.1.23), α-galactosidase (E.C. 3.2.1.22), and β-fructofuronosidase (E.C. 3.2.1.26).[97,109] Other enzymes reported are α-mannosidase,[110] arabinofuranosidase, galacturonisidase, glucosaminosidase, arabinosidase, β-mannosidase, trehalases, and β-glucuronosidase.[44] Some of these are involved in the conversion of extracellular complex carbohydrates or related compounds into monosaccharides for the uptake by the cell. β-glucosidase is shown to be involved in the cleavage of coniferin in spruce during lignification of cell walls.[111] Others may act on the non-reducing terminals of structural polysaccharides or glycoproteins.

A limited number of endoglycanases have been documented in cell walls. The most intensively studied one is endo-β-1,4-D-glucanase ("Cellulase") (E.C. 3.2.1.4). There is no evidence that the natural substrate to this enzyme is cellulose.

Endo-β-1,3-D-glucanase (Callase) (E.C. 3.2.1.39) is also found in cell walls. This enzyme could play an important role in pathology by solubilizing elicitor-active oligoglucans from fungal cell walls.[112,113] It has also been reported in cell walls (mainly middle lamella) (although in low amounts) in ethylene-stressed bean leaves.[114] According to these authors, this enzyme probably degrades invading fungus cells and the oligosaccharides released from the fungal wall are elicitors of phytoalexins.[115]

Although chitin does not occur in higher plants, the presence of chitinase (E.C. 3.2.1.14) in cell walls is a matter of surprise. Chitinase was found in cell walls of many plants including cucumber,[116] probably as a response to fungal infection[117] since chitinase is known to be a "pathogenesis-related" protein.[112] Chitinase is also induced by ethylene.[114]

Pectinase or poly-1,4-α-D- galacturonide glycanohydrolase (E.C. 3.2.1.15) and pectinesterases (E.C. 3.1.1.11) have also been obtained from cell walls.[25,118] Pectinase randomly hydrolyses the α-1→4 linkages of pectin. This is responsible for the softening of fruits; it is also reported to be present in abscission zones probably doing the same function. Cell wall pectin methylesterases might represent a key factor for cell wall expansion processes. These enzymes may function in coordination with pectinase.[25]

Xyloglucan endo-transglycosylase (XET), an enzyme that acts on the cellulose/xyloglucan network (by cleaving the latter) during cell wall loosening and elongation, has been recently characterized.[119-121]

The presence of hydrolytic lysozymes in plant cell walls has also been reported, as a response to fungal infection.

Non-Specific Esterases.

These are a heterogeneous group of enzymes which hydrolyse a range of carboxylic acid esters:

$$R\text{-COOR'} + H_2O \rightarrow R\text{-COOH} + R'OH$$

Thioacetic acid is a substrate normally employed for detecting these enzymes. Esterases are reported in cell walls.[122]

1.2.3 Lipids and Related Substances

Although lipids have been demonstrated biochemically in cell walls of a vast range of plants and fungi, they appear to be almost always associated as glycoconjugates or as part of cutin, suberin, and wax. For example, the sporangiophore wall of *Phycomyces* has a lipid content of more than 25% of the dry weight.[21]

1.2.3.1 *Cutin*
Cutin is present in the cutinized cell wall along with some other wall materials, while it invariably forms the only chemical substance of the non-cellular cuticular

membrane or cuticle[1] which is present in the aerial parts of land plants covering the external surface.[123,124] Since it is laid on the *outer side* of the wall, its precursors have to pass through the wall layers before polymerization.

Cutin is a normally insoluble, high molecular-weight, three-dimensional polyester. It is a substance, that is not a true fat component but is closely related in structure to it. It strongly resists acid attack and decay. However, it readily dissolves in strong alkalis (saponification) to produce a mixture of fatty acids and dihydroxy hexadecanoic acids (mainly the 10,16- and 9,16- positional isomers). The reactive group of these molecules link and cross-link mainly by ester bonds (-CO-O) but also by peroxide and ether linkages.

Cuticle has negative or weak birefringence but strong UV absorption (autofluorescent). The latter character is believed to be due to bound phenolic constituents. The cuticle is also almost isotropic. In the electron microscope, it is completely structureless without any fibrillar or corpuscular component. The cuticles show perhaps the outstanding transpiration defense. As cutin is not a pure lipophilic substance, but contains some quantity of unesterified alcohol and carboxyl groups, it is not completely hydrophobic. Therefore, small amounts of water are continuously lost through the cuticle (cuticular transpiration).

1.2.3.2 Suberins

The suberins are more or less similar to cutins and appear to differ constantly only in one character, i.e., while cutins are deposited on the outside of the cell wall (especially on the surface layer of land plants), suberin is laid on the *inside* of the existing wall. The reader should keep in mind this difference because all the histochemical methods localize both suberins and cutins in the same way. Often the two can be distinguished only by their degree of resistance to saponification: while suberins are saponified in 3% aqueous NaOH, cutins can be saponified only in 5% methanolic KOH.[1]

Suberin is a substance that is commonly present in cork cells, endodermis, root caps of some plants, hypodermis, seed coats, trichomes, and idioblasts (see full literature in Holloway and Wattendorf[125]). Suberin can also be produced in any region and any tissue as a response to injury and infection.[72] It is deposited as definite layers in the form of adcrustation on the inside of the original wall.

Suberin is composed of two-thirds aromatic phenols and one-third long chain aliphatic acids of chain lengths greater than C_{18}. Because of the latter, suberin has more lipid-like properties than cutin.

Suberin is resistant to strong acid treatment but can be dissociated by strong warm alkali treatment. Suberization causes cell wall impermeability and water and solutes are not easily allowed; the same holds true for most pathogens.[126] Since suberin is an excellent thermal insulator, the functioning phloem elements are effectively protected from the solar heat. Suberin has negative birefringence, whereas UV absorption and autofluorescence are strong, and it is also isotropic.

1.2.3.3   Waxes

Waxes are plant surface chemicals that have chemical properties very near to lipids. The major constituents of waxes are C_{27}-C_{31} alkenes (long-chain hydrocarbons), alkyl esters, free primary alcohols, fatty acids, etc. Occasionally phenolics such as flavanoids may also be present. The waxes are embedded in and sometimes exuded over the surface of the cuticle, the latter being called epicuticular waxes. The *epicuticular waxes* may be in the form of granules, crystals, rodlets, needles, corn-flakes, ribbons, or macroni-like, often depending on the plant species. Although the exact role of waxes is not clear, they are believed to perform the functions of protection, water-repelling (thus forming a means of controlling the water balance in plant organs), and disease resistance. Wax is optically anisotropic and birefringent.

1.2.3.4   Sporopollenin

Sporopollenin is a major cell wall material of pollen exines and spore coats and found nowhere else. It has highly polymeric lipoidal esters consisting of carotenoid subunits. Due to the latter, sporopollenin is extremely resistant to microbial and chemical degradation; it is probably the most resistant organic compound in exist-ence. For this reason only, pollen and spores of the geological past have been well preserved as fossils. An exact chemical analysis of sporopollenin is difficult because the products are a function of the hydrolytic method used and may not necessarily represent natural molecular components of the wall. The probable chemical formula for the sporopollenin of rye pollen is $C_{90}H_{134}O_{31}$ and that of pine is $C_{90}H_{158}O_{44}$.

1.2.3.5   Pollenkitt

Pollenkitt is a substance that covers the exine surface in the pollen grains of a number of plants; it is most conspicuously present in insect-pollinated taxa. Pollenkitt is made up of lipoidal materials, flavonoids, carotenoids, and degenerated products of tapetal proteins.[127,128] It is a strongly hydrophobic material. The probable functions of pollenkitt are not clear but it is believed (1) to act as an insect attractant, (2) to help in pollen dispersal, (3) to protect pollen against the damaging effects of UV rays, (4) to control the breeding behavior of the species by playing a role in sporophytic incompatibility, and (5) to help in controlling water loss from the germ pores of pollen grains.

1.2.3.6   Tryphine

It is a complex mixture of hydrophilic substances which include some proteins, all being derived from the breakdown of tapetal cells. Its functions are still not yet clear.

1.2.4 Lignins and Other Phenolic Substances

1.2.4.1   Lignins

Lignins are phenolic polymers present in the secondary cell walls of land plants, especially in woody dead cells as the tracheary elements, xylem and phloem fibers, and sclerenchyma. Approximately 20% of the cell wall materials of the secondary xylem of dicotyledons and 35% of that of the gymnosperms constitute the lignins. Lignins fill up the matrix between the micellae of the cellulose framework of the cell wall, thus welding the cellulose micellae into a single coherent mass, and prevent the bending and buckling of the cellulose strands when subjected to compressional strain. Many hydrogen bonds occur between the polysaccharides at the microfibrillar matrix. Thus, the cell wall becomes a composite in which the polysaccharide polymers are enclosed in a cross-linked lignin cage.[129] The lignin in the matrix is also covalently linked to polysaccharides. The result is a very strong and hydrophobic network which is incapable of plastic extension.

Lignins are complicated macromolecules both structurally and behaviorally and have a molecular weight of *ca* 11,000. On oxidation, lignins yield three simple phenolic aldehydes: *p*-hydroxy benzaldehyde, vanillaldehyde, and syringaldehyde in various proportions depending on the plant. There are several evidences to show that there exist at least three lignin types, sometimes within the same species,[130] perhaps in the same tissue. Sometimes lignification may be induced by wounding or entry of pathogens (either by infusion of phenolics into or the apposition of phenolic-containing materials against the wall) and this renders the wall more rigid and less susceptible to enzymatic degradation.[131] The induced lignin is generally not identical with vascular lignin[132,133] as it possess phenolic polymers with free cinnamaldehyde groups but virtually without syringyl groups.

Because of the presence of double bonds, lignins strongly absorb UV light and so can be distinguished readily from other cell wall components by their strong autofluorescence (pronounced maximum at 282 nm). Lignin has statistically zero birefringence.

1.2.4.2 *Phenolic Acids and Amides*

A phenolic is a compound with an -OH group attached directly to a benzene ring. The plant cell wall often contains phenolic acids as regular constituents. The most important are ferulic acid, diferulic acid, *p*-coumaric acid, truxillic acid (a cyclodimer of the previous), and *p*-hydrobenzoic acid. They are bound by ester linkage to the polysaccharide constituents of cell walls, of especially grasses, Commelinaceae, Philydraceae, Pontederiaceae, and Haemodoraceae.[134] In grasses, the cell walls can contain up to 1 to 2% dry weight, ferulic and *p*-coumaric acids. Ferulic acid has been reported in many cell types of diverse tissues (barring perhaps phloem) including cutinized and suberized cells.[135,136] Ferulic acid is esterified to arabinose and galactose of pectins; two ferulic acid units can be linked by peroxidase activity to form a diphenyl bond. Ether linkages between ferulic acid and pectin may also occur. Thus, ferulic acid has an important role in the crosslinking of pectins. In grasses (especially cereals), which have low amounts of pectin, ferulic acid appears to be linked to the arabinose of arabinoxylans or glucuronoarabinoxylans.[137] The role of these compounds in the cell wall is a debatable topic and several suggestions have been made. These complexes may be involved in alterations in the mechanical properties of the cell wall responsible for growth and development of tissues,[138] since the formation of crosslinks between cell wall matrix polysaccharides such as pectins and hemicelluloses by diferulic acid makes the cell wall mechanically rigid, resulting in limited growth.[87,139,140]

Phenolic acids can also occur in plant cells covalently linked to lignin[141] or may form bridges between polysaccharides and lignin.[142] Wall-bound fungistatic phenolic amides (*phytoalexins*) have been reported in walls of infected host cells.[143] The elicitation of phytoalexin production in response to pathogenic attack is a signaling device exhibited by the cell wall, so that active defense reactions can be triggered in the cytoplasm. Phytoalexins are non-specific toxins; they also have widely varying structures. The elicitation process is initiated by an "elicitor", which is usually an oligosaccharin derived from cell wall carbohydrates.

1.2.5 Mineral Substances

Mineral substances occur in plant cells in diverse forms: soluble, insoluble, or forming part of complex organic substances. They form a fraction of 1% to about 15% or even more of the dry weight of the plant material. More than 40 of the known mineral elements have been identified in some plant or the other but, of these, only 15 are regularly present in appreciable quantities. Among these 15, hardly a few minerals have been detected in cell walls.

In the cell wall, mineral substances are present either in the mineral state or as complexes with organic materials. In the former instance, minerals occur in cell walls either as incrustations or as impregnations.

1.2.5.1   *Silica*

Silicon is deposited in plants as hydrated amorphous silica (SiO_2NH_2O) through the polymerization of monosilicic acid ($Si(OH)_4$) absorbed by roots through soil solutions.[144] Silica is impregnated or incrusted into the matrix of the cell wall such as that of the epidermal cells, stomata and trichomes of grasses, *Commelina* and *Equisetum*.[145] It is also known to be deposited in the cystolith stalk (a wall growth) of several members of Urticaceae and Moraceae.[145,146] Silica is also a characteristic feature of the diatom walls.

The exact function of silica in the cell wall is not very clear. Silica deficiency has been shown to increase transpiration rate and, therefore, silica has a role in checking transpiration rate. Silica is also helpful in preventing grazing by herbivores. The fact that without silica diatoms cannot be cultivated and their cell walls show no evidence of crystalline framework by other materials indicates the possible role of silica as a framework skeletal material[1] in this group of algae.

1.2.5.2   *Calcium*

It is one of the most commonly encountered mineral substances in plants. Although ionic calcium can be present in cell cytoplasm and also likely in cell walls, it often forms complexes with organic materials like pectins (calcium pectate), alginates (calcium alginate), or inorganic materials (calcium oxalate, calcium carbonate, etc.)

in cell walls. Calcium pectate is a common constituent of the middle lamella of land plants, while calcium alginate occurs at the corresponding place in brown seaweeds.

Calcium carbonate deposits usually occur in the form of *cystoliths* in members of families such as Moraceae, Urticaceae, Acanthaceae, Cucurbitaceae, etc.[146] Cystoliths are heavily calcified wall ingrowths that occur in specialized cells called *lithocysts* in leaves, stems, and sometimes in roots. Each cystolith has a matrix of cell wall material, principally cellulose with pectin and callose,[147-150] in which extensive deposits of calcium carbonate are present, in association with calcium silicate. In *Pilea, Ficus,* and other taxa, the peg (stalk), on which the cystolith forms, grows down from the outer wall (before it becomes thickened) of the lithocysts pushing the plasma membrane down into the vacuole.

1.2.5.3   Boron

Boron has been shown to be present in cell walls of a number of plants.[151] Although boron was postulated to form ester systems (didiol-borate crosslink) in the primary cell wall,[152] it is now known that most cell wall borons are associated with pectin and together they affect cell wall expansion. Plants, whose cell walls are poorer in pectins, like grasses, require less boron than plants with greater cell wall pectins.

Other minerals known to occur in cell walls are magnesium (usually as a substitute for calcium in calcium pectate), sodium (as sodium alginate), and iodine (in seaweeds). The presence of aluminium (as aluminium silicate) is not proved beyond doubt in cell walls.

References

1. Frey-Wyssling, A. and Mühlethaler, K., *Ultrastructural Plant Cytology,* Elsevier, Amsterdam,1965, Chap. J.
2. Hooke, R., *Micrographia,* London, 1667.
3. Brett, C. and Waldron, K., *Physiology and Biochemistry of Plant Cell Walls,* Unwin Hyman, London, 1990, Chap. 1.
4. Roberts, K., Structures at the plant cell surface, *Current Opinion in Cell Biology,* 2, 920, 1990.
5. Preston, R. D., Plant cell walls, in *Dynamic Aspects of Plant Ultrastructure,* Robards, A.W., Ed., McGraw-Hill, New York, 1974, 256.
6. Kerr, T. and Bailey, I. W., The cambium and its derivative tissues. X. Structure, optical properties and chemical composition of the so-called middle lamella, *J. Arnold Arbor,* 15, 327, 1934.
7. Moor, H., Ph.D. Thesis, ETH, Zurich (cited from Frey-Wyssling and Mühlethaler, 1965), 1959.
8. Brett, C. and Waldron, K., *Physiology and Biochemistry of Plant Cell Walls,* Unwin Hyman, London, 1990, Chap. 2.
9. Aspinall, G. O., Chemistry of cell wall polysaccharides, in *The Biochemistry of Plants,* Vol. 3, Preiss, J., Ed., Academic Press, London, 1980, 473.

10. Buchala, A. J. and Meier, H., Biosynthesis of ß-glucans in growing cotton (*Gossypium arboreum* L. and *Gossypium hirsutum* L.) fibres, in *Biochemistry of Plant Cell Walls*, Brett, C.T. and Hillman, J.R., Eds., Cambridge University Press, Cambridge, 1985, 221.
11. Delmer, D. P. and Stone, B. A., Biosynthesis of plant cell walls, in *Biochemistry of Plants*, Vol. 14, Priess, J., Ed., Academic Press, New York, 1988, 373.
12. Currier, H. B. and Strugger, S., Aniline blue and fluorescence microscopy of callose in bulb scales of *Allium cepa* L., *Protoplasma*, 45, 552, 1956.
13. Eschrich, W., Kallose, *Protoplasma*, 47, 487, 1956.
14. Smith, M. M. and McCully, M. E., Mild temperature "stress" and callose synthesis, *Planta*, 136, 65, 1977.
15. Stone, B. A., Non-cellulosic ß-glucans in cell walls, in *Structure, Function and Biosynthesis of Plant Cell Walls*, Dugger, W. M. and Bartniki-Garcia, S., Eds., American Society of Plant Physiologists, Rockville, MD, 1984, 52.
16. Blaschek, W., Käsbauer, J., and Franz, G., Analysis of cell-wall polysaccharides from *Pythium aphanidermatum*, in *V Cell Wall Meeting*, Fry, S. C., Brett, C. T., and Reid, J. S. G., Eds., Edinburgh, 1989, 28.
17. Percival, E., The polysaccharides of green, red, and brown seaweeds: their basic structure, biosynthesis and function, *Br. Phycol. J.*, 14, 103, 1979.
18. Waldmann, T., Jeblick, W., and Kauss, H., Induced net Ca+ uptake and callose biosynthesis in suspension-cultured plant cells, *Planta*, 173, 1988.
19. Herth, W. and Hausser, I., Chitin and cellulose fibrillogenesis *in vivo* and experimental alteration, in *Structure, Function and Biosynthesis of Plant Cell Walls*, Dugger, W. M. and Bartnicki-Garcia, S., Eds., American Society of Plant Physiologists, Rockville, MD, 1984, 89.
20. Carlström, D., The crystal structure of ß-chitin, *J. Biophys. Biochem. Cytol.*, 3, 669, 1957.
21. Aronson, J. M., The cell wall, in *The Fungi*, Vol. I, *The Fungal Cell*, Ainsworth, G. C. and Sussman, A. S., Eds., Academic Press, New York, 1973, 49.
22. Selvendran, R. R., Stevens, B. J. H., and O'Neill, M. A., Developments in the isolation and analysis of cell walls from edible plants, in *Biochemistry of Plant Cell Walls*, Brett, C. T. and Hillman, J. R., Eds., Cambridge University Press, Cambridge, 1985, 39.
23. Wilkie, K. C. B., New perspectives on non-cellulosic cell-wall polysaccharides (hemicelluloses and pectic substances) of land plants, in *Biochemistry of Plant Cell Walls*, Brett, C. T. and Hillmann, J. R., Eds., Cambridge University Press, Cambridge, 1985, 1.
24. McCann, M. C. and Roberts, K., Plant cell walls: murals and mosaics, *Agro-Food-Industry Hi-Tech*, 43, 1994.
25. Jarvis, M. C., Structure and properties of pectin gels in plant cell walls, *Plant, Cell and Environ.*, 7, 153, 1984.
26. Baron-Epel, O., Gharyal, P. K., and Schindler, M., Pectins as mediators of wall porosity in soybean cells, *Planta*, 175, 389, 1988.
27. Eberhard, S., Doubrava, N., Marfa, V., Mohnen, D., Southwiek, A., Darvill, A., and Albersheim, P., Pectic cell wall fragments regulate tobacco thin-cell-layer explant morphogenesis, *Plant Cell*, 1, 747, 1989.
28. Fengel, D., Ultrastructural behavior of cell wall polysaccharides, *Tappi*, 53, 497, 1970.
29. Hoffmann, P. and Parameswaran, N., On the ultrastructural localization of hemicellulose within delignified tracheids of spruce, *Holzforschung*, 30, 62, 1976.

30. Parameswaran, P. and Liese, W., Ultrastructural localization of cell wall components in wood cells, *Holz Roh-Werkstoff*, 40, 139, 1982.

31. Stone, J. E. and Scallen, A. M., A structural model for the cell wall of water swollen wood pulp fibers based on their accessibility to macromolecules, *Cell. Chem. Technol.*, 2, 343, 1968.

32. McCandless, E. L., Polysaccharides of the seaweeds, in *The Biology of the Seaweeds*, Lobban, C. S. and Wynne, M. J., Eds., Blackwell Scientific Publications, Oxford, 1981, 559.

33. Varner, J. E. and Lin, L-S., Plant cell wall architecture, *Cell*, 56, 231, 1989.

34. Fry, S. C., The structure and functions of xyloglucan, *J. Exp. Bot.*, 40, 1, 1989.

35. Hayashi, T., Xyloglucans in the primary cell walls, *Annu. Rev. Plant Physiol. Plant Mol. Biol.*, 40, 139, 1989.

36. Acebes, J. L., Lorences, E. P., Revilla, G., and Zarra, I., The cell wall loosening in *Pinus pinaster*. Role of Xyloglucan, in *V Cell Wall Meeting*, Fry, S. C., Brett, C. T., and Reid, J. S. G., Eds., Edinburgh, 1989, 13.

37. Hoson, T. and Masuda, Y., Antibodies and lectins specific for xyloglucans inhibit auxin-induced elongation of Azuki bean epicotyls, in *V Cell Wall Meeting*, Fry, S. C., Brett, C. T., and Reid, J. S. G., Eds., Edinburgh, 1989, 132.

38. Lynch, M. A., Synthesis, secretion and cell localization of secreted complex polysaccharides in the developing root. Ph.D Thesis, University of Colorado at Boulder, Boulder, CO, 1992.

39. Balakrishnan, S., A structural and histochemical study on *Padina boergeseenii* Allender and Kraft. Ph.D. Thesis, Bharathidasan University, Tiruchirappalli, India, 1990.

40. Vreeland, V. and Laetsch, W. M., Identification of associating carbohydrate sequences with labeled oligosaccharides: Localization of alginate gelling subunits in cell walls of a brown alga, *Planta*, 177, 423, 1989.

41. Veluraja, K. and Atkins, E. D. T., Electron microscopic study of guluronate-rich alginate, *Carbohydrate Research*, 187, 313, 1989.

42. Rees, D. A., Structure, conformation and mechanism in the formation of polysaccharide gels and networks, *Adv. Carbohyd. Chem. Biochem.*, 24, 267, 1969.

43. Knox, J. P., Emerging patterns of organization at the plant cell surface, *J. Cell. Sci.*, 96, 557, 1990.

44. Cassab, G. I. and Varner, E., Cell wall proteins, *Annu. Rev. Plant Physiol.*, 39, 321, 1988.

45. Showalter, A. M. and Varner, J. E., Plant hydroxyproline-rich glycoproteins, in *The Biochemistry of Plants*, Stumpf, P. K. and Conn, E. E., Eds., Academic Press, San Diego, CA, 1989, 485.

46. Northcote, D. H., Aspects of vascular tissue differentiation in plants: Parameters that may be used to monotor the process, *Int. J. Plant Sci.*, 156, 245, 1995.

47. Sharon, N. and Lis, H., A century of lectin research (1888–1988), *Trends Biochem. Sci.*, 12, 488, 1987.

48. Vasta, G. R. and Pont-Lezica, R., Plant and animal lectins, in *Organization and Assembly of Plant and Animal Extracellular Matrix*, Adair, W. S. and Mecham, R. P., Eds., Academic Press, San Diego, CA, 1990, 173.

49. Chrispeels, M. J. and Raikhel, N. V., Lectins, lectin genes and their role in plant defense, *Plant Cell*, 3, 1, 1991.

50. Fincher, G. B., Stone, B. A., and Clarke, A. E., Arabinogalactan proteins: structure, biosynthesis and function, *Annu. Rev. Plant Physiol.*, 34, 47, 1983.

51. Gleeson, P. A., McNamara, M., Wettenhall, E. H., Stone, B. A., and Fincher, G. B., Characterization of the hydroxylproline-rich protein core of an arabinogalactan-protein secreted from suspension-cultured *Lolium multiflorum* endosperm, *Biochem. J.*, 264, 857, 1989.

52. Pennell, R. I., Knox, J. P., Scofield, G. N., Selvendran, R. R., and Roberts, K., A family of abundant plasma membrane associated glycoproteins related to the arabinogalactan proteins is unique to flowering plants, *J. Cell Biol.*, 108, 1966, 1989.

53. Baldwin, T., McCann, M. C., and Roberts, K., A novel hydroxyproline-deficient arabinogalactan protein secreted by suspension-cultured cells of *Daucus carota*, *Plant Physiol.*, 103, 115, 1993.

54. Basile, D. V. and Basile, M. R., Changes in cell wall-associated arabinogalactan proteins correlated with experimentally induced altered patterns of cell division and organogenesis, in *V Cell Wall Meeting*, Fry, S. C., Brett, C. T., and Reid, J. S. G., Eds., Edinburgh, 1989, 17.

55. Gell, A. C., Bacic, A., and Clarke, A. E., Arabinogalactan-rich proteins of the female sexual tissue of *Nicotiana alata*, *Plant Physiol.*, 82, 885, 1986.

56. Basile, D. V. and Basile, M. R., The role and control of the place-dependent suppression of cell division in plant morphogenesis and phylogeny, *Memoirs Torrey Bot. Club*, 25, 63, 1993.

57. Knox, J. P., Day, S., and Roberts, K., A set of cell surface glycoproteins forms an early marker of cell position, but not cell type, in the root apical meristem of *Daucus carota* L., *Development*, 106, 47, 1989.

58. Pennell, R. I. and Roberts, K., Sexual development in the pea is presaged by altered expression of arabinogalactan protein, *Nature*, 344, 547, 1990.

59. Kreuger, M. and van Holst, G.-J., Arabinogalactan proteins are essential in somatic embryogenesis of *Daucus carota* L., *Planta*, 189, 243, 1993.

60. Stacey, N. J., Roberts, K., and Knox, J. P., Patterns of expression of the JIM4 arabinogalactan-protein epitope in cell cultures and during somatic embryogenesis in *Daucus carota*, *Planta*, 180, 285, 1990.

61. Adair, W. S. and Snell, W. J., The *Chlamydomonas reinhardtii* cell wall: structure, biochemistry and molecular biology, in *Organization and Assembly of Plant and Animal Extracellular Matrix*, Adair, W. S. and Mecham, R. P., Eds., Academic Press, San Diego, CA, 1990, 16.

62. Roberts, K., Gref, C., Hills, G. J., and Shaw, P. J., Cell wall glycoproteins: structure and functions, *J. Cell Sci.*, Suppl. 2, 105, 1985.

63. Lamport, D. T. A. and Alizadeh, H., Current status of *in vitro* extensin cross-linking, in *V Cell Wall Meeting*, Fry, F. C., Brett, C. T., and Reid, J. S. G., Eds., Edinburgh, 1989, 15.

64. Stafstrom, J. P. and Staehelin, L. A., Cross-linking patterns in salt-extractable extension from carrot cell walls, *Plant Physiol.*, 81, 234, 1986.

65. Stiefel, V., Perez-Grau, L. L., Albericio, F., Giralt, E., Ruiz-Avila, L., Ludevid, M. D., and Puigdomenech, P., Molecular cloning of cDNAs encoding a putative cell wall protein from *Zea mays* and immunological identification of related polypeptides, *Plant Mol. Biol.*, 11, 483, 1988.

66. Stafstrom, J. P. and Staehelin, L. A., A second extension-like hydroxyproline-rich glycoprotein from carrot cell walls, *Plant Physiol.*, 84, 820, 1987.
67. Ye, Z. -H., Song, Y. -R., Marcus, A., and Varner, J. E., Comparative localization of three classes of cell wall proteins, *Plant.*, 1, 175, 1991.
68. Ye, Z. -H. and Varner, J. E., Tissue specific expression of cell wall proteins in developing soybean tissues, *Plant Cell*, 3, 23, 1991.
69. Stafstrom, J. P. and Staehelin, L. A., Antibody localization of extensin in cell walls of carrot storage root, *Planta*, 174, 321, 1988
70. Swords, K. M. M. and Staehelin, L. A., Analysis of extensin structure in plant cell walls, in *Modern Methods of Plant Analysis. N.S.* Vol. 10. *Plant Fibers*, Linskens, H. F. and Jackson, J. F., Eds. Springer Verlag, Berlin, 1989, 219.
71. Staehelin, L. A., Giddings, T. H., Levy, S., Lynch, M. A., Moore, P. J., and Swords, K. M. M., Organization of the secretory pathway of cell wall glycoproteins and complex polysaccharides in plant cells, in *Endocytosis, Exocytosis and Vesicle Traffic in Plants*, Hawes, C. R., Coleman, J. O. D., and Evans, D. E., Eds., Cambridge University Press, Cambridge, 1991, 183.
72. Bacic, A., Harris, R. J., and Stone, B. A., Structure and function of plant cell walls, in *The Biochemistry of Plants*, Vol.14, Stumpf. P. K. and Conn, E. E., Eds., Academic Press, New York, 1988, 297.
73. Benhamou, N., Mazau, D., Grenier, J., and Esquerré- Tugayé, M., Time-course study of the accumulation of hydroxyproline-rich glycoproteins in root cells of susceptible and resistant tomato plants infected by *Fusarium oxysporum* f. sp. *radicis - lycopersici, Planta*, 184, 196, 1991.
74. McNeil, M., Darvill, A. G., Fry, S. C., and Albersheim, P., Structure and function of the primary cell walls of plants, *Annu. Rev. Biochem.*, 53, 652, 1984.
75. Showalter, A. M. and Rumeau, D., Molecular biology of plant cell wall hydroxyproline-rich glycoproteins, in *Organization and Assembly of Plant and Animal Extracellular Matrix*, Adair, W. S. and Mecham, R. P., Eds., Academic Press, San Diego, CA, 1990, 247.
76. Hong, J. C., Nagao, R. T., and Key, J. L., Developmentally regulated expression of soybean proline-rich wall protein genes, *Plant Cell*, 1, 937, 1989.
77. Chen, C. -G., Cornish, E. C., and Clarke, A. E., Specific expression of an extension-like gene in the style of *Nicotiana alata, Plant Cell*, 4, 1053, 1992.
78. Condit, C. M. and Keller, B., The glycine-rich cell wall proteins of higher plants, in *Organisation and Assembly of Plant and Animal Extracellular Matrix*, Adair, W. S. and Mecham, R. P., Eds., Academic Press, San Diego, CA, 1990, 119.
79. Kieliszewski, M. J., Leykam, J. F., and Lamport, D. T. A., Structure of the threonine-rich extension from *Zea mays, Plant Physiol.*, 92, 316, 1990.
80. Bohlmann, H., Clausen, S., Behnke, S., Giese, H., Hiller, C., Reimann-Philipp,U., Schrader, G., Barkholt, V., and Apel, K., Leaf specific thionins of barley—a novel class of cell wall proteins toxic to plant pathogenic fungi and possibly involved in the defence mechanism of plants, *Embo J.*, 7, 1559, 1988.
81. Apel, K., Reimann-Philipp, U., Bohlmann, H., Behnke, S., and Schrader, G., Leaf-specific thionins of barley—a novel class of cell wall proteins toxic to plant-pathogenic fungi: Their possible function and subcellular localization in barley leaves, *Current Topics in Plant Biochem. Physiol.*, 7, 140, 1988.

82. Goldman, M. H., Pezzotti, M., Seurinck, J., and Mariani, C., Developmental expression of tobacco pistil-specific genes encoding novel extension-like proteins, *Plant Cell*, 4, 1041, 1992.
83. Kandasamy, M. K., Paolillo, D. J., Faraday, C. D., Nasrallah, J. B., and Nasarallah, M. E., The S-locus specific glycoproteins of *Brassica* accumulate in the cell wall of developing stigmatic papillae, *Dev. Biol.*, 134, 462, 1990.
84. McClure, B. A., Haring, V., Ebert, P. R., Anderson, M. A., Simpson, R. J., Sakiyama, F., and Clarke, A. E., Style self-incompatibility gene products of *Nicotiana alata* are ribonucleases, *Nature*, 342, 955, 1989.
85. Cosgrove, D. J., Relaxation in a high-stress environment: The molecular basis of extensible cell walls and cell enlargement, *Plant Cell*, 9, 1031, 1997.
86. McQueen-Mason, S., Durachko, D. M., and Cosgrove, D. J., Two endogenous proteins that induce cell wall extension in plants, *Plant Cell*, 4, 1425, 1992.
87. Fry, S. C., Phenolic components of the primary cell wall and their possible role in the hormonal regulation of growth, *Planta*, 146, 343, 1979.
88. Fry, S. C., Gibberellin—controlled pectinic acid and pectin secretion in growing cells, *Phytochemistry*, 19, 735, 1980.
89. Pang, A., Catesson, A. -M., Goldberg, R., Francesch., C., and Rolando, C., Peroxidases and lignification. A re-examination of isoform specificity, in *V Cell Wall Meeting*, Fry, F. C., Brett, C. T., and Reid, J. S. G., Eds., Edinburgh, 1989, 177.
90. Carraro, L., Lombardo, G., and Gerola, F. M., Stylar peroxidase and incompatibility reactions in *Petuma hybrida*, *J. Cell Sci.*, 82, 1, 1986.
91. Hahlbrock, K. and Griesebach, H., Enzymatic controls in the biosynthesis of lignin and flavonoids, *Annu. Rev. Plant Physiol.*, 30, 105, 1979.
92. Olson, P. D. and Varner, J. E., Personal communication, 1993.
93. Kaur-Sawhney, R., Flores, M. E., and Galston, A. W., Polymine oxidase in oat leaves: A cell wall localized enzyme, *Plant Physiol.*, 68, 494, 1981.
94. Smith, T. A., Polyamine oxidase in higher plants, *Biochem. Biophys. Res. Commun.*, 41, 1452, 1970.
95. Angelini, R., Bragaloni, M., Porta-Puglia, A., and Federico, R., Polyamine metabolism, peroxidase acttivity and cell wall modifications in the *Ascochyta rabiei*–chick pea interaction, in *VI. Cell Wall Meeting*, Sassen, M. M. A., Derksen, J. W. M., Emons, A. M. C., and Wolters, A. M. C., Eds., Neijmegen, 1992, 73.
96. Crassnier, M., Noat, G., and Ricard, J., Purification and molecular properties of acid phosphatase from sycamore cell walls, *Plant Cell Environ.*, 3, 217, 1980.
97. Lamport, D. T. A. and Catt, J. W., Glycoproteins and enzymes of the cell wall, in *Encyclopaedia of Plant Physiology* (NS), *Plant Carbohydrates* II, Vol. 13 B, Tanner, W. and Holmes, F. A., Eds., Springer-Verlag, New York, 1981, 133.
98. Gahan, P. B., *Plant Histochemistry and Cytochemistry*, Academic Press, Orlando, FL, 1984, Chap. 5.
99. Gahan, P. B., Dawson, A. L., and Fielding, J., Paranitrophenyl phosphate as a substrate for some acid phosphatases in roots of *Vicia faba*, *Ann. Bot.*, 42, 1413, 1978.
100. Johnson, C. B., Holloway, B. R., Smith, H., and Grierson, D., Isoenzymes of acid phosphatases in germinating peas, *Planta*, 115, 1, 1973.
101. Sheikh, K. L. M. and Gahan, P. B., Soluble proteins and hydrolases during crown gall induction in the tomato, *Lycopersicon esculentum*, *Histochem. J.*, 8, 87, 1967.

102. Gahan, P. B. and McLean, J., Acid phosphatases in root tips of *Vicia faba*, *Biochem. J.*, 102, 47, 1967.
103. Boller, T., Hydrolytic enzymes in plant disease resistance, in *Plant-Microbe Interactions*, Kosuga, T. and Nester, E. W., Eds., Macmillan, New York, 1987, 385.
104. Roldàn, J. M., Verbelen, J. P., Butler, W. L., and Tokuyasu, K., Intracellular localization of nitrate reductase in *Neurospora crassa*, *Plant Physiol.*, 70, 872, 1982.
105. Gross, G. G., Cell wall-bound malate dehydrogenase from horse radish, *Phytochemistry*, 16, 319, 1977.
106. Robinson, D. G., Andreae, M., and Blankestein, P., Plant prolylhydroxylase, in *Organisation and Assembly of Plant and Animal Extracellular Matrix*, Adair, W. S. and Mecham, R. P., Eds., Academic Press, San Diego, CA, 1990, 283.
107. Andreae, M., Blankestein, P., Zhang, Y. -H., and Robinson, D. G., Towards the subcellular localization of plant prolyl hydroxylase, *Eur. J. Cell Biol.*, 47, 181, 1988.
108. Bolwell, G. P., Robbins, M. P., and Dixon, R. A., Metabolic changes in elicitor-treated bean cells. Enzymatic responses associated with rapid changes in cell wall components, *Eur. J. Biochem.*, 148, 571, 1985.
109. Cline, K. and Albersheim, P., Host-pathogen interactions XVII. Hydrolysis of biologically active fungal glucans by enzymes isolated from soybean cells, *Plant Physiol.*, 68, 207&221, 1981.
110. Nagahashi, G. and Seibles, T., Purification of plant cell walls: isoelectric focussing of $CaCl_2$ extracted enzymes, *Protoplasma*, 134, 102, 1986.
111. Schmidt, G. and Grisebach, H., Immunofluorecent labeling of enzymes, in *Modern Methods of Plant Analysis. N.S. Vol. 4. Immunology in Plant Sciences*, Linskens, H. F. and Jackson, J. F., Eds., Springer Verlag, Berlin, 1986, 156.
112. Ham, K. -S, Kaufmann, S., Albersheim, P., and Darvill, A. G., Host-pathogen interactions. XXXIX. A soybean pathogenesis-related protein with ß-1,3-Glucanase activity releases phytoalexin elicator-active heat-stable fragments from fungal walls, *Mol. Plant -Microbe Interaction*, 4, 545, 1991.
113. Keen, N. T. and Yoshikawa, M., ß-1,3-Endoglucanases from soybean releases elicitor-active carbohydrates from fungus cell walls, *Plant Physiol.*, 71, 460, 1983.
114. Mauch, F. and Staehelin, L. A., Functional implications of the subcellular localization of ethylene-induced chitinase and ß-1,3-glucanase in bean leaves, *Plant Cell*, 1, 447, 1989.
115. Benhamou, N., Cytochemical localization of ß-(1→4) -D glucans in plant and fungal cells using an exoglucanase-gold complex, *Electron Microsc. Rev.*, 2, 123, 1989.
116. Boller, T. and Métraux, J. P., Extracellular localization of chitinase in cucumber, *Physiol. Mol. Plant Pathol.*, 33,11, 1988.
117. Benhamou, N., Joosten, M. H. A. J., and De Wit, P. J. G. M., Subcellular localization of chitinase and of its potential substrate in tomato root tissues infected by *Fusarium oxysporium* f. sp. *Radicis-lycopersici*, *Plant Physiol.*, 92, 1108, 1990.
118. Goldberg, R., Durand, L., and Pierron, M., Properties of Pectin-Methyl-Esterases (PME) from mung bean hypocotyls, in *V Cell Wall Meeting*, Fry, S. C., Brett, C. T., and Reid, J. S. G., Eds., Edinburgh, 1989, 11.
119. Fannuti, C., Gidley, M. J., and Reid, J. S. G., Action of a pure xyloglucan endotransglycosylase (formerly called xyloglucan specific endo-(1,4) ß-D-glucanase) from the cotyledons of germinated nasturtium seeds, *Plant. J.*, 3, 691, 1993.

120. Fry, S. C., Smith, R. C., Renwick, K. F., Martin, D. J., Hodge, S. K., and Mathews, K. J., Xyloglucan endotransglycosylase, a new wall-loosening enzyme activity from plants, *Biochemical. J.*, 282, 821, 1991.

121. Nishitani, K. and Tominaga, R., Endoxyloglucan transferase, a novel class of glycosyl transferase that catalyses transfer of a segment of xyloglucan molecule to another xyloglucan molecule, *J. Biol. Chem.*, 267, 21058, 1992.

122. Deising, H., Nicholson, R. L., Haug, M., Hioward, R. J., and Mendgen, K., Adhesion pad formation and the involvement of cutinase and esterases in the attachment of uredospores to the host cuticle, *Plant Cell*, 4, 1101, 1992.

123. Holloway, P. J., Structure and histochemistry of plant cuticular membranes: an overview, in *The Plant Cuticle*, Cutler, D. J., Alvin, K. L., and Price, C. E., Eds., Academic Press, London, 1982, 1.

124. Holloway, P. J., The chemical constitution of plant cutins, in *The Plant Cuticle*, Cutler, D. J., Alvin, K. L., and Price, C. E., Eds., Academic Press, London, 1982, 45.

125. Holloway, P. J. and Watterndorf, J., Cutinized and suberized cell walls, in *Handbook of Plant Cytochemistry*, Vol. II, Vaughn, K. C., Ed., CRC Press, Boca Raton, FL, 1987, 1.

126. Pearce, R. B. and Rutherford, J., A wound-associated suberised barrier to the spread of decay in the sap wood of oak (*Quercus robur* L.), *Physiol. Plant Path.*, 19, 359, 1981.

127. Heslop-Harrison, J., The Cronian Lecture (1974). The physiology of the pollen grain surface, *Proc. R. Sec. London, Ser. B.*, 190, 275, 1975.

128. Wiermann, R. and Vieth, K., Outer pollen wall, an important accumulation site for flavonoids, *Protoplasma*, 118, 230, 1983.

129. Northcote, D. H., Control of cell wall formation during growth, in *Biochemistry of Plant Cell Walls*, Brett, C. T. and Hillman, J. R., Eds., Cambridge University Press, Cambridge, 1985, 177.

130. Wallace, G., Chesson, A., Lomax, J. A., and Jarvis, M. C., Isolation of two different lignin moieties from graminaceous cell walls, in *V Cell Wall Meeting*, Fry, S. C., Brett, C. T., and Reid, J. S. G., Eds., Edinburgh, 1989, 18.

131. Baayen, R. P., Responses related to lignification and intravascular periderm formation in carnations resistant to Fusarium wilt, *Can. J. Bot.*, 66, 784, 1988.

132. Glazener, J. A., Accumulation of phenolic compounds in cells and formation of lignin-like polymers in cell walls of young tomato fruits after inoculation with *Botrydis cinerea*, *Physiol. Plant Pathol.*, 20, 11, 1982.

133. Ride, J. P., Lignification in wounded wheat leaves in response to fungi and its possible role in resistance, *Physiol. Plant Pathol.*, 5, 125, 1975.

134. Harris, P. J. and Hartley, R. D., Phenolic constituents of the cell walls of monocotyledons, *Biochem. Syst. Ecol.*, 8, 153, 1980.

135. Brammal, R. A. and Higgins, V. J., A histological comparison of fungal colonization in tomato seedlings susceptible or resistant to *Fusarium* crown and root rot disease, *Can. J. Bot.*, 66, 915, 1988.

136. Riley, R. G. and Kolattukudy, P. E., Evidence for covalently attached *p*-coumaric acid and ferulic acid in cutins and suberins, *Plant Physiol.*, 56, 650, 1975.

137. Nishitani, K. and Nevins, D. J., Structural analysis of feruloylated glucuronoarabinoxylans, in *V Cell Wall Meeting*, Fry, F. C., Brett, C. T., and Reid, J. S. G., Eds., Edinburgh, 1989, 44.

138. Fry, S. C., Cross-linking of matrix polymers in the growing cell walls of angiosperms, *Annu. Rev. Plant Physiol.*, 37, 165, 1986.

139. Fry, S. C., Feruloylated pectins from the primary cell wall: their structure and possible functions, *Planta*, 157, 111, 1983.

140. Kamisaka, S., Takeda, S., Takahashi, K., and Shibata, K., The content of diferulic acid in cell walls is correlated with mechanical properties of cell walls in oat coleoptiles, in *V Cell Wall Meeting*, Fry, S. C., Brett, C. T., and Reid, J. S. G., Eds., Edinburgh, 1989, 134.

141. Lam, T. B. T., Iiyama, K., and Stone, B. A., Phenolic acid bridges between polysaccharides and lignin in wheat internodes, *Phytochemistry*, 29, 429, 1990.

142. Iiyama, K., Lam, T. B. T., and Stone, B. A., Phenolic acid bridges between polysaccharides and lignin in wheat internodes, *Phytochemistry*, 29, 733, 1990.

143. Niemann, G. J. and Baayen, R. P., Wall-bound phytoalexins. An indication of integration of benzoic and cinnamic acid amides into wall-bound components?, in *V Cell Wall Meeting*, Fry, F. C., Brett, C. T., and Reid, J. S. G., Eds., Edinburgh, 1989, 43.

144. Jones, L. H. P. and Handreck, K. A., Silica in soils, plants and animals, *Adv. Agron.*, 19, 107, 1967.

145. Dayanandan, P., Localisation of silica and calcium carbonate in plants, *Scanning Electron Microscopy 1983/III*, SEM Inc., AMF O'Hare, Chicago, IL, 1519, 1983.

146. Metcalfe, C.R. and Chalk, L., *Anatomy of Dicotyledons*, Vol. II. 2nd ed., Clarendon Press, Oxford, 1983.

147. Ajello, L., Cytology and interrelationships of cystolith formation in *Ficus elastica*, *Am. J. Bot.*, 28, 589, 1941.

148. Eschrich, W., Ein Beitrag Zur Kenntnis der Kallose, *Planta*, 44, 532, 1954.

149. Pireyre, N., Contribution a'l'etude morphologique des cystolithes, *Rev. Cytol. Biol. Veg.*, 23, 93, 1961.

150. Watt, W. M., Morell, C. K., Smith, D. L., and Steer, M. W., Cystolith development and structure in *Pilea cadierei* (Urticaceae), *Ann. Bot.*, 60, 71, 1987.

151. Hu, H., Brown, P. H., and Labavitch, J. M., Species variability in boron requirement is correlated with cell wall pectin, *J. Exptl. Bot.*, 47, 227, 1996.

152. Loomis, W. D., Durst, R. W., Miles, P. W., Kaiser, H. R., and Hirschberg, B., Boron, germanium and the plant cell wall, in *V Cell Wall Meeting*, Fry, F. C., Brett, C. T., and Reid, J. S. G., Eds., Edinburgh, 1989, 46.

Chapter **2**

Light Microscopic Cytochemistry

Contents

2.1   Introduction

Information about the structure of objects comes to us mainly through our sense of sight. What we see is the result of interaction of light with the matter of which the object is made. Since our unaided eye has limitation of perception, microscopes have been developed and perfected to derive more information about objects of interest to us. The majority of cytochemical methods known to us have been developed to suit the use of the light microscope, as it is the oldest of optical gadgets to be designed for visualization of cellular structures. Greater perfection in the design as well as performance of the light microscope and its lens systems and light sources has also contributed substantially to the effective development of many cytochemical methods. Although several other types of microscopes have been developed within the past 100 years, the light microscope is the best suited to visualize coloring reactions of specific dyes with particular chemical components of the cell. Barring the fluorescence microscope, the light microscope is ideally adopted to observe all the visible colors of the electromagnetic spectrum. It is also important to realize that many techniques used in fluorescence and ultrastructural cytochemistry have been (or will be) more or less directly adopted from light microscopic cytochemistry. The light microscope is also the cheapest of all microscopes and is within the reach of any teaching/research institution.

   The light microscope, however, is also beset with difficulties. It cannot be used to perceive colors beyond the visible region of the spectrum. It also has the limitation of lower resolving power (RP). Beyond an RP of 0.2 μm, any more magnification does not make further details visible and only "empty" images can be produced. But it should be borne in mind that benefits outweigh limitations, if we exploit the light microscope in a judicious manner. Light microscopy continues to play a major role in cytochemistry and it will do so in the years to come.

2.2   Tissue Preparation

Tissue preparation is a critical step in light microscopic cytochemistry because it is important that the method chosen should preserve both the chemical and cellular structure. The contradictions reported in literature regarding results obtained after identical cytochemical procedures in the same material are often due to differences in tissue preparation. In other words, the same cytochemical method used on the same tissue may yield very different results, depending solely on how the tissue has been processed.

As Jensen[1] had rightly pointed out, methods in cyto- and histochemistry have many things in common with those in cytology and histology. Techniques of tissue preparation up to staining level are almost the same in both. Therefore, a thorough knowledge of cytological and histological methods is very important for a cytochemist.

2.2.1 Use of Fresh Living Tissue

Cytochemical tests ideally should be carried out upon intact living tissue as, theo-retically, use of living tissue avoids all artifacts (including chemical changes) imposed by tissue processing procedures. For example, it has been shown that, during fixation, covalent bonds are formed with enzymes and this will affect (i.e., reduce) enzyme activity. Thus, it is often found that the better the structural preservation, the lower the enzyme activity and vice versa. This is also true of many other cytochemicals. Living materials are particularly useful in the study of lipids, some cell wall materials, and many enzymes. In living tissue, not only the site of pres-ence/activity of cellular chemicals but the activity itself is maintained without any change, while tissue processing procedures are likely to denature chemicals and affect their *in situ* locus. This is particularly true of respiratory enzymes, such as succinic dehydrogenase. Employing living tissue for cytochemical procedures also avoids, to a large extent,. delay in obtaining the results.

Living materials, however, do pose problems in cytochemistry. The major prob-lem is the difficulty in achieving penetration of the reaction compounds employed during different cytochemical procedures. Unless this is achieved, localization of cellular material is not possible. Another difficulty with living tissues is that there is a reduced retention of water soluble compounds, thereby affecting their localiza-tion. There is also the danger of diffusion of substrates and other reactants into the living cell as well as diffusion of end products of the reaction from their original site. Permanent preparations of living tissues after cytochemical localization are also often difficult to make without affecting either the tissue or the substance localized.

If sections of living tissues are to be used for cytochemical localization studies, additional problems crop up. Hand sections of these living tissues are mostly sug-gested but sections cannot be obtained at the desired thickness (especially at the lower thickness range); neither would the obtained sections be of uniform thickness. Many fresh tissues are not firm enough to be handled during hand-sectioning. To overcome the difficulties in handling fresh tissue for hand-sectioning, support materials are suggested. Pith, thermocole (packing material), or even carrot or similar vegetable materials can be used as supports while taking hand-sections. Other types of supports are suggested for taking thinner sections of fresh materials in a microtome. A matrix material is invariably used for "embedding" the fresh tissue but without actually infiltrating the former into the latter. One of the earliest matrix materials to be used for sectioning fresh shoot pieces was clay.[2] Jensen[3] used paraffin blocks with holes into which the fresh materials to be sectioned were inserted; subsequently the paraffin adjacent to the holes was carefully melted with a warm needle to hold the tissue in position. He successfully used this method to cut root materials for cytochemical analysis. Carbowax[4] or beeswax can be used as a substitute for paraffin wax. In our laboratory, we have successfully used agar and gum arabic as supports for taking sections of diverse types of fresh tissue for cytochemical study.

2.2.2 Use of Frozen Tissue

Freezing is another method that enables us to obtain sections of living materials at a uniform and convenient thickness. This is also one of the best and important alternatives to conventional chemical fixation. Freezing not only brings about prevention of autolysis of cellular structures but also promotes tissue hardening almost simultaneously. These two are very important requirements for proper processing of the material for cytochemical studies.

Both the cell wall and cytoplasm have a significant percentage of water content in them. When a tissue is frozen, this water forms ice crystals; this may cause distortions in histology and can damage cells. One would, therefore, like to have cells with their water frozen in the vitreous (non-crystalline) state by controlled freezing, so as to minimize distortions and damages. Slow cooling results in extracellular ice crystals and these may seriously affect the cell. Rapid cooling, on the other hand, results in intercellular ice crystals, which are, however, very small and much below the resolution of the light microscope (5 nm or less in size).[5] Rapid cooling has, therefore, been suggested for freezing tissues.

Rapid cooling can be achieved by subjecting small materials (preferably less than 1 to 2 mm thick) to treatment with (1) liquid nitrogen (-170°C), (2) isopentane (-165°C) or propane (-185°C) cooled with liquid nitrogen, or (3) liquid helium II. Liquid nitrogen is not suitable because the tissue's latent heat is sufficient enough to cause a gaseous mask around the tissue, thereby decreasing the rate of heat loss. This results in the formation of large crystals, which may fall within the resolution of a light microscope. Liquid helium II causes a very rapid freezing but it is very costly; the pieces of materials should also be extremely small for effective freezing if helium II is to be used. Isopentane is widely used because it is liquid at room temperature and is not flammable.

In a series of experiments, Chayen and associates[6,7] (see also Gahan et al.[8]) found that freezing at -55° to -65°C produces better results than at lower temperatures; at these temperatures, the frozen tissue is comparable to fresh tissue. There is also the absence of ice crystal formation within or outside the cells. They also found that prior treatment of tissues (especially delicate tissues) in 5% aqueous solution of polyvinyl alcohol (PVA) for 10 to 30 min before freezing can stabilize the tissue and yield excellent results in cytochemical localization studies.

2.2.3 Freeze-Substitution and Freeze-Drying

As already detailed, cytochemical procedures are best carried out in tissues that are not subjected to chemical fixation and dehydration. As we shall see subsequently in detail, chemical fixation and dehydration appreciably alter cell chemistry, thus hampering good cytochemical localization. There are a few alternate methods to chemical fixation and the well known among them are freeze-substitution and freeze-drying.

In both methods, the first step followed is the rapid freezing of the tissue. The next step, which concerns tissue dehydration, differs. In freeze substitution, the ice crystals (i.e., water) formed in the tissue by freezing are removed by substitution of cold ethanol or methanol around -30°C. This process is carried out for a period of about 2 days. Two or three changes in alcohol are to be followed before processing the tissue further. In freeze-drying, ice crystals are removed by evaporation at low temperatures (-30 to -40°C) *in vacuo* with a stream of moving gas for a few days. When all ice is removed, the tissue is said to be dry. The third step in both techniques relates to infiltrating the tissue with paraffin or similar material followed by embedding. In freeze-substitution, the dehydrating agent is replaced with toluene and infiltration and embedding with paraffin is done at room temperature as per conventional procedures;[9] sectioning is done in a rotary microtome. In freeze-drying, paraffin infiltration and embedding are carried out *in vacuo* after raising the tissue temperature. This sets the tissue in a hard matrix so as to be handled for sectioning in a conventional microtome. In both procedures, subsequent steps are similar to routine procedures followed in histology.

In freeze-substitution, although proteins of the tissue are reported to be unaffected by alcohol treatment at low temperature and by toluene during clearing[10] and the tissue to remain relatively "unfixed", the probable effects on other cell substances are not clear. Some enzymes (e.g., phosphatases) are reported to survive but many denature (e.g., many dehydrogenases); substances soluble in alcohol and toluene (lipids, pigments, etc.) are lost.[9] Of the two, freeze-drying is definitely more advantageous than freeze-substitution. The advantages include the production of serial and relatively unfixed sections and better preservation is done through evaporation *in vacuo*. The freeze-dried tissue is more closer to fresh tissue than either freeze-substituted or chemically fixed tissue. The following difficulties are, however, encountered in freeze-drying: (1) Different workers have different criteria for deciding when the tissue is "dry" and so may obtain very different results. (2) Although freeze-drying works well with animal tissue,[5,11] plant specimens pose problems. They take a longer time for drying and are also difficult to be infiltrated with wax.

(3) In freeze-drying, there is still a need to infiltrate the tissue with an embedding matrix which calls for the use of a solvent such as xylene. This may have an effect on some chemicals of the cell. (4) Since there is a prolonged exposure of the tissue to low temperature during the drying period, lipid-protein linkages in the cells may split. In spite of the drawbacks mentioned above, freeze-drying can be advantageously used in cytochemical localization studies.[10]

Detailed procedures for freeze-substitution and freeze-drying along with a description of the apparatuses used are provided subsequently. The reader is also advised to see Chapter 6 of this book where freezing methods as applicable to EM are detailed. Many facts mentioned therein are applicable to light microscopy as well.

2.2.4 Chemical Fixation of Tissues

The primary aim of chemical fixation is to convert the dynamic reality of living cells into an immobile and stable state that is as close as possible to the living condition.[12] This process fixes or stabilizes (i.e., prevents diffusion of) the chemical substances in tissues in such a way as to (or at least expected to) reduce their losses from the cell during subsequent tissue-preparative procedures. Fixation often also helps the hastening of diffusion of staining reagents into the tissues better than in living tissue. In the majority of the earlier works, the emphasis in fixation and fixatives has been principally oriented towards the use of chemically fixed materials as objects of morpho-histological rather than cytochemical study. In other words, workers were content when a cytological structure was preserved, without bothering much to know whether the structure preserved retained its earlier chemical identity. Another major function served by fixation in earlier studies was to render the tissue hard and firm so that thin sections could be cut with a microtome. Moreover, all development in the science of fixation and fixatives, at least until three decades ago, have been centered on paraffin-embedment technique of tissue processing.[12,13] Consequently, we did not have adequate information on cytochemical studies. Only in the recent past two to three decades has chemical fixation for cytochemical localization studies been given enough attention.

Most fixation agents used in the past to "kill" tissues produce chemical changes by altering the three-dimensional configuration of bio-molecules (i.e., unfolding them) (*nonadditive fixatives*) or by attaching themselves to chemical groups of the cell (*additive fixatives*) or by altering them chemically in some other way. These reactions may also prevent the subsequent cytochemical demonstration of the chemical groups. So, for cytochemical purposes, the fixative should be such that it should not attach to or otherwise alter the particular chemical group under study, unless attachment or alteration is a crucial part of the cytochemical test itself. In other words, non-additive fixatives are best suited to cytochemistry. However, there is no ideal fixative that will stabilize every type of chemical substance present in the cell wall with no/minimum chemical change. Naturally, therefore, fixatives should be very carefully chosen for the purpose in hand. The inorganic substances present in the cell wall are normally unaffected by fixation unless the fixative actually dissolves them or unless a precipitating agent has been added deliberately to the fixative.

Substances that have a fixative action when used alone are termed *primary fixatives*. On the basis of their action on proteins, these may be classified into *coagulant* and *non-coagulant* fixatives; the former coagulate proteins into fibrous strands. Examples of the former type include ethanol, methanol, acetone, picric acid, mercuric chloride, and chromium trioxide. Formaldehyde, osmium tetroxide, potassium dichromate, glutaraldehyde, acrolein, and acetic acid* are examples of the latter.[12]

Two or more primary fixatives have often been used in mixtures. Such mixtures have been developed in an attempt to combine the advantages of the different ingredients while overcoming their individual disadvantages. This expectation has not been fulfilled by not even one fixative mixture developed thus far even in routine histology, let alone cytochemistry; the major difficulty is imposed by the differential rate of penetration of the different constituents of the mixture, the fastest would reach the target much earlier and fix it in its own way so that slower compounds are deprived of their action.

In addition to the rate of penetration, a number of other factors are involved directly or indirectly in the actual process of fixation. These include duration of treatment, temperature, pH, concentration of the fixatives used, physical state in which the fixative is used (liquid, solution, gaseous state, etc.), osmolality, volume potential, degree of purity, make and manufacture of the fixative, compatibility in mixtures if not used independently, difficulty of elimination, if any, prior to further processing, etc. These features are important either independently or collectively in fixing a material for a particular cytochemical localization study. It is, therefore, advised that each worker should keep all of these points in mind before trying any fixative on his research material.

It is our experience that primary fixatives are better than fixative mixtures in cytochemical studies on cell walls. It is true that fixative mixtures such as FAA, Carnoy's, Farmer's, Navaschin's, etc. do not greatly affect substances such as cellulose and lignin present in the cell walls, but they do have appreciable effect on acidic polysaccharides, enzymes, and lipoidal and similar materials of cell walls.

The most commonly tried primary fixatives in cell wall cytochemical study are formaldehyde, glutaraldehyde, acrolein, acetic acid, acetone, ethanol, chromic acid, potassium dichromate, mercuric chloride, potassium permanganate, picric acid, etc. The most useful among these are formaldehyde and glutaraldehyde.

Formaldehyde is one of the best known and best investigated fixative both in routine histology and in cytochemistry. It is a monoaldehyde available commercially as 38 to 45% (w/v) aqueous solution (more commonly as a 40% solution) under the trade name *Formalin*. Commercially available formaldehyde solution or formalin contains formic acid as an impurity, formed by the self-oxidation of formaldehyde. Fixatives prepared from such samples are often unfit for critical cytochemical investigations and, therefore, it is better that the fixative is prepared fresh from paraformaldehyde powder. Usually a 4% solution of formaldehyde (= 10% formalin) in water (pH around 4) is used but it is advisable to bring its pH around

---

* However, acetic acid is a coagulant of nucleic acids.

neutral range and to prepare it in cacodylate/phosphate buffer if it is to be exploited for cytochemistry.[14]

Fixation in formaldehyde is recommended for cytochemical demonstration of proteins as it does not coagulate them. It reacts with amino, amide, carboxyl, guanidyl, hydroxyl, and imino groups of proteins. In a short-term fixation, reaction of formaldehyde with tissue proteins is reversible and the product is easily hydrolyzable. Therefore, prolonged post-fixation washing of tissues restores the proteins, especially enzymes, in their nearly original state. More than 70% of the total activity of phosphatase, ß-glucuronidase, esterase, etc. is recovered after washing when the tissue is fixed for 2 h at 4°C.[15] As much as 84% of the activity of acid phosphatase is restored if the tissue is washed properly with the cold solvent buffer.[16] More enzyme activity could be restored with 5-norbornene-2,3-dicarboxylic acid-NaOH (NODCA) buffered with cacodylate or phosphate buffers. Other enzymes such as ATPase and glucose 6-phosphatase are cytochemically demonstrable following fixation in neutral buffered formaldehyde for 2 to 4 h.[14] Brief treatment of tissue in cold formalin prior to assay of maleic dehydrogenase helps to prevent it from leaching out from tissue during further processing.[17]

Formaldehyde does not appear to affect cell wall carbohydrates as it has no reaction with them. Although it reacts with lipids giving products containing methylol and cyclic ether linkages, most lipids can be well preserved if calcium with or without cadmium is also present in the fixative;[18] calcium stabilizes lipids and reduces their loss. There are also drawbacks in using formaldehyde. It is known to combine and often crosslink with many reactive groups,[19] so that not all may subsequently be cytochemically demonstrable. But yet it is one of the best fixatives for cytochemical study of cell walls.

Though formaldehyde is historically the earliest and the most commonly used of aldehyde fixatives, in recent years attention, especially by cytochemists, has centered on Glutaraldehyde, a five-carbon dialdehyde. Glutaraldehyde has a molecular weight of 100.12 with relatively low viscosity. It appears to be a better fixative because of its dialdehyde nature. Although for histology, commercial glutaraldehyde, which is a 25% solution, is a good fixative, it has to be made pure for efficient cytochemical fixation properties. In the commercially sold aqueous state, glutaraldehyde has high unsaturated aldehyde content due to aldol condensation. To make the commercial glutaraldehyde pure, distillation or repeated bleaching with charcoal of high surface area such as Norit E should be followed.[20] To decide whether it is pure or not, its absorption maxima is to be tested: pure glutaraldehyde has an absorption maxima at 280 nm, while impure samples have an absorption maxima at 235 nm.

Glutaraldehyde appears to be best suited to localize many constituents of the cell wall. Even enzymes are best localized when this fixative is used. Cytochemical localization of enzymes such as acid phosphatase, alkaline phosphatase, and esterase has been effected even after 1-h fixation in Glutaraldehyde.[17] Sabatini et al.[14] also used this fixative for localizing acid hydrolases which were well preserved.

A mixture of glutaraldehyde and formaldehyde, in the already prescribed dilution percentage, has been found to be quite rewarding in comparison to either one of them used alone.

Acrolein* is yet another aldehyde (monoaldehyde) commonly used for fixation. Like glutaraldehyde, acrolein has also been exploited only in recent years for cytochemical localization studies. However, materials fixed in acrolein should be embedded with Epon, polyester wax, or methacrylate but not with paraffin wax. Acrolein penetrates more quickly and deeply into tissues and so it fixes structural proteins very well, in a manner similar to formaldehyde; enzymes are reported to undergo destruction due to extreme reactivity of acrolein[14] but enzymes such as acid phosphatases, glucose-6-phosphatase, alkaline phosphatase, and ATPase can be localized if a 10% acrolein is used for a very brief period of 1 min at room temperature.[17] Acrolein is not suitable for demonstrating lipids as it extracts them.[21] Acrolein is commercially available as a viscous, transparent to light-yellow liquid, and should be stored in a dark-colored bottle in the freezing chamber of the refrigerator, protected from light.

Acetic acid is one of the earliest-tried fixatives but has never been used alone. It does not fix carbohydrates because while it not only causes swelling and lyophilization of proteins, but also decreases their isoelectric points. It is not recommended individually for any serious cytochemical study.

Acetone is a non-additive fixative, often used ice cold for freeze-dried or cryosections. It is normally used as a 50 to 70% aqueous solution or in 4.2 pH citrate buffer solution. Its greatest asset is quick penetration rate and, therefore, is well suited for enzyme histochemistry. Its drawback lies in its shrinking action.

Ethanol is another primary fixative and is used as a 50 to 70% aqueous solution. It has a rapid penetration property but is known for its high tissue shrinking ability. It coagulates many proteins by a hardening process and dissolves lipids and phospholipids. Although it preserves alkaline phosphatase very well, it is not suitable for many other enzymes because it inactivates them. Even in cases where enzyme activities are well preserved, the cell morphology is greatly affected. It is also not suitable for preserving tissue lipids. Therefore, individually ethanol is not normally fit for cytochemistry; however, in combination with acetic acid and formaldehyde (but not with chromic acid, dichromates, OsO_4, and permanganates) it forms a good fixative mixture (FAA or formalin-acetic-alcohol). FAA is very good for localizing polysaccharides and general proteins.

Chromic acid (Chromium trioxide) is used as a 1% solution in 70% aqueous ethanol for general histological studies, but it is not good for cytochemistry. It hardens the material profoundly and the tissue becomes highly brittle. It denatures the majority of enzymes. It is a strong oxidizing agent, and carbohydrates are oxidized to aldehydes; hence, it should not be used for carbohydrate cytochemistry. It also dissolves the acidic proteins of the cell.[9] It also often alters the general chemistry of the cell. However, lipids are retained fairly well and it is recommended for lipid cytochemistry. Chromic acid with acetic acid and formaldehyde form a good fixative mixture for general histological and few cytochemical investigations, but these mixtures are unsuited for many other cytochemical localizations.

---

* Acrolein is a powerful tear gas (lacrimatory stimulant), has an unpleasant odor and is extremely irritating when inhaled. It should be used in a well-ventilated hood and eyes must be protected during fixation.

Picric acid* was used as a fixative for general histological studies for a long time, especially by animal histologists. Although it fixes non-enzymic proteins of the cell very well, it is not in common use in cytochemistry. When used, it is prepared as a 0.3% solution in 2% aqueous H_2SO_4 or KNO_3 or HCl. It is also often combined with acetic acid or propionic acid and formaldehyde and used as a fixative mixture.

Potassium dichromate, as a 3.6% aqueous solution, has been occasionally used as a useful lipid fixative but because of its strong oxidizing nature is not a favorite of cytochemists.

Mercuric chloride, which was in use in basic histological studies in the past, is not recommended for cytochemical studies as it reacts strongly with proteins and damages enzyme activity.

Potassium permanganate, a favorite of early electron microscopists, is not preferred by a light microscopic cytochemist as it (1) is a strong oxidizing agent, (2) does not fix carbohydrates, and (3) does not preserve fats that get washed away during dehydration.

Osmium tetroxide is a very good lipid fixative. Fixation of glycoproteins by this fixative depends mainly on the stabilization of the protein part. The major drawback of OsO_4 is its ability to block its own penetration as reaction products acccumulate.

2.2.5 Dehydration and Clearing

Dehydration is the process of removing water from the fixed tissue. It is a necessary prerequisite whenever materials are to be infiltrated with and embedded in a material like paraffin wax for obtaining thin and continuous sections in the form of a ribbon. Since paraffin wax and similar embedding media are immiscible in water but are soluble in organic solvents like xylene, water must be gradually removed from tissue (dehydration) and its place must be occupied by the solvent of paraffin (clearing).

Dehydration must be very gradual and should be done carefully; otherwise it may result in shrinkage and distortion of cells. Dehydration consists of treating the tissues with a series of solutions containing progressively increasing concentrations of the dehydrating agent and decreasing concentrations of water. If benzene hydrocarbons are to be used as clearing agents, dehydration must be absolute; if trichloroethylene or its analogs are to be used, dehydration must be only up to 95% dehydrant treatment.

Two types of dehydrants are in use: (1) Tissues are dehydrated in a non-solvent of paraffin and similar embedding media and then are transferred to a solvent (*clearing agent*), e.g., ethanol, isopropanol, glycerin, methylol, methyl benzoate, methyl cellosolve, clove spirit, oil of bergamot (dehydration up to 74% is enough), terpineol, aniline oil (not to be used if fixation is done in osmium tetroxide), carbondisulphide, guaiacol (all dehydration agents), and xylene, chloroform, benzene, toluene, etc. (clearing agents). (2) The dehydrant is also a solvent of paraffin (clearing agent), e.g., N-butanol, tertiary butanol (TBA), dioxan, etc.

---

* It is highly explosive. It should be kept moist and preferably under a layer of water.

Dehydration and clearing have the inherent shortcomings of altering the chemical nature of the tissue and also extract out several metabolites. Enzymes are often denatured and made unlocalizable. The cell wall components that normally withstand the common dehydrating and clearing agents include cellulose, lignin, structural proteins, certain hemicelluloses, and sporopollenin.

2.2.6 Infiltration and Embedding

Embedding media act as a matrix and support for the tissues against the impact of the knife and to hold their parts in proper relation to each other after the sections have been cut. Infiltration is the process by which the embedding media are gradually but increasingly dissolved in the solvent in which the tissues are placed during the clearing process. The solvent is then eliminated by decantation or evaporation or both. Infiltration, therefore, changes the environment to which the tissue was subjected earlier. Factors that control dehydration damage to the tissues also affect the degree of infiltration damage. Among these factors the most important are polarity and viscosity of the infiltrating medium and the temperature of infiltration.

Several embedding media have been tried thus far, of which paraffin wax is the earliest and the most important. Paraffin was first introduced by Klebs,[22] but was popularized by Butschli.[23] Paraffin was preferred in early cytochemical studies (also in histological studies) due to (1) its constant and known melting point (varies from 50 to 60°C depending on the wax), (2) its appropriate hardness, (3) its smooth and even texture with a minimum of crystalline or grainy structure, (4) the absence of dirt including water and volatile or oily substances, and (5) the fact that paraffin by itself does not affect the chemical nature of many cellular metabolites.

Dehydrated tissues permeated with a solvent of paraffin (like xylene) often undergo greater distortions when brought into contact with paraffin wax. These distortions are not only evident on protoplasts but also on cell walls. Therefore, paraffin infiltration should be done very gradually and the dehydration procedure followed should be compatible with the infiltration protocol.

Several substitutes have been tried/used in place of paraffin wax. Some of these have also been used as a mixture with paraffin. These include rubber, beeswax, ceresin (diethylene glycol disterate), colloidin (=celloidine, cellulose nitrate), gelatin added with DMSO or glycerol, agar, Tissuemat, Bioloid, etc. The same problems encountered with paraffin infiltration crop up in these also, although to a variable extent.

In the last two to three decades resin and plastic embedding have made great strides into light microscopic histochemistry, although Jensen[1] had stated that "plastic embedded material is of little value for routine light microscopy". Although more tedious to make, plastic embedded materials have already proved their superiority in light microscopic cytochemical studies.[12] Plastic embedments are homogeneous and hard; therefore, much thinner sections can be obtained than from wax embedment. The very thin sections provide a greater sharpness of image and clarity of details as well as better contrast when cytochemically stained than thicker sections cut from conventionally embedded tissues. Plastic sections are also free from compression defects, which are very common in paraffin sections thinner than 8 μm. Certain plastic media in their pre-polymerized state are miscible in water. Therefore, they obviate the dehydration process which tends to extract tissue metabolites rendering them unlocalized. They also interfere less with the penetration of aqueous staining reagents into the sections during cytochemical localization. The sections adhere to the glass slide easily after drying at 40 to 80°C without any adhesives (since adhesives cause disturbance to cytochemical staining). Whenever adhesion to glass slides is difficult, the materials are placed over a slide coated with ethylene glycol which is then evaporated at 65°C.

Plastic embedding media have their own limitations as well. The major ones are (1) the materials to be embedded in them should be as small as possible, and (2) continuous ribbons cannot be taken.

Details on plastic media and protocols for using them are discussed in Chapter 4. Almost all the protocols mentioned there are directly applicable to light microscopic cytochemistry.

2.2.7 Sectioning of Materials

Protocols and instrumentation for sectioning of plant materials for cell wall cytochemical studies are the same as for basic histological and anatomical investigations and the readers are advised to consult standard books that deal with these aspects.[12,24,25] The same books can be consulted for dewaxing (or removing other embedding media) the specimen before staining.

2.3   Stains and Staining

2.3.1 Introduction

Light microscopic cytochemistry is essentially dependent on specific color development when the metabolites to be localized come in contact with/react with a particular dye/reagent. Color development can be classified into the following categories:[26]

1. *Direct Coloring*—Metabolite to be localized is visualized by its direct combination, either chemical or physical, with the coloring agent. e.g., lignin with safranin, alginic acid with alcian yellow, etc.

2. *Two-Step Reactions*—In the first step, an agent reacts with the metabolites to be visualized and produces a product but does not produce color. The color is developed in the second step when the product of the first step reacts with the dye used. Usually an oxidant is used as an agent in the first step, e.g., PAS reaction to localize cell wall carbohydrates. Here, in the first step the oxidant (periodic acid) produces aldehyde after reacting with cell wall carbohydrates. The second agent, leucofuchsin, forms a highly colored complex with the aldehydes in the second step.

3. *Indirect Coloring*—Here the first reagent (B) combines/reacts with the substance to be localized or with a product obtained from it (A). Then a second reagent (C) is applied and this combines/reacts with B to induce coloring of A, e.g., localization of cell wall acid phosphatases. Here, a substrate for the enzyme (glycerophosphate) is provided to the tissue along with a lead salt (B). If the enzyme is present in the tissue, it will react with the substrate (glycerophosphate) to form phosphates (A) that combine with lead ions to form lead phosphate (AB). Lead phosphate is colorless. Therefore, ammonium sulphide (C) is added, which combines with lead phosphate to produce a black colored lead sulphide. Sometimes colored precipitates may not be formed in Step II but at Steps III or IV when additional substances are sequentially added to produce coloring of the metabolites or their end/byproduct, e.g., staining of alkaline phosphatases.

2.3.2 Stains

The reagents used in coloring various cellular metabolites/substances are called *stains*. A stain is a chemically defined brand of *dye* and is usually certified by the Biological Stain Commission (Geneva, New York) and bears a certificate number.

The stain molecule has two principal groups, viz. *chromophore* and *auxochrome*. The chromophore, usually an aromatic ring, is associated with the coloring property of the stain. It has unsaturated bonds such as -C=C, =C=O, -C-S, =C-NH, -CH=N-, -N=N-, and -N=O. The intensity of color provided by the chromophore to a stain is proportional to the number of such bonds in it. As the degree of unsaturation increases in stains, their emission shifts gradually from violet to red. In addition to the number of unsaturated bonds, their relative location in the stain molecule also substantially influences the color it emits. The auxochrome helps the stain molecule to combine with a metabolite. Dyes without auxochromes are called *chromogens* but these are incorporated, if at all, only superficially by the tissue and by physical adsorption and can easily be removed out of the tissue by mere washing or by treatment with simple solvents. Examples of chromogens are most lipid stains. Auxochromes are, in reality, acid or basic radicals such as amino ($-NH_2$), hydroxyl (-OH), and carboxyl (-COOH) groups.[27]

Stains may be *acidic* or *basic*. Acidic stains have acidic auxochromes like -OH, -COOH, $-SO_3$ while basic stains have basic auxochromes like $-NH_2$, NH, etc. However, it should be understood that stains are not available as acids or bases. Acid stains are made into salts using strong alkalies, while basic stains are combined with acids such as HCl or H_2SO_4. The former are available commercially as salts of ammonium, potassium, or sodium and the latter as salts of bromide, chloride, or sulphate. In commercialy available salts of acidic stains, most parts of the stain are contained in the anion (-) (and so are called *anionic stains*), whereas in the salts of basic dyes, the stain is contained in the cation (+) (and thus are called *cationic stains*). Examples of anionic stains are acid fuchsin, fast green FCF, eosin, erythrosin, etc. and examples of cationic stains are basic fuchsin, safranin, haematoxylin, toluidine blue O, azure B, etc.

Stains used in light microscopic cytochemistry are classified by source and chemical nature. Some stains are derived from mineral sources, e.g., ruthenium red (used for staining cell wall pectins), which is an ammoniated oxychloride of ruthenium. A number of stains are organic dyes, either obtained from some natural source (e.g., haematoxylin), or artificially synthesized (e.g., the aniline dyes or coal-tar dyes). The latter are further classified in accordance with the molecular configuration of their chromophores.

Stains are also classified into *orthochromatic* (Gk: Ortho correct or the first) and *metachromatic* (Gk: Meta-changed or the second) stains. If a stain imparts to a particular cellular component its original color, it is said to be an orthochromatic stain. e.g., safranin, a red-colored stain, imparts the same color to the lignified component of the cell wall. Other examples are Fast green, Orange G, Erythrosin, Ruthenium red, etc. A metachromatic stain is one which, on combining with different cellular components, imparts colors ranging from its own original color (orthochromatic color) to a totally different color (metachromatic color). Toluidine blue O has an orthochromatic blue color (on reaction with lignin), but on combining with acidic cell wall polysaccharides, such as some pectins and hemicelluloses, it turns to a metachromatic magenta to red.

There is still considerable uncertainty as to the exact nature of the metachromatic color development[28] and several hypotheses have been proposed to explain it. The most accepted explantion is as follows: The stain-ions are bound in clusters or aggregates to the strongly charged binding sites of the metabolites. The ions are placed close enough to interact among themselves through hydrogen bonding and van der Wall's forces; this interaction is optically visible as changes in the absorption spectrum of the stain, thus producing metachromasy.[28-31] This view is substantiated by the fact that ethanol, urea, and other substances known to disrupt hydrogen bonds also affect metachromasy.

The property of metachromasy is usually shown by basic rather than acidic stains; but it should be remembered that metachromasy is not fundamentally related to the acidic nature of the stain. The metachromatic color produced by the metabolite conjugate is normally of a lower energy content and possesses a longer wavelength than the original to the stain-solution. This phenomenon is called *bathochromatic* metachromasy. The red metachromasy observed for pectins with toludine blue staining is an instance of the above type of metachromasy. In *hypsochromatic* metachromasy, the metachromatic color is of a higher energy content (consequently of a shorter wavelength) than the original color of the dye.

At this stage, the attention of the reader is drawn to certain problems often encountered in "staining" enzymes cytochemically. One of the major problems in enzyme cytochemistry is the inconsistent staining throughout the tissue. This is due to failure of the penetration of the incubation medium at the same rate into the different regions of the tissue. This penetration depends on the nature of the tissue, its thickness, nature of the incubation medium, and the presence or absence of cut surfaces on the tissue. Also, the different components of the incubation medium penetrate at different rates, posing additional problems. It is better to follow the following precautions: (1) The tissue should be as thin and small as possible. (2) There should be a cut surface on the tissue. (3) If frozen, the tissue should be devoid of large ice crystals. The optimum period for incubation may be defined as the minimum time required to produce recognizable deposits at all sites of activity of enzyme under study. This will vary considerably from enzyme to enzyme and tissue to tissue. To obtain satisfactory staining, it is necessary to incubate several batches of material for different times.

2.3.3 Control Procedures

Proper controls are the most important requirements in cytochemical staining procedures. The cell has innumerable chemical substances, known and unknown, and, therefore, it is likely that more than one substance may respond positively for the same staining schedule followed. It is imperative to find out which is due to the correct substance. Any conclusion from a cytochemical staining should be made only after critical application of control schedules. This is especially important in enzyme localization studies. In any enzyme localization method, there is a possibility of enzyme diffusion. In addition, factors such as diffusion of the reaction medium away from the enzymatic sites and the diffusion of the primary or final reaction products may have significant effects on the validity of the localization observed. Even in the simplest of enzyme localization reactions, where the enzymic product itself is apparently insoluble, the process of precipitation of the resultant product is complex; supersaturation may occur before such precipitation formed allowing time for diffusion of the product to take place. Diffusion of products leads to "stain deposits" at false locations, where enzyme was not initially present ("false positive reactions"). Diffusion may also lead to "false negative reactions", where no product is precipitated at the actual sites of enzyme action. Size of the enzyme sites, the rate of enzyme reaction product, the rate of the reaction involved in localization, and the solubility of the final reaction product are some of the more important factors affecting the enzyme localization process. Therefore, it is essential to run adequate controls to test the validity of the staining reaction. The control procedures used are very diverse and the reader would appreciate this when he peruses the different cytochemial procedures outlined in this book.

2.3.4 Incubation of Materials in Reagents During Cytochemical Staining

Since staining is a chemical reaction/series of reactions involving the use of reagents of diverse nature, proper incubation of the material to be stained is obligatory. Several types of incubating systems are in operation, essentially based on the nature of the material, nature of the reagent employed, the environment under which the reaction is to proceed, availability and costs of the reagents, etc.

The simplest way of incubation is to immerse the material in a *coplin jar*, a vessel of long use in general histology. The advantages of using a coplin jar are

1. The solutions contained in a coplin jar, as measured, can be left in a water bath, an incubator or an ice bath to equlibriate to a constant pre-detemined temperature which can be checked by inserting a thermometer.
2. The materials can be immersed in a constant but larger amount of the solution.
3. There is no danger of drying out of the reagents.
4. The deleterious substances produced by the sections, if any, cannot become sufficiently concentrated as to have inhibitory effects on the reaction.

The obvious disadvantages of using a coplin jar are as follows: (1) it requires a large volume of the incubation reagent, and (2) the greater cost of the reagent (e.g., NAD, ATP, etc.); therefore it would become prohibitively expensive.[32]

A horizontally placed slide kept in a humidity chamber can be used as an incubating vessel. The material is placed on the slide and the incubating reagent is pored over it in one or more drops until the material is covered. This method is not expensive. However, many reagents used in staining schedules easily evaporate leaving portions of material untreated with the reagents, which often leads to negative reactions; the evaporation may also adversely change the concentration of the regents. The material is also not uniformly immersed in the reagent as its drops are of uneven thickness over the material. In addition, the small volume of the drop could allow inhibitory substances released by the cytochemical reaction to reach such a concentration that could affect the result expected of the reaction.

Jones[33] suggested the *micro-cell* or *chamber method* to overcome the difficulties of the slide method mentioned above. However, even here there are problems, but of a different nature. It is difficult to construct micro-cells for cytochemical reactions. The cells are awkward to use and it is virtually impossible to control, to our advantage, the gaseous environment of the micro-cell.

The *open ring method* is suggested wherever costly reagents are to be used. The material is placed on a slide. A perspex ring, to encircle the material and of a depth of about 3 mm, is set around the material and held on to the slide by a thin film of vaseline. The reagent is added to the material and is confined around it by the ring. The gaseous atmosphere can be controlled to our advantage by doing the treatments in specially designed boxes, which can then be placed in locations with appropriate temperature controls.[29]

2.4 Mounting of Cytochemically Stained Materials

The mountants play an extremely important role in light microscopic cytochemistry. They affect the refractive index (RI) and, thus, the visibility of the stained materials. As the colorless cytoplasm is most transparent when it is mounted in a medium with RI between 1.53 and 1.55, it is desirable to use a mountant with RI in this range for advantageously resolving colored materials. Mountants should also be selected in such a way that they should be neutral towards the stained materials and should not affect the stability of the stain-substrate complex. For example, any mountant that impairs hydrogen bonding between the substrate and toluidine blue O (TBO), would affect the metachromatic staining of TBO. The mountant should be colorless/light pale and should not become colored on prolonged exposure to light and with time. It should also be chemically stable and inert. It should be uniform and preferably of a defined composition. It should also be easily available and inexpensive. A permanent mountant should adhere well to the glass, should have quick coverslip setting and drying properties, and should not crack or become granulated on drying.

More than 300 kinds of mountants have been formulated by now and the most cytochemically useful ones are discussed below. Water (RI = 1.336) is one of the best and the cheapest mountants to preserve cell wall constituents soluble in organic solvents, to ensure the stability of the stain-metabolite complex when the stain is soluble in organic solvents, and to sustain the metachromatic color, as mentioned above for Toluidine blue O. Arlex-d-Sorbital syrup (RI = 1.486) added with 10% gelatin (RI = 1.494), fructose syrup (75% aqueous) (RI = 1.476) and commercially available gum arabic mountants such as Abopon (RI = 1.437), Clearcol (RI = 1.404), Paragon (RI = 1.429), and Viscol (RI = 1.417) are ideal for preserving Sudan dye staining of lipoidal/similar constituents of cell walls. Glycerol (RI = 1.467) in a 50% aqueous solution (RI = 1.397) is a general purpose mountant but is unsuited to preserve cell wall calcium carbonates. Glycerin jellies (combination of glycerin and gelatin) are quite unsuited to cytochemistry as they bleach the stained materials. Farrants neutral glycerol-gum arabic medium (RI = 1.44) (gum arabic + 50 ml glycerol + 50 ml water + trace of preservatives such as arsenic oxide or sodium merthiolate + 50 gm potassium acetate) is good for preserving many stained cell wall materials. Mountants like polyvinyl pyrrolidone (PVP) (commercially known as plasdon C) and vinyl acetate in 7:3 proportion are very good mountants to retain metachromasy of cell wall substances as well as cell wall lipid-positive stains.

Non-aqueous mounting media (RI range between 1.4 to 2.55) have long been in use in basic histology and have quickly been adopted to light microscopic cytochemistry. These generally have better coverslip setting and drying properties than the aqueous mountants. Natural resins like canada balsam (from *Abies balsamea*), damar (from *Shorea* species), Oregon balsam (from *Pseudotsuga* species), Styrax (from *Liquidambar* species), Tolu balsam (from *Myrospermum toluiferum*), Diaphane (from *Callitris* species), and Yucatan gum (from *Bursera* species) are dissolved in suitable organic solvents like xylene, toluene, or benzene and are used as mountants. Canada balsam is very good for stained cell wall proteins and cellulose. Staining of acidic polysaccharides is often affected by all these resins.

Several synthetic resins are now available as mountants. These include Diakon (RI = 1.503), Plexiglass (RI = 1.5), Pontalite (RI = 1.5 approximately), Cyanocrylate, Distrene, Lustron, Polystyrene P-1, Gurr's Depex (RI = 1.523), DPX (most commonly used in cytochemistry of cell wall proteins, homopolysaccharides, and lignin), Bioloid (RI = 1.52 to 1.55), Harleco synthetic resin (RI = 1.52 to 1.54), Permount (RI = 1.51 to 1.53), Piccolites (the last four preserve cell wall cationic dyes very well), and Sirax (RI = 1.81) (good mountants for materials stained for cell wall materials like cellulose, callose, lignin, and wall proteins), Naphrax (RI = 1.75 to 1.8), Refrax (RI = 1.6 to 1.8), Technicon (RI = 1.56 to 1.62) preserves PAS and Aniline stains, Caedex (RI = 1.631 to 1.672), Clearax (RI = 1.6 to 1.66) (the last

two preserve aniline blue stains), etc. Many of these have not been adequately tried
in cytochemistry and so deserve wider trials before anything can be said of their
usefulness (see full details in Delly,[34] Gray,[35] and Vijayaraghavan and Shukla[36]).

2.5   Procedures in Tissue Preparation for Light Microscopic Cytochemistry

1. Method for Freezing a Tissue[32]*

1. Carefully remove the tissue from the plant and immediately proceed to chill it. When a delicate plant tissue is involved, immerse it for 10 to 30 min in a 5% aqueous solution of polyvinyl alcohol (PVA) before chilling.

2. Prepare the chilling bath as per the procedure given below:

   A cheap polythene sandwich bath with holes cut in the lid can be used as the chilling bath apparatus. Heat-insulate the box by putting around it expanded polystyrene or other insulating material. Prepare the chilling bath by adding small chips of solid CO_2 to alcohol in the bath until a saturated solution is obtained. The saturation stage is indicated by (a) the viscous appearance of alcohol-CO_2 mixture, (b) the absence of bubbling by the further addition of more solid CO_2, and (c) the indicator in thermometer showing about -70°C.

3. Insert a beaker of about 50 to 100 ml capacity, containing 30 to 50 ml of hexane into the bath through the hole in the lid. (Some workers prefer isopentane instead of hexane.)

4. Add more CO_2 ice to the bath to maintain the temperature.

5. Cut the tissue to be chilled into small pieces (ca 5 mm³) and drop the cut tissue into the hexane when the temperature of the bath is about -70°C. Care must be taken to see that the tissue does not touch the sides of the beaker.

6. Leave the tissue in hexane for at least 30 sec but never longer than 2 min.

7. Transfer the tissue, with the help of pre-cooled forceps, to a dry tube at - 70°C which may then be corked, encased in solid CO_2 in the Dewar flask, and stored for use later on.

2. Method for Freeze-Substitution of a Tissue[10,36,37]

1. Prepare a cold chamber that will maintain a temperature of -30°C or lower. A cold chamber can be arranged either from commercial brands of deep freezers or from a wide-mouthed Dewar flask. The latter is often advisable. Fill up two-thirds of a Dewar flask with powdered dry ice. On top of the metal disc, place one or more cardboard discs and on top of these place an insulated 400 ml beaker. Pour 200 ml of 65% alcohol into the beaker, which will maintain a temperature between -41 and -45°C (temperature can be controlled by varying the number of cardboard discs). Place the vials containing the substitution solvents (such as absolute ethyl or methyl alcohol) and the tissue in

---

* Several modern methods of freezing, freeze-substitution, and freeze drying are discussed in Chapter 6 of this book. The reader is requested to refer to these.

the bath. Cover the beaker with a tightly fitting cover of plastic sheeting. Insert a thermometer through the cover into the vial containing the substitution solvent placed in the bath.

2. Place test tubes or similar vials half full of methanol or ethanol, properly stoppered and labeled, in the cold chamber.

3. Allow 15 min for the alcohol to cool.

4. Place the frozen tissue immediately in the cold alcohol with the help of a cold forceps.

5. After 1 d, replace the alcohol with fresh cold alcohol.

6. After 1 d, replace the alcohol with cold pure toluene.

7. Remove the tubes from the cold chamber after 1 h, place them at room temperature, and change fresh toluene after another hour at room temperature.

8. After 2 to 4 h place the tissue on solid paraffin in the bottom of tubes with a covering layer of toluene and place the tubes in an oven adjusted to 60°C. Change paraffin after regular intervals and embed the tissue once no trace of toluene is detected.

3. Method for Freeze-Drying of a Tissue[10,39-41]

1. Carry out the procedure in Glick apparatus. This apparatus has a dehydration chamber with a well. The chamber has to be connected to a vacuum pump* and all connections should be airtight. Paraffin is to be placed at the bottom of the dehydration chamber and then to be melted until the trapped gas, if any, has been removed (as long as gas is there, paraffin will bubble). Solidify the paraffin with cold water, once all trapped gas is removed.

2. Freeze the tissue as in the previous procedure; however, drop the tissue into a wire basket which is subjected to cooling in the isoteroene.

3. Remove the basket with the tissue from isoterpene and quickly pass it into the dehydration chamber already cooled with a refrigeration unit or in a methyl cellosolve-dry ice mixture at -30 to -40°C.

4. Maintain the temperature of the cold chamber at -30 to -40°C by adding pieces of dry ice at intervals.

5. Seal the system

6. Place a methyl cellosolve-dry ice bath at -65°C or liquid nitrogen around the condenser and in the central well.

7. Start the pump(s)** and keep it (or them) in operation for the whole period in which the tissue would remain in the apparatus (this period ranges from 1 to 6 d depending on the tissue) until dehydration is complete.

8. Remove the cooling bath and melt the paraffin at the bottom of dehydration chamber using warm water.

9. Wait until the tissue completely sinks in the molten paraffin.***

---

* The vacuum pump can be a mechanical pump. Sometimes an additional oil or mercury diffusion pump is also used. If both are to be used, place the diffusion pump between the condenser and the mechanical pump.
** If an oil pump is used, allow the mechanical pump to work for half an hour before starting the diffusion pump.
*** In some cases, materials that do not sink completely are also good enough.

10. Stop the pump and release the seal of the system.
11. Remove the tissue and embed it in paraffin.

4. Procedures for The Preparation Of Fixatives Used in Light Microscopic Cytochemistry

a. Acrolein[14]

A 10% solution buffered with cacodylate or phosphate buffer (0.1 to 0.2 M, pH 6.5 to 7.6) containing an osmotically active substance like sucrose (0.22 to 0.55 M) for making the fixative isotonic with cell sap.

b. Bouin's Fluid [42]

Saturated aqueous picric acid solution	75 ml
Formaldehyde (37 to 40%)	25 ml
Glacial acetic acid	05 ml

c. Carnoy's Fluid [42]

Absolute ethanol	60 ml
Glacial acetic acid	10 ml
Chloroform	30 ml

d. Chrome-Acetic Acid [43]

10% aqueous chromic acid	2.5 ml
10% aqueous acetic acid	2.5 ml
Distilled water	500 ml

e. Clarke's Fluid [43]

Glacial acetic acid	1 part
Absolute ethanol	3 parts

f. Farmer's Fluid [17]

Anhydrous ethanol	75 ml
Glacial acetic acid	25 ml

g. Formaldehyde[17]

Solution A: Dissolve 2 g of paraformaldehyde powder in 20 ml of distilled water by gentle heating (60 to 80°C). Stir well with a glass rod. Add dropwise a 4% solution of sodium hydroxide until the solution becomes transparent, making a total volume of 25 ml.

Solution B: Take disodium hydrogen orthophosphate 0.636 g, potassium dihydrogen orthophosphate 0.194 g and distilled water 25 ml and mix all three.

Mix solutions A and B, adjust pH to 7.2 (50 ml total volume is to be obtained). Use a freshly prepared fixative.

It is ideal to fix at 40°C for 24 h. Material fixed in this fixative can be stored in 1% glycerol in 70% ethanol.

For enzymes, small materials of 1 to 2 mm³ should be fixed at 0 to 4°C for 2 to 4 h.

h. Formalin, Neutral [43]

40% formaldehyde	100 ml
Distilled water	900 ml
NaH_2PO_4, H_2O	4.0 g
Na_2HPO_4 (anhydrous)	6.5 g

i. Formal-Calcium [18]

40% formalin	10 ml
10% calcium chloride in water	10 ml
Distilled water	80 ml
Add chalk to keep the pH neutral	

j. Formalin-AceticAcid-Alcohol (FAA) [43]

Ethanol 50%	90 ml
Glacial acetic acid	5 ml
Commercial formalin	5 ml

k. Glutaraldehyde [17]

Solution A:

Disodium hydrogen orthophosphate	0.636 g
Potassium dihydrogen orthophosphate	0.194 g
Distilled water	25 ml

Solution B:

25% stock solution of glutaraldehyde.

Add B solution to A depending on the desired percentage needed. Normally a 2 to 3% solution is recommended. Adjust the pH to 7.2. Material size should not exceed 2 mm³.

Fixation is for 24 h at 0 to 4° C for routine cytochemistry and for about 30 min for enzyme localization.

Phosphate buffer can be replaced by cacodylate buffer.[44]

A 1.5 to 3% solution of glutaraldehyde in phosphate or cacodylate buffer around 8.00 pH for 2 to 5 h is recommended by Coetzee and Van der Merwe.[45]

l. Lewitsky's Fluid [46]

1% Chromic acid in water	50 ml
10 % Formalin	50 ml

m. Navashin's Solution or CRAF Fixative [43]

Solution A:

Chromic acid	5 g
Glacial acetic acid	50 ml
Distilled water	320 ml

Solution B:

Commercial formalin	200 ml
Distilled water	175 ml

Mix equal volumes of solutions A and B immediately before fixing the material.

n. Osmium Tetroxide [44]

2% solution of Osmium tetroxide in 0.1 M Veronal acetate buffer at pH 7.4.

o. Zenker's Fluid [43]

Potassium dichromate	2.5 g
Mercuric chloride	5.0 g
Glacial acetic acid	5.0 g
Water	100 ml

After mixing all these, add 10 ml of formalin.

p. Zirkle's Fluid [43]

40% Formalin	10 ml
10% Calcium chloride in water	10 ml
Distilled water	80 ml

5. Dehydration and Clearing Procedures

a. Ethyl Alcohol, Acetone, or Isopropyl Alcohol Procedures [47]

1. Begin dehydration of the tissue with a dehydrant percentage having approximately the same percentage of water as the killing (fixing) or storing fluid. For example, if the killing fluid has 30% water, begin dehydration with 30% dehydrant solution.

2. Transfer the tissue to vials containing gradually increasing concentrations of dehydrant with decreasing water content. Decide by trial and error the duration of treatment of the tissue in different concentrations of the dehydrant.

3. Finally give two changes in absolute dehydrant (100%) before proceeding to clearing.

b. Dioxan Procedure [47]

Transfer the tissue through the following three grades of Dioxan (= diethylene dioxide) at 4 to 12 h intervals.

(a)   1/3 dioxan + 2/3 water
(b)   2/3 dioxan + 1/3 water
(c)   Pure dioxan (3 changes)

Then pass on to infiltration of paraffin.

c. n-Butyl Alcohol Procedure [47]

1. Wash the tissue in water and dehydrate it in ethyl alcohol up to 30% if the tissue is unfixed. If tissues fixed in ethyl-alcohol-containing fixatives are used, wash them in ethyl alcohol of the same concentration as used in the fixative.
2. Treat the tissue in the grades of solutions shown in Table 2.1. Decide the timing of treatment under each grade by trial and error for the tissue in question.
3. Make two to three changes in absoloute n-butyl alcohol and proceed with wax infiltration.

TABLE 2.1
N-Butyl Alcohol Procedure for Dehydration and Clearing

Grade No.	n-Butyl alcohol (ml)	Ethyl alcohol (ml)	Distilled water (ml)
1.	10	20	70
2.	15	25	60
3.	25	30	45
4.	40	30	30
5.	55	25	20
6.	70	20	10
7.	85	15	-
8.	100	-	-

Source: From Berlyn, G. P. and Miksche, J. P., *Botanical Microtechnique and Cytochemistry*, The Iowa State University Press, Iowa, 1976. With permission.

d. Tertiary Butyl Alcohol (TBA) Procedure [47]

1. Dehydrate the tissue in ethyl alcohol up to 50%
2. Transfer the tissue in the following grades of solutions shown in Table 2.2. The time under each grade should be decided for the selected tissue by trial and error.
3. Make three changes in absolute TBA and proceed to wax infiltration.

TABLE 2.2
Tertiary Butyl Alcohol Procedure for Dehydration and Clearing

Grade No.	95% ethyl alcohol (ml)	Absolute ethyl (ml)	TBA (ml)	Distilled water (ml)
1.	50	-	10	40
2.	50	-	20	30
3.	50	-	35	15
4.	50	-	50	-
5.	-	25	75	-

Source: From Berlyn, G.P. and Miksche, J.P. *Botanical Microtechnique and Cytochemistry*, The Iowa State University Press, Iowa, 1976. With permission.

e. Clearing Dehydrated Tissues in Solvents of Paraffin and Other Embedding Media

The clearing agents that can be used are xylene (or xylol), chloroform, benzene, toluene, etc. Transfer the tissue from the 100% dehydrant (after 1 to 3 changes) to a series of vials containing the dehydrant-clearing fluid mixtures in which the percentage of dehydrant is slowly reduced and that of the clearing fluid increased gradually. Alternatively, add the clearing fluid in drops at regular intervals into the container having the tissue in the 100% dehydrant until the volume of clearing fluid added reaches twice the volume of the dehydrant. In both cases, immerse the tissue in 100% clearing agent at the end. Paraffin can then be infiltrated slowly into the tissue.

f. A Suitable Dehydration, Clearing, and Embedding Procedure[48,49]

1. Transfer fixed tissue directly to 100% methyl cellosolve.
2. Transfer successively to 100% ethanol, n-propanol, and n-butanol.
3. Finally transfer the tissue to an embedding medium (waxes or glycol methacrylate).

Notes: All operations should be done at 0°C. Dehydration is done in methyl cellosolve solution.

6. Removal of Resins from Sections[50]

a. Preparation of Solution to Remove Araldite and Spurr's Resin

Add 10 single pellets of NaOH (approx. 1.0 g) to 10 ml absolute ethanol and 10 ml propylene oxide. Stir this mixture on a magnetic stirrer (about 5 min) until at saturation a cloudy white precipitate is formed.

b. Preparation of Solution to Remove Epon Resin

Add 20 single pellets of KOH (approx. 2.0 g) to 10 ml absolute methanol and 5 ml propylene oxide. Stir this mixture in a magnetic stirrer (about 5 min) until dissolved.

c. Removal of Resin

Attach the sections from aldehyde and osmium fixed blocks to the slides with 1% egg albumin in distilled water and flood with appropriate solution (1 or 2 above). Allow the solution to act at room temperature for about 2 min in the case of araldite and spur resins and 5 min for Epon resin. Remove any residual resin by rinsing sections in absolute ethanol (araldite and spur) and absolute methanol (Epon). (Thicker sections would obviously require a slightly longer application.) Use fresh solutions because storage for more than 4 to 5 h reduces the resin removing properties. Wash the sections well in running water for 5 to 10 min. Subject the sections to treatment in postfixation mixture (potassium dichromate 3 g, formalin 25 ml, acetic acid 5 ml, and distilled water 70 ml) for 1 h at room temperature; then treat the sections in oxidation solution (2.5% potassium permanganate in distilled water) for 15 min at room temperature. Bleach the sections in 1% oxalic acid for 15 min at room temperature. Wash well in tap water. Stain in any desired dye.

2.6. Procedures for the Cytochemical Localization of Different Cell Wall Substances.

2.6.1 Carbohydrates

2.6.1.1   Insoluble Polysaccharides (with 1, 2-glycol groups) Periodic Acid-Schiff's (PAS) Method[200,203,204]

Tissue Preparation

Use fresh, feeeze-dried, freeze-substituted, or chemically fixed (Farmer's fluid ideal) tissue. Tissue embedded in paraffin or other substances can also be used. Avoid fixatives that contain aldehydes (such as formaldehyde, acetaldehyde, gultaralde-hyde, and acrolein) because proteins tend to bind to aldehydes (more so to acrolein) in a manner that allows them to react with Schiff's reagent.[48]

Procedure

1. Wash the material in water; if embedded, bring the material down to water after removing the embedding medium.

2. *Block the tissue aldehydes in a saturated solution of 2,4dinitrophenyl hydrazine (DNPH) for 30 min (DNPH 0.5 g in 10 ml of 15% aqueous acetic acid, agitated for 1 h and then filtered). Dimedone can be substituted for DNPH. Blocking can also be done with 0.5% (w/v) sodium borohydride.[53]

3. Oxidize the materials in 0.5 to 1% aqueous periodic acid** for 5 to 30 min at 17 to 22°C. Hotchkiss[51] recommended a solution consisting of 10 ml of distilled water, 35 ml of 70% alcohlic solution bufferd with 5 ml of 0.2 M sodium acetate (27.2 g of the hydrated salt in 1000 ml) and 0.4 g of periodic acid. A 1% periodic acid in anhydrous DMSO (dimethyl sulphoxide) has been recommended by others.[28] Keep the periodic acid in the dark at 17 to 22°C. Discard if a brown color appears.

4. Wash in running water for 5 min.

5. Place the materials in Schiff's reagent for 10 to 30 min.

    > Prepare Schiff's reagent according to the following procedure:
    > Add 1 g basic fuchsin in 200 ml boiling distilled water, followed by shaking for 5 min. Upon cooling to 50°C, filter the solution and add 30 ml HCl and 3 g sodium or potassium metabisulphite. Store the solution in the dark for 24 to 28 h. During this time, the red color should disappear.*** To the decolorized solution, add 0.5 g vegetable charcoal, and shake for 1 min. Filter the mixture rapidly under vacuum and then store at 5°C in the dark until use. Schiff's reagent can keep for several weeks to a year, but the only way to know if it is good is to try it.

6. ****Transfer the material quickly and directly to 3 successive baths of 0.5% sodium sulphite, 2 min each.

7. Rinse the material in running water for 5 to 10 min, mount in water and observe. If the material is to be made permanent, then dehydrate, clear, and mount.

---

* Step 2 is necessary to prevent aldehydes already existing in the tissue from reacting with Schiff's reagent.
** Periodic acid is claimed to be advantageous over other oxidizing agents ($KMnO_4$, H_2CrO_4, H_2O_2, etc.) in that the aldehydes that it forms are stable in its presence and are not further oxidized. Since periodic acid oxidizes many other substances, it is important that, for each tissue, after standardization, it is used at the suitable concentration and for particular duration (within the limits mentioned above), which conditions are conducive to the oxidation of only -CHOH-CHOH- group of carbohydrates.
*** The loss of color is due to the following reason: In acid solution (HCl), with excess SO_2 (supplied by potassium metabisulphite), fuchsin is transformed into colorless N-sulphinic acid, also called leuco-fuchsin.
**** The step was avoided in the procedure outlined by McManus.[52]

8. If desired, counterstain in celestian blue (dissolve 2.5 g ferric ammonium sulphate in 50 ml distilled water by stirring overnight at room temperature. Add 0.25 g celestian blue B and boil for 3 min. Filter after cooling and then add 7 ml glycerol). Counterstaining can also be done in Orange G (add 2 g orange G in 100 ml of 5% phosphotungstic acid. Allow the solution to stand for a day and collect the supernatent for use).

Result

Cell wall polysaccharides stain a purplish red to magenta. This reaction is shown by cellulose, chitin, carboxylated polysaccharides (like pectins and some hemicelluloses), and alginic acid. Sulphated polysaccharides such as fucoidan do not get stained.[54,278] Although Nanda and Gupta[55] and Albertini et al.[56] have claimed a positive staining of callose of microspore mother cells and microspores with PAS schedule, it has been shown to be an erroneous conclusion by the results obtained by Helslop-Harrrison[57] as well as by our long experience of the use of this schedule on a variety of plant materials. Laminarin, related to callose, is also not stained by PAS procedure.[54]

Rationale of the Color Reaction

The 1,2-glycol(vicinal) linkage within the sugar molecule of carbohydrate is acted upon by the oxidant, periodic acid (HIO_4) and two free aldehyde groups are produced. The aldehydes, so formed, react with leucofuchsin producing highly colored (pink red to magenta) complexes.

For this reaction to occur, the hydroxyl groups of the sugar must be free or substituted with an amino group, an alkylamino group, or a carbonyl group.[58] This is the reason why chitin is also stained with PAS procedure.

Controls

Employ any one of the following controls:

1. Acetylation: Treat the control materials at 22°C for 1 to 24 h (depending on tissue) in 16 ml acetic anhydride in 24 ml dry pyridine.* Wash in water and then subject to PAS procedure as detailed earlier. A negative reaction will indicate that 1.2-glycol groups are responsible for the original color reaction observed.

2. Benzoylation: Fix the materials in absolute ethanol for 8 min and change in the same for another 8 min. Then place the materials in acetonitrile* in a desiccator containing phosphorus pentoxide for 3 min. Subsequently, immerse the materials in the benzoylation solution for 2 h. (Prepare this solution by adding 50 ml acetonitrile, 4.2 ml benzoyl chloride* and 2.2 ml dry pyridine.) Remove the materials from the desiccator, wash them well in absolute ethanol, and then stain according to PAS procedure. A

---

* Pyridine, acetonitrile, and benzoyl chloride are harmful chemicals and should be handled in a fume cupboard. They should not be pipetted by mouth.

negative reaction will indicate that 1,2-glycol groups are responsible for the original color reaction observed.

3.   Omit periodic acid treatment.

Comments

1.   Many, including us, have often been puzzled by PAS negativity (or very feeble staining) of cellulose in intact cell walls,[59] although theoretically it should not be the case. We have encountered this difficulty while studying the gelatinous and non-lignified layer of these fibers of tension wood of Angiosperms. The gelatinous and non-lignified layer of these fibers possesses a highly crystalline form of cellulose as detected through x-ray diffraction studies, but PAS technique has failed to stain this layer. It seems likely that the high crystallinity of cellulose might be the causative factor for this negativity; prolonged treatment with periodic acid (60 min to overnight depending on the tissue) enables cell wall cellulose PAS positive as this treatment affects the crystalline nature of cellulose. As O'Brien and McCully[12] have remarked, the same may be the reason for the PAS-positivity of alkali-treated cell walls (see also Jensen,[60] pages 190-192).

2.   The PAS reaction is highly suitable[60] because (a) it breaks the polysaccharide chains to result in diffusion, (b) it is specific, (c) the basis of the reaction is known, (d) it offers little or no scope for interference or of false localizations, and (e) it gives an intense and stable reaction.[58]

Application of the Procedure

PAS technique has been one of the most widely used methods in botanical research. Hundreds of research papers have used this technique to localize insoluble polysaccharides in different materials ranging from algae to angiosperms and it is difficult to cite all of them here. The presence of wall ingrowths in, and transfer-cell morphology of, antipodals, central cell, synergids (the filiform apparatus), suspensor cells, endosperm haustoria, and endothelial cells have been essentially brought to light by using the PAS technique (see Vijayaraghavan and Bhat[61] and Vijayaraghavan et al.[62] for full literature). The chemotactic nature of the synergid and its filiform apparatus and the carbohydrate nature of its secretory product which is released into the micropyle and which attracts the pollen tubes at the time of fertilization were proved beyond doubt by the application of the PAS technique.[61-65] This technique has also been exploited to study the cells at the leaf abscission zone,[66] where it has been shown that the cells separate along the line of the middle lamella.

Another instance where PAS technique has solved beyond doubt a long-existed controversy is that of the nature of the matrix and stalk of cystoliths (calcium carbonate crystal bunches) of taxa belonging to families such as Acanthaceae, Moraceae, Urticaceae, Cucurbitaceae, etc. The stalk or peg was proved to be a wall ingrowth by employing this technique.[67,68]

The differential extraction procedure followed by a PAS staining was first designed by Jensen[69] to study the distribution of the different wall materials indirectly, since many carbohydrate components of cell walls reacted equally well in PAS. The details of the procedure, as outlined in Jensen,[60] are as follows: Four adjacent sections of the same material are mounted on four different slides to facilitate differential extraction of adjacent sections of the same tissue. The first slide is kept as a control. The other three slides are treated with warm ammonium oxalate to extract and remove both the water-soluble and pectic carbohydrates. One of these slides is set aside while the other two are treated with 4% NaOH to remove hemicelluloses. One slide is set aside and this has tissue minus pectins and hemicellulose. The other slide is extracted with 17.5% NaOH to remove all carbohydrates except cellulose from the cell wall. Finally all four slides are stained in PAS. A comparison of the intensities of color of cell walls after the different extraction procedures would give an indication of the amount of the different cell wall carbohydrates present. Jensen[69] first used this method to successfully study the cell wall composition of onion root cells. Subsequently this procedure was used to study cell wall composition in developing fruits,[70] the liverwort *Geothallus*,[71] wood,[72] tension wood fibers,[73,74] rhizoids of the gametophytes of *Polypodium vulgare*,[75] etc. In the last two instances, modifications were introduced in Jensen's procedure not only in the duration of treatment with different extraction chemicals but also by introducing additional/modified methods to extract other polysaccharides.

2.6.1.2 Cellulose

Procedure 1. Chlorazol Black E Method[76,77]

Tissue Preparation

Use fresh, freeze-dried, freeze-substituted, or FAA fixed materials. Materials embedded in paraffin or similar materials can also be used.

Procedure

1. Bring the material down to water.
2. Place it in distilled water for 5 min.
3. Dehydrate in ethanol.
4. Stain in 1% chlorazol black E in methyl cellosolve for 5 min and observe.
5. If necessary, wash it in absolute alcohol very briefly (less than 2 min), clear in xylene, and mount in any suitable mountant.

Result

Cellulose gets stained to a black or greyish black tinge.

Rationale of the Color Reaction

The basis of the color reaction is not very clear.

Controls

Use any one of the following control procedures:

1. Acetylation (see Page 55).
2. Benzoylation (see Page 55).
3. Treat the material with the enzyme cellulase. The enzyme is available in various commercial names such as Cellulase Onozuka, Meicelase, Cellulysin, Driselase, etc. The last one is a mixture of cellulase, hemicellulase, and polygalacturonase. The optimum treatment mixture of the enzyme is as follows: Cellulase 5 mg/ml in 0.05 M phosphate buffer, pH 5.5.[78] Standardize the duration of application of the mixture for each material.
4. Cuprammonium Treatment: Treat the material in cuprammonium (also called Schweitzer's reagent) for 15 min to 2 h depending on the material and its thickness. Prepare cuprammonium as follows: To 15% aqueous cupric sulphate add sufficiently strong ammonia until the precipitates of cupric hydroxide formed initially are dissolved. You can also prepare it either by adding 15% ammonia water to copper turnings or filings or by adding aqueous solution of $CuSO_4$ to NaOH solution, collecting the precipitate of cupric hydroxide formed by the reaction of the above two chemicals by centrifugation, and by dissolving the collected cupric hydroxide in minimal amounts of 28% ammonia water.

Comments

Chlorazol black E gives a good contrast with lignin pink, which stains lignin to a pink color. If this staining is desired, it should be carried out after Step 2 in the procedure outlined. Use a 1% aqueous solution of lignin pink for 5 min to stain the material.

Application of the Procedure

One of the major applications of this procedure has been to identify and study tension wood (reaction wood) in dicotyledons. Robards and Purvis[76] were the first to use this technique for this purpose. The tension wood is characterized by a special type of fiber, the gelatinous fiber (G-fiber), in which the inner layer of the secondary wall is non-lignified and gelatinous. This layer contains cellulose and carboxylated polysaccharides but no lignin. When chlorazol black E is applied, the cellulose of the inner, gelatinous and non-lignified layer gets stained black. Subsequently this technique was extensively used by many others for studying G-fibers of tension wood[73,74,79-85] and has now become one of the standard techniques to identify tension wood.

Procedure 2. Zinc-Chlor-Iodide or Chlor-Zinc-Iodide Method[60,86]

Tissue Prepration

Preferably use fresh, freeze-dried, or freeze-substituted tissue. Ethanol fixed (and subsequently paraffin embedded) material may also be used. Avoid fixatives with chromic acid as a constituent.

Procedure

1. Bring the material down to water.
2. Place the material in the zinc-chlor-iodide solution. The solution may be prepared in any one of the following ways:
    (a) Dissolve commercial chlor-iodide of zinc in about its own weight of water and add enough or even excess of metallic iodine* to give the solution a deep brown color.
    (b) Dissolve 50 g zinc chloride and 16 g potassium iodide in 17 ml distilled water. Add iodine to make a saturated solution. Allow this solution to stand for several days. Collect only the supernatant in brown bottles and use.
    (c) Dissolve 10 g zinc chloride in 7 ml water. To this viscous solution, add 2 ml I_2-KI solution (1.0 g KI, 0.1g I_2 and 10 ml of water). Centrifuge and use the supernatant solution, or
    (d) Add 30 g zinc chloride, 5 g potassium iodide, 1 g iodine, and 14 ml distilled water.[87]
3. Observe the material in the solution itself.

Result

Cellulose turns a dichroic blue to violet.

Rationale of the Color Reaction

The basis of the reaction is not very clear.

Controls

Use any control procedure which can selectively remove cellulose prior to staining with zinc-chlor-iodide.

---

* Iodine is poisonous.

Comments

1.   Cellulose gets stained only if present in large amounts in the cell wall.
2.   The test is non-specific in that hemicelluloses are also colored blue. Therefore, the procedure should be carried out after extracting hemicelluloses from the cell wall.
3.   If cellulose occurs with lignin, chitin, cutin, suberin, etc. blue color is not obtained but the walls appear yellow to orange.[88] Therefore, remove these wall substances to enhance the specificity of the reaction.
4.   Naylor and Russel-Wells[89] consider this test as unreliable.

Application of This Procedure

This is one of the classic and often-used methods to test the presence of cellulose in cell walls. Several earlier research papers have relied exclusively on this procedure for verifying the presence of cellulose in cell walls of algae to angiosperms.

Van Wisselingh[90] used this procedure to demonstrate the absence of cellulose in *Plasmodiophora*, a plasmodial fungus. These results have been confirmed by modern analytical methods for the sporangia and resting spore of *Sorodiscus*, *Plasmodiophora*, and *Woronia*.[91]

The presence of cellulose in the cell walls of red algae and brown algae have been doubted for some time (see discussion in Percival[92]), but its presence has been demonstrated by this technique (see, for example, Matty and Johansen[93]).

Watt et al.[67] used this technique, in addition to others, to prove the wall nature of the stalk of cystoliths of *Pilea cadierei*.

Procedure 3: Potassium Iodide-Iodine-Sulphuric Acid Method[86,94]

Tissue Preparation

Fresh, freeze-dried, or freeze-substituted materials are preferable.

Procedure

1.   Stain the material in potassium iodide-iodine solution for 15 to 60 min. The solution is prepared by mixing 0.2% iodine in 2% aqueous KI solution.
2.   Mount in the staining solution itself and observe. Cellulose will stain yellow at end of this step.
3.   Slowly add 60 to 75% H_2SO_4 through the sides of the coverslip until it diffuses into the material. Then observe once again.

Result

Cellulose walls swell and take a bright blue to violet color. The different layers of the cellulose microfibrils would be evident in many cases.

Rationale of the Color Reaction

The native structure of cellulose is disrupted and its glucose strands are separated by the removal of hydrogen bonds by the action of sulphuric acid. Iodine is then believed to accumulate within the disrupted cellulose molecule and impart the color.

Controls

Use any control method which can selectively remove cellulose.

Comments

The color will not develop if cellulose occurs along with lignin as the latter often masks the color. Hemicellulose interferes with the staining reaction and also gets stained blue.[88] But, after extracting the non-cellulosic components, the cellulose can be localized excellently. For extracting non-cellulosic polysaccharides, adopt the following procedure: Heat the materials in a 2.5% solution of sodium carbonate in a boiling water bath for 1 h. Then wash the material until free from alkali. Again heat the materials for 1 h in a boiling water bath with 25% H_2SO_4 and finally wash in water until free from acid.

Application of This Procedure

This procedure is one of the most common that had been in use for a long time to demonstrate cellulose in a wide variety of plants. Naylor and Russel-Wells[89] used this technique extensively to demonstrate the presence of cellulose in a number of brown and red algae where there was dispute about the occurrence of cellulose. Subsequently, it was used by many others to show the presence of cellulose in many algal taxa.[93,95,96]

Watt et al.[67] used this procedure to show cellulose in the cystolith stalks of *Pilea cadierei* and, thereby, proved conclusively the wall-nature of the stalk.

Procedure 4 : Cellobiohydrolase-Colloidal Gold Procedure[97]

Fix specimens in 2.5% glutaraldehyde in 0.1 M cacodylate buffer, pH 7.4 for 2 h. Wash the specimens in cacodylate buffer. Post-fixation in 1% osmium tetroxide for 1 h is optional. Dehydrate in ethanol series and embed in a mixture of epon-araldite epoxy resins or in LR white. Take sections at the desired thickness and stick them to a glass slide. De-embed the plastic as per the procedure of Maxwell.[50] Conventional as well as cryosections can also be used.

1. Incubate the sections on 0.05 M citrate-phosphate buffer pH 4.9, for 10 min.
2. Treat the section with a 1/30 dilution of the enzyme cellobiohydrolase (E.C. 3.2.1.91)-colloidal gold solution with the incubation buffer for 30 min. Cellobioydrolase enzyme can be commercially obtained or can be isolated and purified from cellulose degrading fungi like *Trichoderma* species. The preparation of colloidal gold solution is detailed on Pages 63 to 65. The preparation of the enzyme-gold complex is given on Page 65.
3. Wash several times in buffer.
4. Wash thoroughly in distilled water.
5. Air-dry the slide.
6. Cover the sections with a few drops of silver enhancement mixture for 6 to 10 min. Silver enhancement mixture can be commercially obtained (Intense M. Janssen Kit, Belgium) or can be prepared in the laboratory as per the procedure outlined on Page 265.
7. Wash the sections threefold in distilled water.
8. Counterstain, if necessary, in 2% fast green in 50% ethanol and wash in ethanol.
9. Air-dry the slide.
10. Mount in any suitable mountant such as Eukitt.
11. Observe under a microscope, preferably in the dark.

Result

Regions of the tissue containing cellulose show bright silver particles.

Rationale of the Reaction

Cellobiohydrolase provided to the tissue first reaches its substrate-containing (i.e., cellulosic) regions in the tissue. The enzyme is first tagged to the marker, i.e. gold. The gold particles bound to the enzyme provide a signal too weak to be clearly seen at the light microscopic level. Silver reduction amplifies this weak single. As a result, a cloud of reduced silver is seen around the gold particles, the whole becoming distinctly visible.

Controls

Run controls with an enzyme-gold complex solution to which carboxymethyl cellulose (CMC) has been added, at 1.0 mg ml^{-1} final concentration, 1 h before use.

Comments

1. If gold-labeling is done with the resin-embedding, background problems arise during silver intensification. Therefore, gold-labeling should preferably be done on plastic-de-embedded sections.

2. Adsorption of macromolecules such as enzymes to gold particles is not based on chemical covalent cross-linking but rather complex electrochemical interactions. Therefore, the bound cell wall molecules essentially retain their biological activity.

3. Silver enchancement of colloidal gold is a very sensitive method capable of detecting low levels of reactivity on conventional thin sections, cryosections, semithin resin sections, and paraffin sections. This method also allows correlative localization studies with light microscope and electron microscope.

Application of this Procedure

Benhamou[98] studied cellulose in plant and fungal cell walls (host-parasite interaction) using this techinique. Vian and Roland[97] used this protocol to study the location of cellulose in the cultured tissues of *Cucumis melo* raised from seedling epicotyl. They have also used this procedure to study the cells of the elogating zone of the mung bean (*Vigna radiata*) hypocotyl. They found that the primary cell walls were labeled very intensely, especially the outer thick walls of the epidermal cells and the walls between parenchyma cell corners. The young cell walls separating recently divided cells were weakly labeled.

Procedure 4a: Preparation of Colloidal Gold[99-104]

Colloidal gold can easily be obtained from commercial sources. But those who are desirous of preparing it themselves can do so by following any one of the three methods outlined below:

Colloidal gold (CG), can be prepared by reduction of gold chloride (also called chloroauric acid, HAu_3Cl_4) with sodium citrate (CG cit), sodium ascorbate (CG ase), phosphorus-ether (CG phos), tannic acid (CG tan), or sodium thiocyanate (CG thio). To avoid flocculation of the colloid during its peparation, filter the solutions (0.45 µm pore size, millipore) prior to use and manipulate it in siliconized glassware or polystyrene tubes. Wrap the glassware in aluminium foil, as otherwise the gold solution will be affected by light.

CG cit

1. Add 0.5 ml of 1% sodium citrate (w/v in water to 50 ml of deionized water heated under reflux).

2. Immediately add 0.5 ml of 4% (w/v in water) gold chloride.

3. Boil the solution under reflux for 15 min, until a light red (absorbance at 526 nm) colloid is obtained.

4. Cool the colloid.

5. Adjust its pH to 4.8 with 0.2 M K_2CO_3, using pH paper.

The gold particle size will be around 24 nm diameter and there will be 33 particles/ml.

CG ase

1. Add 1 ml of 1% gold chloride and 1 ml of 0.2 M K_2CO_3 to 25 ml of deionized water, at 4°C.

2. Stir well and quickly add 1 ml of 0.7% sodium ascorbate. The color should become purple–red.

3. Adjust the volume to 100 ml with deionized water.

4. Heat the solution under reflux until the color becomes orange–red (absorbance 523 nm).

5. Cool and adjust the pH to 7.5-8.00 with 0.2 M K_2CO_3, using pH paper.

The gold particle size will be around 13 nm diameter and there will be 160 particles/ml.

CG phos

1. Add 2.5 ml of 0.6% gold chloride and 0.7 ml of 0.2M K_2CO_3 to 60 ml of deionized water. This is stock solution 1.

2. Dilute at 1:4 anhydrous ethyl ether saturated with phosphorus (also termed yellow phosphours)* with ethyl ether. This is stock solution 2.

3. Mix the entire volume of solution A with 1 ml of solution B and shake for 15 min at room temperature.

4. Heat the solution under reflux until its color turns to brown to red (which approximately takes 10 min) (absorbance 518 nm).

5. Immediately cool the solution on ice.

6. Adjust the pH to 7.5-8.0 with 0.2 M K_2CO_3, using pH paper.

The gold particle size will be around 5 nm diameter and there will be about 3500 particles/ml.

Stabilized colloidal gold, prepared by any of the above three methods, can be stored for months at 4°C, without any adverse effect.

---

* White phosphours is highly toxic and flammable. So, it must be handled with caution, stored and manipulated under water, and transferred rapidly to the ether. Excess phosphorus-containing solutions should be decomposed before being discarded by adding excess aqueous copper sulphate.

Comment

The size of gold particle ranges from 2 to 150 nm and this can be adjusted by varying the reductant used. Large size (20 to 45) particles are ideal for light microscopy, while smaller ones (5 nm) are good for EM work. However, the relative labeling efficiency and selectivity by colloidal gold particles decrease with an increase in their size.

Procedure 4b: Preparation of Cellobiohydrolase-Colloidal Gold Complex

The enzyme-gold complex can be prepared as per the following procedure.[105,106]

1. Adjust the pH of the 10 ml of 15 nm colloidal gold solution to 4.5.
2. Add a 100 μl aliquot of a stock solution of cellobiohydrolase (2 mg.ml[1]) under stirring to the gold solution.
3. After 5 min add 500 μl of 1% polyethylene glycol (PEG) to the enzyme-gold mixture.
4. Centrifuge the latter at 14000 rpm for 1 h at 4°C.
5. Collect the mobile pellet in 5 ml of 0.05 M citrate-phosphate buffer, at pH 4.9 to which 0.02% PEG was added. (Discard the supernatant as it will contain free enzymes that may compete with the enzyme-gold complex during labeling.)
6. Centrifuge the complex again under the same conditions and collect the pellet a second time in the same buffer.
7. Store the complex at 4°C.

2.6.1.3   Chitin

Procedure 1: Potassium Hydroxide-Iodine-Potassium Iodide Procedure.[90,107-109]

Tissue Preparation

Fresh material is preferable.

Procedure

1. Coat the material in celloidin. Treat the material in autoclave at 15 psi and 121°C in 23 M KOH for 10 to 15 min. A 21.4 M KOH was used by Kaminskyj and Heath.[108]
2. Gently remove the KOH through a syringe.
3. Air-dry the slide, coat again with celloidin, and hydrate.
4. Stain in iodine-potassium iodide solution in 1% suphuric acid.
5. Drain off excess solution and mount in 1% aqueous H_2SO_4.

Result

A violet or red-violet color is indicative of chitin. Brown color indicates a negative reaction.

Rationale of the Color Reaction

Potassium hydroxide converts chitin into chitosan (= alkalinated chitin). Chitosan is stained a violet or violet-red by the iodine-potassium iodide solution in 1% H_2SO_4.

Control

After autoclaving in potassium hydroxide, the material is treated in 2% v/v aqueous acetic acid in which the chitosan is soluble. Subsequently, proceed with the staining procedure outlined above.

Application of this Procedure

Van Wisselingh[90] first used this technique to study the chitinous cell walls of fungi. Subsequently it has been used by many investigators in several fungi (see details in Aronson[91]). Now it has become the routine method by which chitin is localized.

A fine case of the use of this procedure is the study of cell walls in the fungus *Rhizidiomyces*.[110] This procedure was used in conjunction with differential extraction procedures, specific staining for cellulose and use of bright field and polarization optics. Based on these, the location of cellulose and chitin in the cell walls of this fungus was determined very successfully (also see Jensen[60] for details).

This test has been used by Pearlmutter and Lembi[96] to show the unequivocal presence of chitin in the cell walls of the green alga, *Pithophora*; the presence of chitin in yeast cell walls has been demonstrated[109] on the basis of this procedure.

Procedure 2: Chlor-Zinc-Iodide Method[28]

Tissue Preparation

Use fresh, freeze-dried, freeze-substituted, or chemically fixed (and embedded) tissue.

Procedure

1.   Treat with warm 50% KOH or diaphanol (chlorodioxyacetic acid).
2.   Wash with alcohol.

3. Treat with chlor-zinc-iodide reagent (iodine 6.1 g; KI 10 g; zinc chloride 60 g; distilled water 14 ml).

Timing in Steps 1 and 3 is to be standardized for the tissue to be studied.

Result

Chitin stains reddish violet in KOH-treated tissue and violet if treated with diaphanol.

Rationale of the Color Reaction

The basis of the reaction is not very clear.

Control

As in Procedure 1 for chitin detailed earlier.

Comment

This reaction works much better in the test tube than under histochemical conditions.[28]

Procedure 3: PAS Reaction[58,60]

Tissue preparation and procedure similar to PAS technique (Page 53). Chitin becomes a red color. The only precaution to be taken is that the materials should not be air-dried before mounting. It is to be mentioned, however, that arguments concerning the presence or absence of positive staining in chitin by the PAS have not been resolved in a satisfactory manner.[28]

2.6.1.4   Callose

Procedure 1: Lacmoid* blue method[111,112]

Tissue Preparation

Fresh, freeze-dried, freeze-substituted, or chemically fixed tissue can be used. Materials embedded in paraffin or other media can also be used. CRAF III was found to be the most successful fixing fluid.

---

* Lacmoid is very often confused with resorcin blue or resorcinol blue. Lacmoid is really the non-brominated homolog of resorcin.

Procedure

1. Bring the material down to water.
2. Stain for 15 to 72 h in 0.17 to 0.25% lacmoid blue in 30% ethanol.
3. Wash thoroughly in tap water.
4. Place the material in 1% sodium bicarbonate in 50% ethanol for a few seconds to 10 min.
5. Dehydrate, clear, and mount if necessary or otherwise it can be examined in the medium mentioned in (4) above.

Result

Callose stains a greenish blue.

Rationale of the Color Reaction

The mechanism of the reaction is not very clear.

Controls

1. Treat materials in 5% sodium hydroxide, in saturated calcium chloride,[112] stannous chloride,[113] or in glycerine heated to 280°C[113] for variable periods depending on the material. These treatments will remove callose from the material. Then, proceed with the protocol.
2. Subject the materials to Lyticase, a β-1-3-glucanase enzyme, 5 mg/ml before staining with lacmoid blue. β-1-3-glucanase from other sources (e.g., helicase from the gut of the snail *Helix pomata*) can also used.[112]

Comments

1. This method is more specific and dependable than the aniline blue technique.[112]
2. O'Brien and McCully[49] have suggested that for bright field microscopy, the material should be stained in resorcinol blue rather than in lacmoid. Prepare this by dissolving 3 g pure white resorcinol in 200 ml distilled water. Add to this 3 ml of 0.88% NH₃. Heat the mixture for 10 min without boiling in a steam bath. Store the reddish brown solution in a bottle stoppered with cotton until a dark, bluish color appears. Heat on steam bath for a further 30 min, filter hot solution into evaporating vessel, and heat until no significant NH₃ escapes by checking with moistured red litmus paper. Store stock solution in dropping bottle. Stain by mounting materials in drops of stock solution freshly diluted with 10 ml of water. Callose stains to a cobalt blue. Lignified walls also

stain, but to a red color when sections are mounted in buffer at pH 3.2, whereas callose remains cobalt blue. For photography, mount in aqueous mountants.

3. Cheadle et al.[111] used this method in combination with tannic acid and ferric chloride schedule for studying sieve elements.

Application of This Procedure

Lacmoid blue and the related resorcinol blue have been primarily used to localize callose in the sieve tubes and sieve plates of a variety of vascular plants.[111,114] It has not found a wider application, although rightly deserves so. Reynolds and Dashek[110] found that lacmoid blue was more specific and dependable in localizing pollen tube wall callose than the other callose localizing techniques they had used. Indra[115] had used it to study the pattern and timing of callose deposition and depletion in meiocytes during micro- and mega-sporogenesis in *Ottelia alismoides*. Senthil Kumar and Krishnamurthy[116] have used this method to identify the insulating material that separates the fungal pelotons (hyphal balls) from the remaining part of the host cell as callose in the mycorrhizae of the orchid *Spathoglottis plicata*.

Procedure 2: Aniline Blue Method[86]

Tissue Preparation

Fresh tissue is preferable; if fixed in FAA, the material should be washed several times in water.

Procedure

1. Bring the material to water.
2. Place the material for about 1 h in a dilute aqueous solution of aniline blue (0.005%) or for 4 to 24 h in a 0.005% solution of aniline blue in 50% ethanol.
3. Extract the excess stain by treatment with glycerine.
4. Mount in gelatin; if embedded tissue has been used, dehydrate, clear, and mount in DPX.

Result

Callose stains a clear blue.

Application of This Procedure

For a very long time aniline blue has been used routinely by light microscopists to localize callose in sieve plates and pollen tubes.

Procedure 3: Rosalic Acid Method[113]

Tissue Preparation

Fresh tissue is preferable.

Procedure

1. Stain the material in 1% solution of rosalic acid in 4% aqueous sodium bicarbonate.
2. Mount and observe.

Result

Callose stains a red color.

2.6.1.5 Sulphated and Carboxylated Polysaccharides (Including Pectins, Some Hemicelluloses, Alginic Acid, Fucoidans, Carrageenans, and Ascophyllans)

Procedure 1: Alcian Blue 8G* Method[117-122]

Tissue Preparation

Fresh, freeze-dried, freeze-substituted, or chemically fixed and embedded tissue can be used. Scott and Dorling[122] recommended neutral phosphate-buffered 4% formaldehyde or 95% ethanol at 4°C as fixatives. If cryostat sectioned material is to be used, fix it in picrate-formalin for 5 min before proceeding further.

Procedure

1. Bring the material down to water.
2. Stain for 45 min at room temperature in the staining solution. Scott and Dorling[122] recommended staining for at least 8 h and insisted that less staining time would give

---

* Formerly 8GX. Use only Alcian Blue 8G and not Alcian Blue 8GS, 5GX, or 7GX, as the former is more soluble and less resistant to decoloration than the others. Commercially available lots of alcian blue contain about 49% dye, the remaining constituents being boric acid sulphates, dextrin, and other unknown organic substances. It can be purified by the method detailed below:
1 g Alcian blue is placed in a flask to which is added 100 ml of a 9:1 mixture of acetone and distilled water. This mixture is stirred magnetically for 1 h and then filtered through No. 2 filter paper. Quantities of additional acetone are used to wash the dye from the flask and down the sides of the filter paper. The filtrate has a faint yellow green color. The residue is recovered from the filter paper and dried. A small quantity of the dye should be purified as it is needed. This method yields alcian blue containing 78 to 84% dye depending on the commercial source.

innaccurate results. However, our experience with a wide variety of tissues from algae to angiosperms showed that good results can be obtained by staining with the given dye concentration for about 45 min. The staining solution should be prepared as follows: Dissolve 3 g alcian blue 8GX in 100 ml of 0.1 N sodium acetate buffer, pH 5.6. Solution of alcian blue at pH 5.6 is unstable at room temperature, but the stability can be increased by the method of dye preparation by Scott and Dorling[122] who suggested that 0.05% alcian blue, pH 5.8 in 0.025 M sodium acetate buffer added with 0.05% magnesium chloride. Smith[226] used a 0.2% dye solution at pH 2.5 to study only sulphated polysaccharides. Cole et al.[123] used a 3% dye solution in 0.01 M HCl at pH 1.0 to stain sulphated polysaccharides (result is an aqua color) and a 3% solution in 3% acetic acid at pH 2.5 for both sulphated and carboxylated polysaccharides (both stain to an aqua color).

3. Wash in tap water for 5 min, dry, mount, and observe.

Results

Sulphated and carboxylated polysaccharides get stained to an aqua blue color.

Rationale of the Color Reaction

Carboxylated and sulphated polysaccharides are negatively charged either due to carboxyl groups of uronic acids or to sulphate esters or both. Therefore, they bind to this cationic dye by salt linkages at the isoindole-conferred cationic sites.[124] The staining reaction is due to electrostatic linkages between the polyvalent basic dye molecules and the anionic sites on the substrate. At a very strongly acidic pH, only sulphate moieties appear to retain their negative charge and bind to alcian blue.[125]

Controls

The control procedures depend upon the type of carboxylated or sulphated polysaccharides to be localized. For sulphated fucoidans, Smith[75] followed methylation as the control step while O'Colla[126] boiled the materials for 10 min in distilled water before staining. La Claire II and Dawes[127] did methylation with 4% methanolic $SOCl_2$ for 4 to 6 h. This process esterified carboxylated polysaccharides, but the carboxyl groups can be restored by subsequent saponification. For alginic acid, the material should be treated with 1 to 3% sodium carbonate solution at 20 to 30°C for 2 to 3 h in a sodium salt of EDTA for a similar duration under similar conditions.

Comments

Results with angiosperm tissue and brown algae indicate that some phenolic compounds were also stained by alcian blue.[128]

Application of This Procedure

Smith[75] used this procedure to investigate the presence of sulphated polysaccharides in the rhizoidal wall of the gametophytes of *Polypodium vulgare*. Heslop-Harrison[129] and Shivanna et al.[130] used this method to study pectins in the stigma cells. Watt et al.[67] used this technique to detect acidic polysaccharides in the cystoliths of *Pilea cadierei*. Simpson[131] used this technique to demonstrate pectins in the developing pollen walls of *Xiphidium coeruleum* (Haemodoraceae).

Procedure 2: Toluidine Blue O Method[21,132,133]

Tissue Preparation

Fresh, freeze- dried, freeze -substituted, or chemically fixed and paraffin-embedded (or in other embedding media) tissue can be used. If acrolein is used as a fixative, it may induce the formation of aldehyde groups in the tissue; therefore, materials must be immersed in chlorous acid[134] for 30 to 60 min to block the tissue aldehydes.

Procedure

1. Bring the material down to water.
2. Stain for 1 to 5 min in the dye solution [0.05% w/v toluidine blue O in benzoate buffer, pH 4.4 (benzoic acid 0.25 g and sodium benzoate 0.29 g in 200 ml water)]. O'Brien et al.[133] suggested a 0.05% toluidine blue O in 0.1 M phosphate buffer, pH 6.8. At higher pH, the stain can localize both carboxylated and sulphated polysaccharides. But for specific metachromasy (pink color) (Y metachromacy) of sulphated polysaccharides, a lower pH is usually preferred. The pH of dye solution has to be adjusted to 1.0 to 1.5 using 1 N HCl and benzoate, phosphate, or phosphate-citrate buffer.
3. Wash in running water until most of the excess stain has washed out.
4. Preferably air-dry and mount. Dehydration is not advised. If desired, it should be carried out very fast; otherwise, ethanol or its substitutes will remove the metachromatic effects of toluidine blue O.

Results

Sulphated and carboxylated polysaccharides stain a metachromatic pink to reddish pink. An orthochromatic blue color is taken by lignin and a green color by phenols, if present.

Rationale of the Color Reaction

The production of red and purple color with toluidine blue O is thought to be due to the formation of staked dye ions. Water is essential for the formation of these dye aggregates. If treated with ethanol, stacking is inhibited strongly as the material becomes dehydrated and loses water.[12]

Controls

1. Methylation: Mowry[135] has demonstrated that methylation esterifies carboxyl groups. So, for demonstrating sulphated groups, methylation is a control step. The carboxyl groups can be restored by subsequent saponification as they become demethylated, thus restoring *basophilia.*[75,127]
2. Watt et al.[67] used a treatment in 2% calcium chloride before toluidine blue O staining. This treatment removes metachromasy.

Comments

1. Cellulose is not stained by this method. In fact, the negative reaction is a strong proof for the presence of cellulose.[136]
2. The pink or red metachromasia shown by sulphate groups is ethanol resistant[75] while that of carboxylated polysaccharides is not resistant.

Application of This Procedure

This is one of the very extensively used techniques to study cell walls, especially those of brown and red algae and those of the gelatinous fibers of tension woods. Dawes et al.[137] used this technique to study the cell walls of the hapterons of the brown seaweed *Laminaria*; Evans and Holligan[95] studied *Dictyota*; Cole et al.[123] studied *Bangia atropurpurea*; McCully[138] studied *Fucus*; Quatrano and Crayton[139] and Quatrano et al.[140,141] studied *Fucus distichus* zygotes (to study sulphated polysaccharides); LaClaire II and Dawes[127] studied *Eucheuma nudum* (to study carrageenans); Balakrishnan[128] studied *Padina boergesenii* (to study alginic acid and fucoidans); and Waterkeyn and Bienfait[142] studied the diatom *Pinnularia* to study sulphated polysaccharides, etc.

The chemical nature of gelatinous fibers of tension wood of dicotyledons was elucidated mainly by employing this technique.[73,74,85] The inner gelatinous layer showed a characteristic metachromatic pink or red color with this dye at a higher pH (4.5 to 6.8) indicating the presence of carboxylated polysaccharides in it (of course, in addition to cellulose), while the outer layer of the fiber wall showed an orthochromatic blue indicating its lignin nature. A number of other investigators have also used this method to study tension wood fibers.[79-81]

Watt et al.[67] have used the procedure to localize acidic polysaccharides and, in combination with ethanol solubility, non-methylated pectins (carboxylated polysaccharides) in the cystolith matrix and stalk of *Pilea cadierei*.

Procedure 3: Alcian Blue–Alcian Yellow Method[143-145]

Tissue Preparation

Fresh, freeze-dried, freeze-substituted, or chemically fixed tissue can be used. Chemically fixed tissue can be embedded in paraffin or other substances and the microtome sections can be used. For marine forms, fixation in 10% formalin in sea water can be used. If ribboned materials are used, do not employ adhesives on the slide.

Procedure

1. Bring the material down to water
2. Stain for 30 min in filtered alcian blue 8GX (0.5 g in 100 ml acid water; pH adjusted to 2.5 with 1 N HCl).
3. Wash for 10 min in acid water (pH 0.5) and then in distilled water.
4. Stain for 30 min in filtered alcian yellow (0.5 g in 100 ml acid water; pH adjusted to 2.5 with 1N HCl).
5. Wash in distilled water
6. Dehydtrate, clear, and mount.

Results

Sulphated and carboxylated polysaccharides are, respectively, stained blue and yellow.

Rationale of the Color Reaction

The exclusive reaction of alcian blue with sulphated polysaccharides at pH 0.5 is explainable by the fact that carboxyl groups are no longer dissociated below pH 1.0. Thus, only the dissociated sulphate groups complex with alcian blue. Subsequently at pH 2.5, alcian yellow complexes with the dissociated carboxyl groups but does not react with the sulphate groups as they have already been bound to alcian blue.

Controls

See Page 71.

Comments

For studying sulphated and carboxylated polysaccharides separately, the concerned parts of the procedures alone can be followed.

Application of the Procedure

This procedure has been used thus far only in some algae where alginic acid (carboxylated polysaccharide) occurs along with fucoidan or carrageenan (sulphated polysaccharide). LaClaire II and Dawes[127] used this procedure to study cell wall polysaccharides of *Eucheuma nudum*, Evans and Holligan[95] looked at *Dictyota*, and Dawes et al.[137] looked at *Laminaria*. Recently, Balakrishnan[128] has used this procedure to study the relative distribution of alginic acid and fucoidan in the thallus of *Padina boergesenii*. He found that the intercellular matrix is rich in sulphated polysaccharides while the wall proper contained both alginic acid and sulphated polysaccharides or predominantly alginic acid.

Procedure 4: Iron Diamine–Alcian Blue Method[28,146]

Tissue Preparation

Fresh, frozen,or chemically fixed tissue can be used. The last can be paraffin/resin embedded.

Procedure

1. Bring the material to water.
2. Hydrolyze the material for 10 min in 1 N HCl at 60°C to overcome interference by nucleic acids.
3. Wash for 5 min in running water.
4. Oxidize in 1% aqueous periodic acid for 10 min.
5. Stain in the dye reagent for 18 h. Use either high iron diamine (HID) or low iron diamine (LID) methods. HID specifically demonstrates sulphated polysaccharides, while LID demonstrates both. Prepare HID reagent in the following way:

N,N- Dimethyl-m-phenylenediamine (HCl)$_2$	120 mg
N,N- Dimethyl-p-phenylenediamine HCl	20 mg
Distilled water	50 ml

    Add 1.4 ml of 40% (w/v) ferric chloride to the above solution.
    Adjust the pH of the solution to 1.3 to 1.6.

Prepare LID reagent as per the following procedure:

N,N-Dimethyl-m-phenylenediamine	30 mg
N,N-Dimethyl-p-phenylenediamine	5 mg
Distilled water	50 ml

Add 0.5 ml of 40% (w/v) ferric chloride to the above solution.
Adjust the pH to 2.5 by adding sufficient 0.2 M Na$_2$HPO$_4$. Prepare the solution fresh before use.

6. Rinse very rapidly in water.
7. Stain in 1% alcian blue in 3% acetic acid for 30 min.
8. Observe. If permanent slides are desired, dehydrate, clear, and mount in synthetic resin.

Results

This procedure stains sulphated polysaccharides black and carboxylated polysaccharides aqua.

Control

Incubate the control specimens for about 18 h at 22°C in MgCl$_2$ solution which is prepared by adding 0.5 ml of 40% MgCl$_2$ to 50 ml of distilled water and adjusting the pH to 1.8 with HCl.

Application of This Procedure

Cole et al.[123] have used this procedure on the cell walls of *Bangia atropurpurea* (red alga) in order to identify the nature of their polysaccharide components.

Procedure 5: Astra Blue Method[121,147]

Tissue Preparation

Fresh, freeze-dried, freeze-substituted, or chemically fixed and embedded tissue can be used; neutral phosphate-buffered 4% formaldehyde was very effective among the fixatives.

Procedure

1. Bring the material down to water.
2. Stain for about 1 h at room temperature in the staining solution. A 1% aqueous solution was found to be good, although for more critical localization, dissolve 1 g dye in 100 ml 0.1 N sodium acetate buffer, pH 5.6. For sulphated polysaccharides, the pH of the dye solution is to be less than 2.5.
3. Wash in tap water and observe. If permanent slides are desired, dehydrate the material, clear, and mount.

Results

Acidic polysaccharides give a brilliant blue color.

Comments

Best results are obtained when used along with safranin in vascular plants. For cryptogamic materials, the dye works very well singly.

Application of This Procedure

This procedure was used by Höster and Liese[147] to study the nature of the walls of gelatinous fibers in tension woods. We have extensively used this method to study G-fibers of tension woods in a wide variety of dicots.

Procedure 6: Toluidine Blue O–Aluminium Sulphate Method[148]

Tissue Preparation

Fresh or chemically fixed tissue can be used. Avoid fixation in Zenkler's fluid. If necessary, materials can be embedded in paraffin or similar media and processed for microtomy.

Procedure

1. If embedded, section the material at about 6 to 10 μm thickness, deparaffinize and bring down to water. Other materials are also to be brought to water.
2. Stain the material in the staining solution for 5 to 30 min. (Dissolve 0.1 g of toluidine blue O in boiling 5% aluminium sulphate solution. This is the stock solution. Dilute this 1 to 2 with further 5% aluminium sulphate solution before use.)
3. Differentiate very rapidly in 70% ethanol, dehydrate, clear and mount. If differentiation is done very slowly the metachromatic dye color is removed by ethanol.

Result

Sulphated polysaccharides like fucoidan get stained pink to reddish purple.

Rationale of the Color Reaction

It is likely that the dye forms a coordination compound with the metallic cation. This coordination compound and the polysaccharides interact to produce the color in a reaction similar to mordanting in the textile dyeing.

Controls

Methylate the materials prior to staining.

Comments

1. Toluidine blue can be substituted by methylene blue, neutral red, or nuclear fast red, and equally good results can be obtained.
2. If pink or red coloration developed by the material is removed by ethanol treatment, it is a good evidence for the presence of sulphated polysaccharides. Ethanol has a solubilizing activity on toludine blue O.[67]

Application of This Procedure

This technique has been used in the rhizoids of the gametophytes of *Polypodium vulgare* to find out the chemical nature of their cell walls.[75] A metachromatic staining reaction was not found indicating the absence of sulphated polysaccharides in the walls.

Procedure 7: Krajcinovic Amine Test[28]

Tissue Preparation

Fresh or chemically fixed tissue can be used. Materials can be dehydrated, cleared, and embedded and the sections of the embedded tissue can be used.

Procedure

1. Bring the material down to water.
2. Treat for 2 min in 1% HCl.
3. Wash in three changes of absolute ethanol.
4. Teat with 0.1 M benzidine* for 45 min.
5. Wash in four changes of absolute ethanol.
6. Diazotize by immersion for 1 min in 0.1 N HCl with 0.1 N sodium nitrite (3:1).
7. Wash in water.
8. Couple with 0.1M ß-naphthol.
9. Wash in water and mount in glycerol.

Results

Pectins stain red.

---

* It is a carcinogen and adequate care should be taken while using it.

Rationale of the Color Reaction

An addition compound between pectic acid and benzidine is formed. After diazotization of the free amino group, which remains at one end of the molecule, an azodye was formed by treatment with ß-napthol in alkaline solution. This gives a red color.

Controls

1. Dissolve pectin in alkaline carbonates or in ammonia before proceeding to stain.
2. Subject the material to ammonium oxalate extraction before staining. The treatment should be done for 12 h at 25°C. This procedure may not be very effective if pectin is very tightly bound with other polymers such as lignin.
3. Pectinase treatment prior to staining.

Application of This Procedure

Although Pearse[28] did not recommend this method, it was found that this is a very useful method to localize cell wall pectins of not only wood cells,[74] but also of soft parenchymatous cells of fruits.[149] The latter author, by using this method, was able to record precisely the changes in cell wall pectin in ripening banana fruit tissues.

Procedure 8: Ruthenium Red Method[86,136]

Tissue Preparation

Fresh, freeze-dried, freeze-substituted, or FAA fixed materials can be used. Materials embedded in paraffin or other embedding media can also be used.

Procedure

1. Bring the material down to water.
2. Place the material in aqueous ruthenium red solution (1 g in 5000 ml of water). Jensen[60] and Smith[75] used a 0.0001% aqueous solution. We have used a 0.01% solution in 0.1% ammonium chloride, pH 9, adjusted with ammonium hydroxide. The solution has to be kept in the dark until use.
3. Observe.

Results

Non-esterified pectins get stained to a red or pink color.

Rationale of the Color Reaction

Ruthenium red (ammoniated ruthenium oxychloride) is an inorganic, synthetically prepared, intensively colored, crystalline compound. It has long been used as a standard pectin stain in plant tissue. Ruthenium red is a hexavalent cation (diameter 1.13 nm) and reacts with a large number of polyamines having high charge density. The basis of the coloring reaction is not very clear.[60] According to Sterling,[150] the staining group is the ruthenium ion and the associated complex of four ammonia molecules. Staining is effected when the pectin molecule has two negative charges 0.42 nm apart, so as to accommodate the staining group. The staining sites are intramolecular and the staining can be intensified if the pectins are de-esterified, since strongly negative carboxyl groups are reestablished. At low pH, the carboxyl groups are dissociated and the coordination ion of the stain may form a salt linkage (i.e., binding is electrostatic) with the anionic groups.[136,151]

Controls

1. Ammonium oxalate extraction prior to staining: Coat the sections with celloidin and hydrate them. Place the material on a slide kept horizontally flat in a petriplate, which is filled subsequently with 0.047 M ammonium oxlate solution. Incubate at 90°C for 12 h and remove the solution carefully through syringes. Air-dry the slide, coat again with celloidin, and keep it as control.

2. Pectinase treatment before proceeding to stain. Pectinase is a commercial name for polygalacturonase. It is present as a mixture in commercially available Driselase and Macerozyme, but it is also available in pure form as pectinase and pectolyase. Prepare a 1% w/v solution of pectinase in 0.005 M phosphate buffer at pH 5.8. To extract pectin, immerse the hydrated and celloidin coated section in this solution at 37°C overnight.

3. Treatment with 4% NaOH solution at room temperature for a period from 10 min to a few hours before staining.

Comments

1. Ruthenium red stains many other substances besides pectin (but cellulose and sulphated polysaccharides do not respond). On this basis, O'Brien and McCully[49] considered this technique as non-specific to pectins. Smith[75] also found that after treatment with ammonium oxalate or pectinase, staining of the walls of the rhizoid of *Polypodium vulgare* gametophytes was not eliminated. Therefore, he inferred that some other wall material other than pectins also get stained with ruthenium red. Certain lipids are also known to react with the stain.[151]

2. Pectic substances are stained only if they are present in high concentrations.

3. The staining reaction is interfered by the presence of other substances. Therefore, failure to get stained does not confirm the absence of pectins.

4. This technique is relatively unsuitable to plant tissues.

Application of This Procedure

This is one of the most widely used techniques to localize pectins especially in the early works but in view of the above, the results of such studies should be viewed with reservation.

Procedure 9: Alkaline Hydroxylamine Hydrochloride Method or Hydroxylamine-Ferric Chloride Method[60,152-154]

Tissue Preparation

Fresh tissue is preferable.

Procedure

1. Treat materials placed on a slide with a few drops of fresh alkaline hydroxylamine solution for 5 min or longer. [The solution is prepared by mixing equal volumes of sodium hydroxide (14 g in 100 ml of 60% ethanol) and hydraxylamine hydrochloride (14 g in 100 ml of 60% ethanol).]

2. Add to the slide 5 to 10 drops of a solution of a 1 part concentrated HCl and 2 parts 95% ethanol mixture. This acidifies the reaction mixture.

3. Remove the solution from the tissue and treat the materials with 10% ferric chloride solution in 60% ethanol containing 0.1 N HCl.

4. Observe. If need be, the materials can be dehydrated, cleared, and mounted in DPX.

Results

Esterified pectins stain red. Intensity of staining can be increased by methylation of tissue (i.e., by placing the tissue in a hot solution of absolute methanol containing 0.5 N HCl).

Rationale of the Color Reaction

Alkaline hydroxylamine hydrochloride reacts with the methyl esters of pectins to produce pectin hydroxamic acid. The latter reacts with ferric ions to produce red complexes that get deposited as precipitates.[153] The intensity is proportional to the degree of methyl esterification in pectin polymer. Hence, methylation of existing unesterified carboxyl radicals by treatment in 0.5 N HCl in absolute methanol (prepared by mixing 4.36 ml of con. HCl with 95.64 ml of absolute methanol) may be used for obtaining more intense coloring.

Controls

Any control procedure that can remove pectins can be used, but the best is deme-thylation, since this staining procedure is based on localizing mainly the methyl esterified pectin; demethylation (or saponification) may be used as an effective blockade of the reactive group of pectins. Demethylation is done by treating the celloidin-coated and hydrated sections with 15 w/v solution of KOH in 70% aqueous ethanol, for 20 to 30 min. It can also be done by incubating the sections in 1:4 v/v ethanol and commercial ammonia for 24 h.

Comments

1. This staining reaction permits a distinction between nonesterified and esterified pectins.
2. This method permits quantification of the pectin from stained materials.
3. It can be extended to the ultramicroscopic level.

Application of This Procedure

Watt et al.[67] used this method to detect methylated pectins in the cystoliths of *Pilea cadierei* but found negative results. Smith[75] used this method in the rhizoid of the gametophytes of *Polypodium vulgare*. This has also been used to study changes in cell wall pectin in the abscission zone of leaves[161] and in the ripening fruit tissue of banana.[149]

Quantification (through microspectrophotometry) of esterified pectin stained with this procedure has been successfully made by Gee et al.[155] and Reeve.[153] Using this, the degree of esterification of pectic substances has been found in fruits.

Procedure 10: Tannic Acid–Ferric Chloride Method[156]

Tissue Preparation

Fresh, freeze-dried, freeze-substituted, or chemically fixed (with or without embed-ment in wax or similar materials) tissues can be used.

Procedure

1. Bring the material down to water.
2. Place the material in a 1% aqueous solution of tannic acid for 10 min. (To this solution is usually added 1 g of sodium salicylate or a few drops of phenol to prevent the growth of fungi.)
3. Wash in running water for 10 min.

4. Place in 3% aqueous solution of ferric chloride for 3 to 10 min.
5. Wash and observe. If the cell walls are stained black or blue–black, the staining can be stopped. If not, repeat from the beginning until the above colors are obtained.

Results

Pectate of calcium is stained black or blue black.

Rationale of the Color Reaction

The exact mechanism of the staining reaction is not yet known.

Application of This Procedure

It is an excellent stain combination for middle lamella as well as for pectic materials of the cellulosic primary walls. Hundreds of research papers dealing with various aspects of developmental anatomy of vascular plants have used this procedure. Special mention must be made of its use in studying shoot and root apical meristems, developing leaves, and embryogeny. Tectonics of wall formation in developing embryos of dicots and monocots have been fully brought to light exploiting this method, leading to the correct understanding of the true morphology of cotyledon and shoot apex in monocot embryos.[157,158]

Procedure 11: Colloidal Iron Method[58,159]

Procedure

1. Bring the material down to water.
2. Flood the tissue with dialyzed iron solution for 10 min. The solution can be prepared by taking 1 volume of pharmaceutical grade of dialyzed iron (5% Fe_2O_3) and adding 1 volume of 2 N acetic acid. If not readily available, iron solution can be prepared by heating ferric chloride to boiling and then adding acetic acid to make a solution of pH 1.8.
3. Wash in water.
4. Immerse in acid-ferrocyanide solution for 10 min. This is prepared by adding 85 ml of 1% solution of potassium ferrocyanide and 15 ml of 1 N HCl.
5. Wash in water and observe. If needed, the material can be dehydrated and mounted in Euparal or DPX.

Result

Pectins (and many other acidic polysaccharides) take a blue color.

Rationale of the Color Reaction

Pectins, being acidic, are absorbed by colloidal Fe^{3+} which are positively charged. The iron is then stained blue by the acid ferrocyanide treatment.

Application of the Procedure

This method has been used by Smith[75] to study the cell walls of the rhizoids of *Polypodium vulgare* gametophytes. It has been found that the wall is positive for pectins.

2.6.2 Proteins

2.6.2.1   Non Specific Cell Wall Proteins

Procedure 1: Coomassie Brilliant Blue R 250 Method[160-162]

Tissue Preparation

Fresh or frozen tissue preferred. However, tissue fixed in 2.5% glutaraldehyde can be used.

Procedure

1.   Hydrate the material and wash thoroughly in distilled water.
2.   Treat the materials in a staining medium of 0.1% coomassie brilliant blue in methanol:water:acetic acid (25:73:2 v/v) with a sucrose concentration in the range of 10 to 20% chosen to balance the osmolarity of the material. A 0.25 g stain in enough of 7% aqueous acetic acid to make a volume of 100 ml was used as the staining solution.[162,163] A 0.02% solution of the dye in Clarke's solution, pH 2.0 was advocated by Cawood et al.[160] (see also Gahan[164]). Stain at room temperature for 3 min. The staining solution recommended by Cawood et al.[160] needs to be used for 24 h.
3.   Briefly immerse in 70% acetic acid to remove excess stain adsorbed to the tissue. Destaining in fresh Clarke's solution for 20 min is recommended if the stain was prepared in that solution.
4.   Blot the material and air dry. The material can also be dehydrated in ethanol.
5.   Mount in 5% v/v acetic acid solution in glycerol. Cawood et al.[160] suggest mounting in Euparol.

Result

Proteins appear blue.

Rationale of the Color Reaction

It is an anionic stain. It stains free end groups of proteins at low pH (similar to aniline blue black).

Control

1. Deamination: Treat the material in the deamination reagent (15 g of sodium nitrite in distilled water + 0.75 ml of glacial actic acid to make a total volume of 100 ml) at room temperature for 1 to 24 h.
2. Protein digestion using enzymes.

Application of This Procedure

This method was used to study stigma surface proteins of *Hypericum calycinum*.[130] It was originally used for the same purpose with a lot of success by Heslop-Harrison[129] in *Crocus* species. Lu et al.[165] emloyed this technique to study structural proteins of the stigma cell walls in groundnut.

Procedure 2: Mercuric Bromophenol Blue Method[166,167]

Tissue Preparation

Fresh, frozen, or fixed tissue can be used. Fixation in neutral buffered formaldehyde or FAA is ideal.

Procedure

1. Hydrate the materials. If sectioned materials are to be used, preferably employ sections less than 10 μm.
2. Stain in the staining mixture for 15 to 30 min. (Prepare the staining solution by dissolving 10 g mercuric chloride and 100 mg of bromophenol blue in 100 ml of distilled water or 95% ethanol.)
3. Wash in 0.5% acetic acid for 1 to 20 min. It removes the excess dye.
4. Wash in distilled water for not more than 3 min, otherwise the dye is removed.
5. Dip in absolute tertiary butanol for 1 min.
6. Place in absolute tertiary butanol for 2 min.
7. Clear and mount in DPX.

Result

Proteins bind the stain and give a blue color. A higher degree of dye binding due to more concentrated proteins may produce a metachromatic reddish blue color.

Rationale of the Color Reaction

Proteins are amphoteric compounds. A protein becomes positively charged (cationic) at a particular pH that is below its isoelectric point, and becomes negatively charged (anionic) at a pH that is higher than its isoelectric point. In the mercuric bromophenol blue staining process, the protein is brought into contact with the acid (anionic) stain in the first step. Rinsing in a 5% acetic acid subsequently brings the pH around tissue proteins down, so that all proteins with isolectric point well above the pH of 0.5% acetic acid become positively charged (or cationic) and form electrovalent salt with anionic bromophenol blue. Sulphydryl, phenolic, and carboxyl groups of proteins bind with bromophenol blue through the assistance of mercuric chloride.

Controls

1. Deamination (See Page 85).
2. Dinitrofluorobenzene treatment. (Dissolve 0.4 g of sodium hydroxide in distilled water to make a total volume of 10 ml. Mix 0.5 g 2,4 nitrofluorobenzene in 90% aqueous ethanol enough to make a volume of 99 ml. Add 1 ml of NaOH prepared earlier.) This treatment blocks all free amino groups, sulphydryl groups, hydroxyl groups, and imidazole groups.

Comments

It stains a wide variety of reactive protein groups; one can stain a desired group by blocking the rest.

Application of This Procedure

This procedure was used to study the rhizoids of *Polypodium vulgare* gametophytes whose cell walls stained pale blue indicating the moderate presence of proteins.[75]

Procedure 3: Naphthol Yellow S Method[164,168,169]

Tissue Preparation

Frozen material was used by Knox and Heslop-Harrison.[170] They also tried fixation in 3% glutaraldehyde in 0.1 M phosphate buffer at pH 7.0 at room temperature for

5 to 8 h or in acetic-methanol (1:3 v/v) at room temperature for 3 h or at 3 to 5°C overnight to prevent loss of wall proteins by diffusion. However, there was a loss of proteins from the walls. Tas et al.[169] and Gahan[164] suggest fixation for 2 h in absolute ethanol:glacial acetic acid (5:1) or in absolute methanol:40% formaldehyde:glacial acetic acid (17:2:1).

To prevent diffusion of wall proteins, pre-soak the material for 5 min at room temperature in half-saturated $(NH_4)SO_4$ before rinsing quickly in distilled water and staining.

Procedure

1. Bring the material down to water.
2. Stain in the staining medium for 30 min. (1% naphthol yellow in water; 0.1% solution in 1% acetic acid, pH 2.8 was recommended by Tas et al.[169] and Gahan.[164])
3. Differentiate the material in 1% acetic acid with constant agitation for 30 min, changing the bath at 0.5, 5, and 15 min, respectively.
4. Dehydrate in ethanol series; air drying was suggested by Tas et al.[169] and Gahan.[164]
5. Mount in balsam or euparol; in the latter case, brief rinsing in TBA three times is required.

Result

Sites of basic proteins (arginine rich) will stain yellow.

Comments

Aldehyde fixation and dehydration would considerably alter the stainability of the proteins;[169] in many cases, staining intensity was considerably reduced.

Application of This Procedure

Knox and Heslop-Harrison[170] used this method to study pollen grain wall proteins in several angiosperms.

Procedure 4: Amido Black 10 B Method[107]

Tissue Preparation

It is preferable to use fresh tissues.

Procedure

1. Section the material, if necessary.
2. Stain for 30 min in the staining mixture. (Prepare the mixture by adding amido black 0.5 g, mercuric chloride 5 g, glacial acetic acid 5 ml, and distilled water 100 ml. Filter the stain before use.)
3. Wash in 2% acetic acid for 5 min three times.
4. Wash in distilled water.
5. Mount in glycerine.

Result

Cell wall proteins produce a blue color.

Application of This Procedure

This method was used to study the fungal cell wall proteins by Heath and Perumalla.[107]

Procedure 5: Aniline Blue–Black Method[16]

Tissue Preparation

Use fresh or fixed tissue. In the case of the latter, fix in neutral buffered formaldehyde, FAA, or 3% glutaraldehyde. If sectioned materials are tried, use sections preferably below 10 μm thickness.

Procedure

1. Bring the material down to water.
2. Wash thoroughly in distilled water.
3. Stain the material in the staining mixture (1 g of aniline blue–black in enough of 7% aqueous acetic acid to make a total volume of 100 ml) at 50 to 60°C for 10 to 12 min.
4. Immerse the material in 7% aqueous acetic acid to remove excess dye adsorbed to the tissue.
5. Blot the material and mount in 5% v/v acetic acid in glycerol (v/v).

Result

Proteins stain deep blue to bluish black.

Controls

Although any control method may be used, it is preferable to use the deamination method.

Rationale of the Color Reaction

At low pH of the staining mixture, free amino groups of proteins become positively charged (or cationic) by acquiring protons from the solution. Cationic protein moiety then binds with the anionic aniline blue–black forming electrovalent salt. Since the process involves amino groups of proteins, deamination control is the most usually applied control for this staining.

Application

This protocol was used to stain cell wall proteins of fungi.[171] Vijayaraghavan et al.[62] used this method to show the absence of wall proteins in filiform apparatus of the synergids of *Aquilegia formosa*.

2.6.2.2   Enzymic Proteins and Glycoproteins

Peroxidases.   Peroxidases catalyze the reduction of H_2O_2 to produce water in the presence of a suitable electron donor. A broad range of electron donors are involved in this: mostly phenols, amines, and alcohols; some donors are oxidized to partially insoluble color products. The following are some of the most used donors: Benzidine was one of the first to be tried.[172] Its oxidation produces a blue product and so is useful in light microscopy. However, oxidation product of benzidine is neither osmiphilic nor of substantial electron density. The products are also somewhat soluble.[173] Therefore, it is not useful in electron microscopy; it is also of very limited use in light microscopy.

Diaminobenzidine (DAB) or 3, 3', 4, 4'-tetra aminobiphenyl is a diamino derivative of benzidine. It is a more efficient and more widely exploited reagent (electron donor) for light as well as electron microscopic studies on peroxidases[174] since its debut as a detector of horse radish peroxidase.[175] Essner[176] has remarked about DAB as follows: "By almost any criterion the DAB reaction emerges as the most sensitive and versatile cytochemical staining method available" for peroxidase demonstration. The reaction involved[177] is shown in Figure 2.1.

Reliable specificity of DAB is noticed in most cases but with judicious use and proper interpretation of controls. The DAB procedure has been successfully used to detect peroxidases in a variety of plant tissues.[178-185]

The common alternatives to DAB are N, N' bis (4-aminophenyl)-1,3, xylene diamine (BAXD), N, N' bis (4-aminophenyl)-N, N'-dimethyl ethylene diamine (BED),[186] tetramethyl benzidine (TMB),[187] and a combination of *p*-phenylenediamine and pyrocatechol (PPE-PC).[188,189] Other electron donors tried include ferulic acid-iso propylamine salt (FIS), syringaldazine (SYR), and guaiacol (GUA).

The common problems encountered in peroxidase localization include: (1) diffusion artifacts of the oxidized DAB, (2) diffusion of enzymes, (3) auto-oxidation of DAB oxidase to produce a brown product (this can be prevented if DAB is kept in the dark), and (4) endogenous generation of peroxide by hydrogen peroxidase systems. (This can be prevented by including a commercially purified catalase or sodium pyruvate in the medium.)

Materials for peroxidase localization preferably are to be fresh. Wachstein and Meisel[190] have used fresh frozen sections which they spread on lukewarm water and then put directly onto incubation medium. However, in general, peroxidase is relatively unaffected by fixatives. The effect of fixatives on peroxidase activity has been investigated by Al-Azzawi and Hall.[191] 1% and 5% glutaraldehyde and formaldehyde for 2 h at 4°C, respectively, had an inhibition percentage of 14, 12, 7, and 8. The use of fixation seems to help in minimizing the effects of diffusion of enzymes and in improving the rate of penetration of DAB into tissues which is normally low.[192] For the

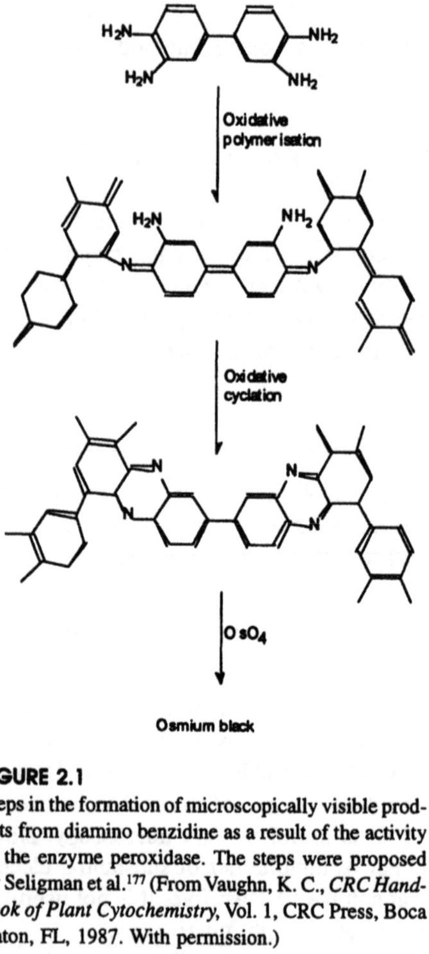

FIGURE 2.1
Steps in the formation of microscopically visible products from diamino benzidine as a result of the activity of the enzyme peroxidase. The steps were proposed by Seligman et al.[177] (From Vaughn, K. C., *CRC Handbook of Plant Cytochemistry*, Vol. 1, CRC Press, Boca Raton, FL, 1987. With permission.)

benzidine method, Straus[193] has used a 10% formalin plus 30% sucrose, 0 to 4°C for 18 h, but not less than 15 h. The tissue should be as small as possible since DAB and other electron donors do not easily penetrate into tissues.

Procedure 1: Benzidine Method[19]

Incubation Medium

50 to 70% ethanol 50.0 ml
Benzidine 0.1 to 0.2 g
Hydrogen peroxide 0.015 ml.

Procedure

1. Treat the materials in ice cold incubation medium until peroxidase is distinctly blue. Stop reaction before blue pigment crystals form.
2. Transfer to nitro prusside solution A, ice cold for a few seconds (sodium nitroprusside 4.5 g + 70% ethanol 50.0 ml).
3. Transfer to nitroprusside solution B, ice cold, in refrigerator, 1 to 2 h (30% aqueous sodium nitroprusside 15.0 ml + absolute methanol 35.0 ml + 0.2 m acetate buffer, pH 5.0, 2.5ml).
4. Observe. If permanent slides are required, dehydrate, clear, and mount in resinous mountant.

Result

Peroxidase sites stain grey green to black.

Rationale of the Reaction

Peroxidases catalyze the transfer of two electrons from substrate to hydrogen peroxide to form water and oxidized dyes. Benzidine serves as an acceptor in the transfer of O_2 from H_2O_2. A blue product is formed and in a short time fades and diffuses. Treatment with nitroprusside stabilizes this reaction product by forming a more stable salt.

Controls

Incubate without H_2O_2 or incubate heat-killed materials.

Procedure 2: Benzidine Method[190]

Incubation Medium

Sodium ferricyanide	600 mg
Benzidine	600 mg
25% ethanol	100 ml

Retain the incubation medium in the icebox for about 30 min. Filter. Before use, add 6 drops of 30% solution of H_2O_2.

Procedure

1. Quench the materials in isopentane or petroleum ether for a very short time if fresh tissue was used. If the tissue was already frozen, there is no need of the above treatment.
2. Bring the materials to 4°C for staining.

3. Subject the slides to the incubation medium for 5 to 30 min, at 4°C.
4. Wash in distilled water and mount in glycerogel.

Result

Sites of peroxidase activity are stained to a blue color, which become brown immediately.

Rationale of the Reaction

As in Procedure 1.

Controls

Use boiled sections or subject the materials to incubation medium minus the substrate.

Procedure 3: BAXD and BED Method[186]

Tissue Preparation

Fix in 4% depolymerized paraformaldehyde (dissolved in 0.05 M phosphate buffer, pH 7.0) at 0 to 4°C for 30 min in phosphate buffer (0.5 m, pH 7.0), which contains 5% sucrose.

Incubation Medium

0.05 M phosphate buffer, pH 7.0, BAXD or BED 1 mg/ml.

Procedure

1. Incubate the material in incubation medium at 37°C for 50 to 60 min.
2. Rinse tissue for 30 min in cold phosphate buffer (0.05 M, pH 7.0) containing 5% sucrose.
3. Observe.

Result

Black to brown deposits indicate places of peroxidase activity.

Controls

Treat tissue at 80°C in 4% formaldehyde for 20 min or use medium without substrates.

Application of This Procedure

This method has been used to study the peroxidase distribution in the root tips of corn.[186]

Procedure 4: Diaminobenzidine (DAB) Method[174,178-185,192,194,195]

Tissue Preparation

Use fresh, frozen, or fixed tissue. Fix for 4 h in cold 5% glutaraldehyde in 0.1 M phosphate buffer, pH 7.5. Wash at least overnight in 0.1 M phosphate buffer containing 5% sucrose. Alternatively, fix for 5 h at room temperature containing 4% formaldehyde and 5% glutaraldehyde in 0.1 M cacodylate buffer, pH 7.2, added with 25 mg of $CaCl_2$. Wash tissue at least overnight in 0.1 M cacodylate buffer, pH 7.2.

Incubation Medium

10 ml of 0.05 M (or 50 mm) tris-HCl buffer, pH 7.6, 5 mg of DAB* (it is necessary to warm the obtained solution in order to increase the solubility of DAB. This should be cooled and filtered prior to the addition of H_2O_2) and 0.1 ml of freshly diluted 1% H_2O_2** (prepared by diluting 30% stock before use).

Procedure

1. Treat the material in the incubation medium for 3 to 10 min at room temperature.
2. Wash in water and mount in any aqueous medium such as glycerine jelly.

Result

Brown black deposits are observed at the site of peroxidase activity, which should be absent from the controls.

---

* DAB is carcinogenic and the powder scatters readily; great care is needed.
** H_2O_2 concentration can be between 0.002 to 0.02%.

Controls

1. Use incubation medium without H_2O_2.
2. Use heat treated (15 min in boiling water) sections or materials.
3. Add 2 mm sodium pyruvate to the incubation medium to suppress peroxidase.[196]

Application of This Procedure

It is one of the most widely used procotols for studying peroxidase activity in a variety of plant systems.

Procedure 5: p-Phenylenediamine and Pyrocatechol Method[188]

Incubation Medium

20 ml of 0.1 M tris-HCl buffer, pH 7.6

10 mg p-phenylenediamine

20 mg pyrocatechol

0.2 ml of freshly diluted 1% H_2O_2

Procedure

1. Treat the material in the incubating medium.
2. Wash in water and mount in any aqueous mountant such as glycerine jelly.

Results

Blue deposits are found at the sites of peroxidase activity.

Controls

As in the previous procedure.

Rationale of the Reaction

In this procedure, DAB is replaced by pyrocatechol and p-phenylenediamine in the reaction mixture. The peroxidase coupling reactions of p-phenylenediamine are greatly accelerated in the presence of pyrocatecol and a deeply colored synthetic melanin like co-polymeric compound is formed. This is water insoluble and blue in color.

Procedure 6: Homovanillic Acid Method[197]

Tissue Preparation

Use fresh or fixed tissue. Fixation is done either in 95% ethanol for 1 min or in 1.5% glutaraldehyde in 0.1 M cacodylate buffer, pH 7.4, for 10 min. If fixed, wash in 0.2 M imidazole buffer, pH 7.4, for 10 min and then in 0.2 M acetate buffer, pH 6.0, for 5 min.

Incubation Medium

10 ml of 0.2 M acetate buffer, pH 6.0.
   Mixture 1: 51 mg of homovanillic acid (adjust pH to 6.0 using 0.2 M NaOH).
   Mixture 2: 10 ml of 0.2 M acetate buffer, pH 6.0, 132 to 134 mg of lead nitrate.
   Slowly add Mixture 2 to Mixture 1, with stirring. To the mixture of 1 and 2 add 0.2 ml of a freshly prepared 1% H_2O_2.

Procedure

1. Incubate for up to 5 to 15 min at 37°C. Incubation of fresh tissue is advised; if fixed in ethanol (95%) or formaldehyde, incubate for 30 min.
2. Wash well (preferably at least twice) in acetate buffer 0.2 M, pH 6.0.
3. Immerse in aqueous saturated H_2S or ammonium sulphide solution for 2 min.
4. Wash well in distilled water.
5. Mount in Farrants medium or, if desired, dehydrate, clear, and mount in euparal.

Result

Brown–black deposits are formed at the sites of peroxidase activity.

Controls

1. Incubate heat-inactivated materials in the reaction mixture.
2. Incubate materials in the reaction mixture without H_2O_2 (substrate).

Rationale of the Reaction

The basis of the reaction is the conversion of homovanillic acid from its monomeric to its dimeric form by the action of H_2O_2 and peroxidase. The dimer is then precipitated by the lead ions (supplied by lead nitrate). The lead complex is treated with hydrogen sulphide or ammounium sulphide to yield lead sulphide, which forms the brown–black deposits at the sites of enzyme activity (Figure 2.2).

Monomeric   **Dimeric**

FIGURE 2.2
The conversion of monomeric homovanillic acid to dimeric acid as a result of the activity of the enzyme peroxidase.

Procedure 7: p-Phenyleneamine Method[198-200]

Incubation Medium

10 mg of p-aminodiphenylamine (variamine blue RT base) +10 mg of m-methoxy - p-aminodiphenylamine (variamine blue PG base)

0.5 ml of absolute ethyl alcohol

35 ml of distilled water

15 ml of 0.2 M Tris buffer, pH 7.4

0.25 ml of 3% H_2O_2

Procedure

1. Place tissue in incubation medium for 15 to 60 min.
2. Transfer the material to a solution of 100 ml of 10% cabaltous acetate in 10% formalin, plus 5.0 ml of 0.2 M acetate buffer at pH 5.2 for 60 min.
3. Wash in distilled water.
4. Mount in glycerine jelly.

Results

Peroxidase activity is manifested as a bluish precipitate.

Procedure 8: Syringaldazine Method[201]

Tissue Preparation

Fresh or frozen tissue is preferred.

Incubation Medium

Syringaldazine 0.1% in ethanol, 1 to 2 droplets
Aqueous H_2O_2 0.03%, 1 to 2 droplets

Procedure

Add the components of the incubation medium on to the material in the same order.

Result

Intense purple color indicates the sites of peroxidase activity.

Application of This Procedure

This method has been used extensively to study the involvement of peroxidase during lignification in several dicot and conifer trees.

Procedure 9: Guaiacol Method[201,202]

Tissue Preparation

Fresh free hand sections or sections of tissue fixed in 2.5% (w/v) glutaraldehyde buffered with 50 mm sodium cacodylate, pH 7.0, for 1 h at 4°C followed by washing for 2 to 3 h in running tap water.

Incubation medium

Guaiacol 48 mM
Sodium cacodylate, pH 7.0, 50 mm
H_2O_2 0.02%
Incubation for 5 min

Result

Peroxidase containing sites are marked by precipitates.

Controls

Media without substrate or use of boiled sections.

Application of This Procedure

Fielding and Hall[202] have used this procedure to study the distribution of peroxidase in the root tip region of *Pisum sativum*. They found that the activity was high in root cap cells at the root surface, epidermis, inner cortex, phloem, and pith. Goldberg et al.[201] used this method to know peroxidase profiles in the cells which are at the final stages of lignification.

Polyamine Oxidase.

Procedure 1: Starch Reagent Method[203]

Tissue Preparation

Tissue sections, peeled leaf, or stem segements or, if small, whole plant material can be used. Fresh material is ideal.

Incubation Medium

Soluble starch	1.3%
Potassium iodide	20 mm
Spermine or spermidine (substrate)	10 mM
Phosphate buffer solution	1 mM, pH 5.8

Procedure

1. Float material on the incubation medium for a period varying from 30 min to several hours.
2. Observe.

Result

The enzyme sites turn blue.

Rationale of the Reaction

The development of blue color is the result of I^- oxidation to I^+ by enzymatically (by polyamine oxidase) generated H_2O_2 formed following addition of selected polyamines such as spermine.

Application of This Procedure

Kaur-Sawney et al.[203] have proved by experimentation (including cytochemical localization) that the enzyme is located in the cell wall. They have localized the enzyme in the cell walls of stomatal guard cells, vascular tissue cells, etc. in oat leaves.

Hydrogen Peroxide.

Procedure 1: Starch–KI method[204]

Tissue Preparation

Use only fresh hand-cut sections or materials.

Procedure

1. Prepare the starch–KI solution as follows: Dissolve 4 g of potato starch in 100 ml of 0.2 M aqueous potassium iodide solution and heat to boiling. Allow the colorless solution to cool to room temperature.
2. Apply this solution to freshly cut sections of the plant material on a slide.

Result

Location of H_2O_2 is indicated by the development of dark starch-I_2 complexes on the cut surfaces within 5 to 30 min after treatment.

Rationale of the Reaction

Starch–KI is oxidized in the presence of H_2O_2 and forms dark complexes.

Controls

1.   Fresh sections coated with starch–KI reagent under anaerobic conditions.
2.   Sections soaked only in pure water.

Comments

The starch-KI assay is relatively specific and does not require peroxidase and the conditions conducive for the activity of that enzyme to detect H_2O_2, as the method employing guaiacol required.[201] Further, this method is of low toxicity and low cost. However, as with other possible H_2O_2 assays, a negative test result is difficult to interpret. Absence of substrate oxidation could result from several different factors:

1. H_2O_2 production might in fact be absent in the tissue.
2.   Catalase may be decomposing the H_2O_2 in tissue as quickly as it is evolved.
3.   A negative result could result from the presence of both H_2O_2 and some endogenous reducant(s) such as ascorbic acid, which could reduce H_2O_2 and/or oxidized substrate back to colorless starch–KI before enough of the complex had accumulated to be reduced.

Therefore, the starch–KI histochemical reagent serves as an assay for H_2O_2 but not as an assay for H_2O_2 non-production.

Application of This Procedure

1.   This test has been used as a valid assay for the involvement of H_2O_2 in the lignification process in *Zinnia*. It can, thereby, identify possible zones of ongoing lignification.
2.   The test has also added further support to the proposed role of H_2O_2 in wounding and disease responses.
3.   This test can also be used to evaluate the induced oxidative bursts and detection of polyamine oxidases.

Malate Dehydrogenase.

Tissue Preparation

Fresh tissues preferable. As the enzyme in animal tissue could be localized from 10 min to 24 h after fixation in formaldehyde,[205] the same fixative can be used for plant tissues as well; but, the effect of fixatives on bound malate dehydrogenase in plant tissues has not yet been studied.

Procedure 1[194]

Incubation Medium

10 ml of 0.05 M tris-HCl buffer, pH 7.4

40 mg of NBT (nitro blue tetrazolium chloride)

2.2 g of polyvinyl alcohol

20 mg of NAD (or NADP)

30.3 g of sodium malate

pH adjusted to 7.4 with sodium hydroxide

Bubble oxygen-free nitrogen through medium prior to use.

Procedure

1. Incubate materials for 15 to 60 min at room temperature in the dark.
2. Wash in warm water.
3. Mount in Farrants medium.

Result

Formazan deposit sites indicate enzyme locations.

Controls

1. Test solution lacking NAD(P).
2. Test solution lacking substrate (sodium malate).
3. Heat treated (boiled water 10 min) materials.

Procedure 2[206]

Incubation Medium

Triethanolamine-HCl-buffer, pH 7.6, 50 mM

EDTA, disodium salt 5 mM

D,L-malate (sodium malate, Na_3 salt) 60 mM

NAD, 1-8 mM

NBT, 1.5 mM

PMS (phenazine methosulphate) 0.1 mM

Procedure

Incubate for 10 min at 37°C and examine immediately.

Result

Sites of enzyme turn to a blue color.

Controls

1. Omission of substrate.
2. Addition of p-chloromercuribenzoate (PCMB) 0.1 mM.

Procedure 3[207]

Incubation Medium

0.2 ml	substrate (1M)
0.25 ml	phosphate buffer (1/15M, pH 7.0)
0.25 ml	NBT (2 mg/ml distilled water)
5 mg	NAD
0.3 ml	distilled water

Substrates include D, L-isocitrate (Na$_3$-salt), 2H$_2$O; L-malic acid; D, L-glyc-erophosphate (Na$_2$ salt); ethanol (1M); and D, L-lactate (Na salt).

Procedure

Materials are incubated for 10 to 60 min at 37°C, washed briefly with buffer and fixed in 4% formaldehyde for at least 30 min. After washing with buffer the samples are examined.

Result

Sites of enzyme activity are stained to a blue color.

Controls

Omit the substrate or add PCMB (0.1 mM) or incubate without NAD or NADP.

Procedure 4[208,209]

Tissue Preparation

Fix in 3% formaldehyde for 10 min just before incubation.

Incubation medium

Stock solution	Final concentration (mM)
0.1 Sorensen's phosphate buffer, pH 7.6, 1ml	80
0.5M K-Na tartarate, 30 ml	300
0.3M CuSO$_4$, 0.35 ml	21 adjust pH to 7.0
0.1M D, L-malate 0.5 ml	10
NAD or NADP 1 to 5 mg/ml of medium	Same as in stock
0.05M K-ferricyanide, 0.15 ml	1.5

Procedure

Incubate for 2 h at 3°C.

Result

Specific precipitation of bluish green color indicates the loci of the enzyme.

Controls

1. Heat the samples at 70 to 80°C during fixation.
2. Omit the substrate from the incubation medium.

Acid Phosphatase.

There are two general methods for cytochemically localizing phosphatases:

1. Capture of the "R" group to which the phosphate was originally attached (the azo dye method).
2. Capture of the liberated phosphate group via lead ions, as shown below:

Phosphate ester of "R"

Inhibition by 0.01 M NaF → ↓ ← Acid Phosphatase, pH 5.2

R-OH   +   phosphoric acid

| |

"R"capture via   or Phosphate capture

Naphthol derivatives   via Gomori reaction

Tissue Preparation

Acid phosphatase activity can be localized in materials which are preferably fresh or frozen. 10% acrolein fixation for 2.5 h at 4°C inhibited the enzyme activity in maize roots by about 25%, while in 3% glutaraldehyde for 2 h at 4°C, 16% inhibition was shown by the same tissue[210] (see also Shnitka and Seligman[211]). The effect of other fixations has been studied only in animal tissues. In the lead salt method, 10% formalin containing 1% calcium chloride, pH 7.2, can be used as a fixative. Fixation at 4 to 10°C is better. The material can be embedded in paraffin or resin.

The enzyme exhibits optimal activity at pH below 7.5, mostly between 3.8 and 6.0. This fact must be taken into account while adjusting the pH of the incubation medium. The fact that the enzyme is readily made soluble and released from the tissue is a probable cause of failure in past methods.

Many histochemical methods have been tried to localize acid phosphatase in biological materials. But it is difficult to recommend one method over another, because the commonly used substrates can demonstrate a multiplicity of acid phosphatases and, in addition, do not necessarily demonstrate the enzyme.[194] Thus, it is possible to arrive at a situation where one substrate may show the presence of acid phosphatase activity, while another used on the same tissue will not.[212] In consequence, when starting to work with a new tissue, it is of value to try more than one substrate. For example, the method of Livingstone et al.[213] has worked satisfactorily on root apices of *Vicia faba*,[214] but has not seemed adequate when used with strawberry receptacles.[215] One further point is that the reactants often failed to penetrate the tissues adequately. Size of the material, therefore, is an important factor.

Procedure 1: Lead Salt Method[216,192,194,199,217-226]

Incubation Medium

Sodium ß-glycerophosphate*	10 mM (pH adjusted to 5.0 or 5.5 before mixing with acetate buffer)
Acetate buffer	50 mM (pH 5.0 or 5.5)**
$(PbNO_3)_2$	3.6 mM

Barka and Anderson[217] favor a slightly modified medium which consists of the following:

Distilled water	10 ml
1.25% sodium ß-glycerophosphate (pH 5.0 aqueous)	10 ml

---

* 0.1 M *p*-nitrophenyl phosphate was recommended by Halperin[227] and Poux;[228] Cytidine monophosphate by Marty;[229] Glucose-6-phosphate by Tice and Barnett.[226] Poux[228] included 5% sucrose for helping in penetration but this was not recommended by others.

** 2.85 ml glacial acetic acid in 500 ml of distilled water and 6.85 g of sodium acetate in 500 ml of distilled water.

0.1 M tris-maleate buffer* (pH 5.0)   10 ml

0.2% $(PbNO_3)_2$ 20 ml

The lead nitrate solution is added dropwise with continuous stirring and the medium is used immediately after preparation. This medium stabilizes the lead and keeps it in solution. A slight precipitate may form, which is filtered off before use. The medium is recommended to be used immediately after preparation by Holt and Hicks[230] while a storage up to 16 h at 37°C was recommended by others[195, 231] to reduce artifacts.

Procedure

1. Incubate materials at 25 to 37°C in the medium. Standardize the incubation time and temperature for each tissue.
2. Briefly rinse in distilled water.
3. Place in 1% aqueous acetic acid.**
4. Briefly rinse in distilled water.
5. Immerse the material for 2 to 10 min in distilled water in which H_2S or ammonium/sodium sulphide is bubbled through.
6. Wash in distilled water and mount in Farrants medium.
7. Examine immediately.

Result

Brown-black deposits of lead sulphide indicate the sites of enzyme activity.

Rationale of the Reaction

In this enzyme reaction, the product is soluble and in order to localize the enzyme cytochemically, the product must be precipitated immediately by a so-called simultaneous capture mechanism:

$$\text{Substrate} \xrightarrow[\text{reaction}]{\text{enzyme}} \begin{array}{c}\text{Primary}\\(\text{soluble})\\\text{reaction product}\end{array} \xrightarrow[\text{reaction}]{\text{Capture}} \begin{array}{c}\text{Final}\\(\text{insoluble})\\\text{reaction product}\end{array}$$

The incubation medium includes a suitable phosphate ester as substrate and soluble lead nitrate. Inorganic phosphate released by enzyme activity is immediately precipitated as insoluble lead phosphate and so marks the site of enzyme activity. In light microscopy, lead phosphate is difficult to see and so is converted into another

---

* Tris (hydroxy methyl) aminomethane 1.212 g in 100 ml water, maleic acid 1.16 g in 100 ml water, sodium hydroxide 800 mg in 100 ml water. Make 100 ml buffer solution with the three above components in the following proportion: 10 ml tris solution, 10 ml maleic acid solution, 2 ml NaOH solution, and 78 ml water.
** This step is not recommended by Goldfischer et al.[232] and Sexton and Hall[192] as it might remove the reaction product especially from an enzyme site of low activity.

insoluble substance, lead sulphide (black deposit) by exposing the tissue to weak hydrogen sulphide or ammonium/sodium sulphide.

Controls

1. Omit the substrate in the incubation medium.
2. Use substrate with the addition of 0.42g/l NaF which inhibits the enzyme activity.
3. Incubate boiled sections/materials as per schedule.
4. Trypsin digestion (enzyme trypsin in Tris-HCl buffer).
5. Add Zn^{2+}, Cu^{2+}, Mg^{2+} or Fe^{2+} ions to the incubation medium.[226,233]
6. Add D-glucose, ammonium molybdate, or the oral antidiabetic drug orinase to the incubation medium.[334]

Application of This Procedure

This method is one of the widely used methods of localizing acid phosphatase activity. For example, it was used by Catesson and Roland[235] to study the implication of this enzyme in cell formation during cambial activity. Moore et al.[224] have used this to demonstrate the enzyme activity in the root cortical cells of *Zea mays*. Knox and Heslop-Harrison[170,236] have used the lead salt method to study the distribution of acid phosphatase in the pollen grain wall of many angiosperms. They found its regular presence in the cellulosic intine. Smith[237] found high phosphatase activity on the rhizoidal walls of young gametophytes of *Polypodium vulgare* by employing this procedure.

Procedure 2: Azo Dye Method[194,238,239]

Incubation Medium

5 to 6 mg ASBI, ASMX, AS-TR, or synthetic hydroxy carbazole phosphate

0.2 to 1 ml dimethyl formamide (or acetone)

50 to 60 ml 0.1 M - 0.2 M citrate, phosphate, or tris-maleate buffer, pH 5.0 to 5.2 (shake well)

30 to 50 mg fast blue BBN, fast red violet salt LB, or fast garnet GBC (shake well and filter)

Tissue Preparation

Fix at room temperature for 30 sec in 60% acetone buffered to pH 4.2 - 4.5 with 0.03 M citric acid-sodium citrate solution; unfixed material can also be used.

Procedure

1. Incubate materials at 37°C* in the medium. (Length of incubation time needed will depend upon the tissue to be studied. Hence, a range of time from 1 min to 2 h should be tried.)
2. Rinse in distilled water for 20 to 30 sec.
3. Mount in Farrants medium (glycerine jelly or polyvinyl alcohol can also be used). In all cases, color fades in course of time.

Result

Blue deposits indicate the sites of enzyme activity in fast blue BBN and red deposits in fast red violet LB.

Rationale of the Reaction

The naphthol AS series of phosphates produced an insoluble substituted naphthol at the site of hydrolysis, which when coupled to a suitable diazonium salt (like fast blue BB) produced a brightly colored insoluble dye (as shown below):

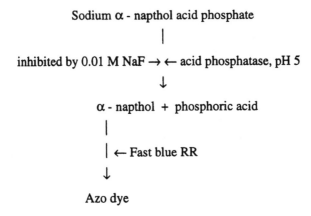

$$Sodium\ \alpha - napthol\ acid\ phosphate$$

$$|$$

$$inhibited\ by\ 0.01\ M\ NaF \rightarrow \leftarrow acid\ phosphatase,\ pH\ 5$$

$$\downarrow$$

$$\alpha - napthol\ +\ phosphoric\ acid$$

$$|$$

$$| \leftarrow Fast\ blue\ RR$$

$$\downarrow$$

$$Azo\ dye$$

Controls

1. Incubate materials in solution mentioned, from which the substrate has been omitted.
2. Incubate materials in solution mentioned above to which 0.01 M NaF or 0.001 M sodium molybdate has been added.
3. Incubate boiled materials in solution mentioned above.

---

* 25°C, if the synthetic hydroxy carbazole phosphate is used in the incubation medium.

Comments

This method cannot be easily modified for EM, as the dye product is electron-opaque; it obscures cytological detail and is often soluble in the dehydrating and embedding media.[213] Furthermore, diazonium salts themselves can inhibit phosphatases in plants.

Application of This Procedure

Lu et al.[165] have studied the activity of cell wall-bound acid phosphatase in the stigmatic papillae of groundnut.

Procedure 3: Azo Dye Method—Hexazonium Pararosanilin Complex[217]

Tissue Preparation

Fix thin tissue slices (1 to 3 mm thick) for 24 to 36 h at 4 to 5°C in formal calcium (1 g anhydrous calcium chloride in about 60 ml distilled water and 10 ml of 40% formaldehyde; adjust pH to 7.0 to 7.2 with 1N NaOH; add distilled water to make 100 ml). Take frozen sections of 3 to 6 μm thickness and air dry them at room temperature for 2 h before use. Sections may also be preserved for 24 h in a refrigerator without appreciable loss of activity.

Preparation of Stock Solutions

1. Pararosanilin solution: 1 g pararosanilin hydrochloride in 20 ml distilled water + 5 ml concentrated HCl with gentle warming. Filter after cooling and store at room temperature.
2. Sodium nitrite solution: 4% in distilled water and store it in refrigerator.
3. Michaelis veronal acetate buffer solution: 9.714 g of sodium acetate. $3H_2O$ (or 5.85 g of the anhydrous salt) + 14.714 g of sodium barbiturate in CO_2^- free distilled water to a final volume of 500 ml.
4. Substrate solution (naphthyl phosphate): 400 mg sodium α-naphthyl phosphate in 100 ml buffer stock solution. Store it in cold.
5. Substrate solution (naphthol AS phosphate): 100 mg of naphthol AS-TR or AS-BI phosphate in 10 ml N,N-dimethyl formamide. Store in cold.

Working Solutions

1. Naphthyl phosphate–hexazonium pararosanilin: Mix 0.8 ml of (1) and (2) above at room temperature in a test tube. Pour this mixture into 5 ml of (4) above previously diluted with 10 ml of distilled water. Adjust pH to 6.0 with 1 N NaOH. Add distilled water to a final volume of 20 ml.

2. Naphthol AS-TR phosphate–hexazonium pararosanilin: Dilute 5 ml buffer stock (3) above with 12 ml of distilled water and add 1 ml substrate solution (5) above. Mix 0.8 ml of (1) and (2) above in a test tube and pour into buffered substrate solution. Adjust pH to 5.0 with 1 N NaOH.

Incubation

1. α-naphthyl phosphate technique: Incubate sections for 10 to 30 min at room temperature. Rinse in distilled water, dehydrate quickly in alcohol,* clear in xylol, and mount in synthetic mountant.

2. Naphthol AS-TR phosphate technique: Incubate section for 30 to 90 min (if time beyond this is required, use new incubation medium) at room temperature or for a shorter period at 37°C. Rinse briefly in distilled water, dehydrate quickly in alcohol,* clear in xylol and mount. Counterstain, if needed, with methyl green, celestian blue, or haematoxylin.

Results

α-naphthyl-phosphate technique: A reddish brown precipitate indicates the enzyme site. The background is pale yellow.

Naphthol AS-TR phosphate technique: A vivid red azodye color with an unstained or pale pink background indicates enzyme site.

Application

Knox and Heslop-Harrison[170,236] have used this method to study acid phosphatase distribution in the pollen grain wall of several angiosperms. They found that it was regularly present in the cellulosic intine. Heslop-Harrison[129] studied this enzyme on the stigmatic receptive surface of *Crocus*, while Smith[237] found its high activity on the walls of the gametophytic rhizoids of the fern *Polypodium vulgare*.

Procedure 4: Azo Dye Method with Coupling by Mercury Substituted Diazotate[240]

Tissue Preparation

Tissue fixed in (1) formal-calcium, pH 7.4 for 18 to 20 h at 4°C, (2) cold phosphate buffered (pH 7.4) 10% formaldehyde containing 0.1% chloral hydrate for 18 to 20 h, or (3) cold 1.5% glutaraldehyde containing 1% sucrose and buffered at pH 7.4 with phosphate or 0.067M sodium cacodylate.

---

* Dehydration may cause a slight reduction in staining intensity. Therefore, water mounts or aqueous media can be used.

After fixation, wash and store at 4°C in 0.1 M cacodylate or veronal acetate buffer, pH 7.4, containing 7% sucrose, for varying periods of time up to 1 week.

Fixed tissue are to be frozen first. Frozen materials are then to be precoated with gelatin and hardened with 10% formalin. Dry at room temperature for 2 h. Materials can be stored in refrigerator until use.

Incubation Medium

100 mg naphthol AS-BI phosphate dissolved in 10 ml of N,N dimethyl acetamide. The buffer is to be made as follows: (1) 5 ml stock veronal-acetate buffer , (2) 7 ml 0.1 NHCl, and (3) 13 ml distilled water. The working incubation medium consists of 15 ml buffer, 35 ml distilled water, and 0.6 ml stock substrate solution. If necessary, adjust pH to 5.0 with 0.1 N HCl.

Procedure

1. Incubate in medium for 15 to 45 min in an oven at 37°C.
2. Wash in two changes of ice-cold distilled water containing 7% sucrose or in 0.1 M veronal acetate wash buffer, pH 7.4, containing 7% sucrose for less than 15 min.
3. Couple in coupling agent, pH 7.25, for 2 to 3 min at room temperature. Prepare the coupling agent as follows:

    a) Make a 4% p-(acetoxymercuric) aniline (AMA) in 50% acetic acid at room temperature with constant agitation. Plunge the prepared solution into ice. An ice-cold solution is stable for some hours but it is preferable to prepare fresh.

    b) Prepare 4% aqueous sodium nitrite solution by dissolving 2 g of sodium nitrite in 50 ml of water. The solution is kept cold by putting its container in an ice box. The solution is stable for 2 d in cold condition but it is better to prepare it fresh.

    c) After keeping solution (a) and (b) above in an ice box for at least 45 min, mix equal parts of them. Diazotization takes place in 15 min when kept in the box; diazotization is indicated by change of color from orange to red–brown, with a brown precipitate. Filter through Whatman 1 filter paper and keep the filtered solution in the ice-box. The solution is a clear yellow–orange. Use it within 12 h of its preparation.

    d) Dilute 0.7 ml of solution (c) above to 50 ml with 1 N phosphate buffer, pH 7.6. Adjust final dilution solution to a pH of 7.25 to 7.3. Do not refrigerate this. Use this solution within 15 min and at room temperature.

4. Rapidly pipette off the coupler and wash in ice-cold veronal acetate buffer, three times, 5 min, each.
5. Mount the material in polyvinyl pyrollidine.
6. If permanent preparations are needed, omit Step 5, and treat the material for 5 to 10 min with a 1% solution of thiocarbohydrazide containing 2% sucrose at 25°C to room temperature.

7. Dehydrate, clear, and mount in permount.

Result

Red colored deposits indicate the locations of enzyme activity.

Controls

1. Use of heated sections.
2. Substrate mixed with saccharolactone, 20 mg/ml H_2O; 0.01 ml saccarolactone to 0.99 ml substrate.
3. Lack of substrate in the incubation medium.

Esterases.
These are a heterogenous group of enzymes which hydrolyze a broad range of carboxylic acid esters.[241,242] A few procedures have been suggested to cytochemically localize non-specific esterases, of which the following are the most followed:

Procedure 1[214]

Incubation Medium

2×10^{-4} M naphthol AS-D acetate

0.1 M tris maleate (pH 6.5) or tris-HCl buffer (pH 7.1)

Procedure

1. Incubate for periods up to 60 min at 37°C.
2. Transfer to a post-coupling medium containing 10 mg lead phthalacyanin diazotate[213] in 1 ml 0.1 M acetate buffer at pH 4.8 for 10 min at 2°C.
3. Dehydrate in water soluble Durcapan mountant and observe. If permanent slides are required, mount the material in araldite, section, and remove the resin.

Results

Sites of enzyme activity are colored.

Controls

Heat killed materials.

Rationale of the Reaction

The substrate (Naphthol AS-D acetate) is hydrolyzed to give H_2S and acetic acid as products. Since the material is then subjected to lead salt treatment, lead sulphide will be formed as a precipitate, thus indicating enzyme sites.

Application of This Method

Livingston et al.[213] and Gahan and McLean[214] have used this method to study the activity of esterase in the cells of *Vicia faba* root tips.

Procedure 2[243]

Incubating Medium

> Naphthol AS-D chloroacetate 10 mg dissolved in 0.5 ml acetone
>
> Distilled water 25 ml
>
> Sodium barbiturate buffer 25 ml (sodium barbiturate 1.03 g/50 ml water, 50 ml + 0.1 N HCl 31 ml + enough distilled water to make 100 ml, pH 7.4)
>
> Fast blue RR or Garnet GBL salt 30 mg
>
> Shake for 30 sec before use

Procedure

1. Hydrate the materials.
2. Incubate in substrate at room temperature for 10 to 30 min.
3. Rinse in running water, 1 min.
4. Mount in glycerine jelly, Farrants medium, or Gelva.*

Result

Dark blue deposits indicate enzyme sites.

Comment

Fast blue RR is preferable as Garnet GBL salt produces an undesirable yellow background.

---

* Gelva is prepared as follows: 20% solution of polyvinyl acetate in 80% alcohol.

Application

This method was used by Lu et al.[165] to study esterase activity in the stigmatic papillae of groundnut.

Procedure 3: Indoxyl Acetate Method[244]

Tissue Preparation

Unfixed frozen tissue should be used.

Incubation Medium

Tris buffer (0.2M), pH 8.5	2 ml
Oxidant (potassium ferrricyanide 210 mg + potassium ferrocyanide 155 mg + distilled water 100 ml)	1 ml
1 M calcium chloride	0.1 ml
2 M sodium chloride	5 ml
Distilled water	2 ml

Add this solution rapidly, with agitation, to 1 ml of absolute ethanol containing 1.3 mg of 5-bromo indoxyl acetate.

Procedure

1. Incubate in the reaction medium at 37°C for 15 to 30 min.
2. Wash in 30% ethanol containing 0.1% acetic acid.
3. Wash in distilled water.
4. Mount, preferably in Farrants medium.

Result

Blue precipitates indicate sites of enzyme activity.

Rationale of the Color Reaction

Barrnett and Seligman[244] first suggested that when indoxyl acetate is used as an ester substrate, esterase would liberate the indoxyl group which would become oxidized to indigo. Holt[245] developed on this idea and proposed the reaction sequence as follows: Indoxyl acetate is the ester substrate and on activity with esterase, it is oxidized by potassium ferricyanide-potassium ferrocyanide into indigo blue.

Comments

Many alternative indoxyl acetates have been tested. Of these, 5-bromo 4-chloroindoxyl acetate provides the most precise staining. The concentration of the oxidant is a vital factor; if not carefully decided, products of enzyme activity (therefore, color) are diffused to artifactual sites.

Application of This Procedure

Deising et al.[246] have used this method to localize esterase activity on the spore surface of the germinating uredospores on the host cuticle. They found that the spore showed esterase activity upon exposure to water.

Ribonuclease[247].

Tissue Preparation

Fresh, frozen, or fixed tissue could be used. In the last instance, formalin vapor fixation for 10 min is recommended.

Incubation Medium

Ribonucleic acid (yeast derived)	15.0 mg
Acid phosphatase	5.0 mg
0.2 M acetate buffer	12.5 ml
0.2 M lead acetate	1.0 ml
Distilled water	50.0 ml

Procedure

1. Incubate the material for 30 min in the medium.
2. Rinse in distilled water.
3. Place the material in dilute ammonium sulphide solution for 10 min.
4. Again rinse in distilled water.
5. Counterstain, if necessary, in 1% safranin for 30 sec.
6. Rinse in distilled water, air dry, and mount in glycerin jelly .
7. Observe.

Result

Enzyme sites appear as small dark granules.

Control

Omission of RNA from the incubation medium.

Comments

Knox and Heslop-Harrison[170] reduced the lead ion concentration to 0.12% to get a clearer contrast without background colors. They also used H_2S-saturated water to convert lead phosphate into black lead sulphide instead of ammonium sulphide.

Application of This Procedure

This method was used by Knox and Heslop-Harrison[170,236] to study the distribution of ribonucleases in the cellulosic intine of the pollen grains of several angiosperms.

Proteases.

Procedure 1[248]

The hydrolysis of amino acid derivatives of ß-naphthylamine by amino peptidases has been used for the localization of these enzymes.

The substrate (e.g., L.leucyl-ß-naphthylamide) is hydrolyzed by corresponding amino acid (e.g., leucine) and ß-naphthylamine. The latter yields a colored azodye precipitate in the presence of a suitable diazonium salt. Unfortunately, considerable diffusion of the product occurs due to the slow rate of coupling of the β-naphthy-lamide to the diazonium salt. The procedure has not yet been put to rigorous practice.

2.6.3 Lignins

Procedure 1: Calcium Hypochlorite–Sodium Sulphite Method[249-251]

Tissue Preparations

Fresh, freeze-dried, freeze-substituted, or chemically fixed (with or without embedding) tissue can be used.

Procedure

1. Bring the material down to water.
2. Place the material in a fresh, saturated, acidified calcium hypochlorite solution (acidified with a few drops of 1 N HCl or with 1 ml of a 56% v/v HCl)[163] for 5 min. Chlorax can be substituted for calcium hypochlorite.

3. Place the material in a fresh and cold 1% aqueous solution of sodium sulphite cooled to 10ºC, and observe.

Result

A bright red color or a brown color will develop within a few minutes but will gradually fade in 45 min. These colors indicate the presence of lignin.

Rationale of the Color Reaction

The basis of the reaction is not very clear, although it is believed that syringyl groups of lignin are the sites of reaction in angiosperms and the coniferyl or guaiacyl groups of lignins in gymnosperms.[252]

Control

Ethanolysis: Mix 8.5 ml of concentrated HCl in enough of rectified spirit to make 100 ml solution. Boiling of tissue in this solution solubilizes lignin by a combination of solvolysis and depolymeriaztion.

Comments

1. The red and brown colors are believed to be due to the presence of different species of lignins in angiosperms and gymnosperms rich in syringyl and guaiacyl moieties, respectively.
2. Even where an initial red color is obtained, the color changes to brown in about 30 min.[49] This needs some explanation.

Procedure 2: Maule's Reaction[253]

Tissue Preparation

Fresh tissue is preferred, although chemically fixed and paraffin embedded ones can also be used. Even very old herbarium specimens have been used successfully after revival with warm water (with some detergents added).

Procedure

1. Bring the material down to water.
2. Treat for 5 to 20 min in 1% neutral potassium permanganate solution at room temperature.
3. Wash in distilled water.

4. Decolorize with dilute HCl (2% is ideal). A 5.65% solution is suggested by Vijayaragha-van and Shukla.[163] A 12% HCl treatment for 1 to 3 min duration is suggested by Iiyama and Pant.[254]

5. Wash thoroughly in water.

6. Treat with a few drops of ammonium hydroxide or sodium bicarbonate solution and observe.

Results

Lignin stains red to purple–red in some cases, brown in others, and colorless in yet others.

Rationale of the Color Reaction

Meshitsuka and Nakano[255-257] have shown by model experiments that methoxy cat-echol was produced from "end syringyl nuclei" with free phenolic hydroxyl groups in hard wood (dicot) lignin by demethylaton with permanganate treatment. It is then chlorinated with chlorine generated from HCl by permanganate oxidation and this was later oxidized to chlorinated methoxyquinone structures in ammonium hydrox-ide solution. These quinones give an intense red or purple–red color.

However, this interpretation has been refuted recently by Iiyama and Pant,[254] according to whom the purple–red coloration obtained from methylated hardwood lignin with the Maule's color test is generated by reaction of syringyl groups which were liberated by ß-ether cleavage under the permanganate oxidation conditions.

Control

Ethanolysis (see Page 116).

Comments

1. Baayen[258] claimed that this method stained mainly syringyl lignins.

2. Decolorization time in HCl depends on the materials. Our experience shows that some get decolorized within 10 to 15 min, while others take as long as 12 h. This time should be standardized for each material.

3. Our experience shows that in some plants the lignin does not respond to produce a red or brown color, but remains colorless. This indicates the presence of the grass-type lignin.

Application of This Method

This method was used by Baayen[258] to study the responses related to lignification and intravascular periderm formation in carnations resistant to *Fusarium* wilt.

This was used earlier by many researchers to distinguish conifer lignin from hardwood lignin.

Procedure 3: Schiff's Reagent Method[60,257]

Tissue Preparation

Fresh or chemically fixed tissue can be used. Fixatives containing heavy metals are not suitable.

Procedure

1. Bring the material down to water.
2. Stain directly in Schiff's reagent for 15 min to 4 h (prepare Schiff's reagent as per procedure given on Page 53).
3. Wash and observe. Slides can also be made permanent.

Results

Lignin stains pink or magenta, indicating the presence of natural aldehydes in it.

Rationale of the Color Reaction

The aldehydes in lignin get colored with Schiff's reagent as detailed on Page 55.

Controls

1. Bromination: Treat materials after rinsing them in a 2.5% v/v solution of bromine in carbon tetrachloride for 1 h. This treatment impairs the charcteristics of lignin to react positively to this test. This charcteristic may be restored, if desired, by treating the brominated sections in boiling hydroiodic acid.
2. Ethanolysis (see Page 116).

Comments

1. Cuticles[49] and waxes are also stained. We have observed in several cases a positive staining of certain phenolics also.
2. Staining is permanent.

Procedure 4: Toluidine Blue O Method[133]

Tissue Preparation

Fresh, freeze-dried, freeze-substituted, or chemically fixed (with or without paraffin embedding) tissue can be used.

Procedure

1. Bring the material down to water
2. Stain in aqueous toluidine blue O solution, pH 4.4 [0.05% stain in benzoate buffer (benzoic acid 0.25 g and sodium benzoate 0.29 g in 200 ml water)].
3. Wash in running water and mount in any aqueous mountant.

Result

Lignin stains to a blue color.

Rationale of the Color Reaction

The precise nature of the staining reaction is not clear.

Application

Fisher[80] has used this method to stain lignins of tension xylem. This method has been extensively used by Ranjani and Krishnamurthy[74,85] to study the gelatinous fibers of tension woods. The G-layer and the lignified layer of the secondary walls of these fibers get stained differently; the former showing a metachromatic pink color, while the latter shows an orthochromatic blue color. This indicates that the G-layer is not only devoid of lignins but also that it is made of acidic polysaccharides in addition to cellulose.

Procedure 5: Aniline Sulphate Test[87,94]

Tissue Preparation

Fresh, frozen, or chemically fixed tissue can be used.

Procedure

1. Mount the material in a 2:1, v/v, mixture of aniline and sulphuric acid.*
2. Observe.

Results

Lignin appears deep yellow.

Controls

1. Mount in aniline alone.
2. Ethanolysis (see Page 116).

Procedure 6: Potassium Iodide-Iodine-Sulphuric Acid Method[86]

Tissue Preparation

Fresh tissue is preferable.

Procedure

1. Stain in potassium iodide-iodine (Lugol's iodine) solution.
2. Transfer the materials to 60 to 70% sulphuric acid solution and observe.

Result

Lignin becomes yellow, yellowish orange, or brown.

Comment

This method is not very specific because it stains to the same colors cuticle and suberin.

---

* Purvis et al.[94] employed a saturated solution of aniline sulphate to which a few drops of HCl were added, while Gahan[87] used a solution consisting of 1 g aniline sulphate mixed with 10 ml of 0.1 N H_2SO_4 and 89 ml of aqueous ethanol.

Procedure 7: Chlor-Iodide of Zinc Method[86]

Tissue Preparation

Fresh tissue is preferable.

Procedure

1. Place material in chlor-iodide of zinc solution (see Page 59 for preparation of this solution).
2. Observe.

Result

Lignin turns yellow to orange.

Comment

Cutin and suberin also get the some color with this reagent and, so, this method is not specific.

Procedure 8: Pholoroglucinol-HCl Method or Weisner's Reaction[60,86,87,94,251]

Tissue Preparations

Fresh, freeze-dried, freeze-substituted, or chemically fixed (with or without embedding) tissue can be used.

Procedure

1. Bring the material down to water.
2. Place the material in a saturated aqueous solution of phloroglucinol in 20% HCl for 1 min to a few minutes. Chamberlain[86] suggested a simultaneous application of a 5% dye solution (aqueous or ethanolic) and HCl to the material. Purvis et al.[94] suggested a 1% solution of phloroglucinol in 95% ethanol for 1 to 2 min, followed by a treatment with a few drops of concentrated HCl. Gahan[87] recommended a 10% solution of dye in 95 ml of absolute ethanol for 1 to 3 min, followed by a treatment with a few drops of concentrated HCl. Vijayaraghavan and Shukla[163] prefer a 1% solution of phloroglucinol in 95% ethanol followed by a treatement of 90% HCl through the sides of the cover glass.
3. Observe in the acid medium itself or transfer the material to a weak glycerin and then observe. Gahan[87] used a 75% glycerin as a mounting medium.

Results

Lignin will stain red to red-violet.

Rationale of the Color Reaction

The basis of the staining reaction is not clear, although phloroglucinol appears to react with the cinnamaldehyde and coniferyl end groups of lignin to yield a cationic chromatophore.[259]

Comments

1. The preparation is not permanent.
2. Jensen[60] opined that this method is reasonably specific but not too sensitive. However, Reeve[260] and Shah and Babu[261] stated that this test is not very specific for lignin as it was also positive to hemicelluloses and suberin. Other lignin tests should also be carried out before coming to a definite conclusion.
3. Our experience shows that this stain does not give positive results to the lignin of some monocots and that of the protoxylem of some dicots. Probably it stains lignins that are rich in some component(s) only.

Application

Baayen[258] used this method to study lignification and intravascular periderm formation in carnations resistant to *Fusarium* wilt.

Controls

1. Ethanolysis: See Page 116.
2. Bromination: See Page 118.
3. Treatment with chlorous acid.
4. Chlorination followed by extraction with organic solvents such as ethanol containing 3% ethanolamine.

Procedure 9: Azure B Method[60]

Tissue Preparation

Fresh, freeze-dried, freeze-substituted tissue, or tissue fixed in Carnoy's fluid can be used. The last may be embedded in a suitable embedding medium.

Procedure

1. Bring the material down to water.
2. Stain for 2 h in azure B solution at 50°C [0.25 mg/ml stain in McIlvaine buffer (24.6 ml of 0.1 M citric acid + 15.4 ml of 0.2 M disodium phosphate, adjusted to pH 4.0)].
3. Rinse in TBA from 30 min to variable time depending on the material (but not more than 3 h).
4. Observe. If need be, the material can be made permanent after customary procedures.

Results

Lignin stains a clear blue-green color, whereas the other components of the wall do not react.

Rationale of the Color Reaction

Not clear.

Comments

This test is not very specific. Our studies on the ovules of *Ottelia* (Hydrocharitaceae) showed that a wall material, which is not lignin (as tested by all lignin tests), was also positive (stained green) to azure B. It is probably a wall rich in phenolic substances.[115]

2.6.4 Sporopollenin

There is no specific method that will selectively indicate the presence of sporopollenin. The best way to detect sporopollenin is as follows:[163]

1. Acetolyze the material. (For acetolysis, treat the material in 1:9 v/v concentrated sulphuric acid and acetic anhydride at 70 to 80°C for 15 to 30 min. If desired, bleach the acetylosed material by treating it in a mixture of 20 ml glacial acetic acid, 2 ml concentrated hydrochloric acid, and 1 g potassium chlorate.) All wall substances other than sporopollenin, cutin, suberin, lignin, and some lipids are eliminated by acetolysis. Now do the following combinations of tests to specifically localize sporopollenin.
2. Sudan dye test (see Page 124): Stains sporopollenin, cutin, suberin, and lipids but not lignins.
3. Ferric salts test. (Treat materials either in 2% W/V solution ferric sulphate in 10% aqueous solutions of formalin or in 0.5 to 1% w/v solution of ferric chloride in 0.1 N HCl.) Stains sporopollenin (turned bluish green); cutin and suberin not generally stained; lipids do not get stained.
4. Ethanolysis (See Page 116 for procedure): This treatment dissolves lignin but not others.

5. Methanolic potash extraction: 5% (w/v) KOH in 95% methonol, dissolves cutin, suberin, and lipids but not sporopollenin and lignin.

6. Ethanolamine treatment. Treat in fresh 2-4, amino ethanol heated to 90°C for 5 min. Then wash in distilled water after cooling. Dissolves sporopollenin.[262]

7. Schiff's (see Page 118 for procedure) and azure B (see Page 122 for procedure) tests: Lignin, sporopollenin, cutin, and suberin are positive, but not lipids.

2.6.5 Cutin and Suberin

Procedure 1 : Sudan Dyes Method[263,49,264]

The sudan dyes, which can be used, are sudan III, sudan IV, and sudan black B, either independently or as a mixture.

Tissue Preparation

Fresh, freeze-dried, freeze-substituted, or chemically fixed tissues can be used. Material embedded in paraffin or acrylic resins and sectioned can also be used.

Procedure

1. Bring the material down to water.
2. Stain in a saturated solution of any of the sudan dyes or dye combinations in 70% ethanol for up to 60 min.
3. Remove excess stain by rinsing in ethanol.
4. Mount in dilute glycerin (1 volume glycerol:2 volume water).

Result

Sudan III and IV independently and together give a red/orange color to suberin and cutin, while sudan black B gives a black–blue black color.

Rationale of the Color Reaction

Sudan dyes are lipophilic. Therefore, reactivity is probably due to the physical association rather than a chemical reaction.

Comments

1. To remove lignin staining, prior treatment in aqueous sodium hypochlorite solution can be followed.
2. The staining reaction is not affected even if the waxes are extracted out of the cuticle.

3.   This procedure is best suited and gives a greater contrast.

Applications

This method has been one of the most commonly used to study plant cuticles and cork tissue. Baayen[258] has used it to study the cell wall responses (especially formation of suberins) of carnations to *Fusarium* wilt infections. Brammal and Higgins[265] have also used this procedure to study pathogenesis-related cell wall changes.

Procedure 2: Caustic Potash–Chloriodide of Zinc Method[86,163]

Tissue Preparation

Fresh tissue is preferable. Coat the material (placed on a slide) in celloidin.

Procedure

1.   Place the material in an aqueous solution of caustic potash. The time of treatment is to be standardized for each tissue by trial and error. Vijayaraghavan and Shukla[163] suggested placing the slides horizontally flat with the section side facing up in a deep and heat resistant petriplate to which a 28% aqueous solution of KOH is passed until the sections are about .5 cm below the upper layer of the solution. Treatment time is 2 to 8 h. By this treatment, color of cutin and suberin changes yellow to deep yellow. Put the petriplate with slides gently in an oven at 90°C for 10 to 20 min. Yellow granules will appear in suberized/cutinized cells. Withdraw KOH with the help of a syringe. Air dry the slides and coat the slides again with celloidin. Hydrate in cold water. Wash the material repeatedly with cold water.
2.   Treat the material in the chloriodide of zinc (see Page 59) for 2 to 3 min and mount in the same.
3.   Observe.

Result

Cutin and suberin turn reddish violet.

Rationale of the Color Reaction

Suberin consists of phellic acid; whereas cutin does not. This gives potassium phellonate when treated with KOH. Potassium phellonate appears as yellow granular substances in KOH treated suberized cells. These granules become reddish violet after reacting with iodine of chlor-zinc-iodide. The reason why cutin turns reddish violet is not clear, although a reaction similar to the one mentioned for suberin is likely to be involved here.

Comments

The method can be stopped at the end of Step 1 itself, when cutin and suberin get stained yellow.

Application of This Procedure

Pearce and Rutherford[266] have used this procedure to study suberization under wounding in the sap wood of *Quercus*.

Procedure 3: Azure B Method[261]

See Page 122 for detailed procedure.

Result

Cutin and suberin stain red.

Procedure 4: Potassium Hypochlorite–Cyanin Method[86]

Tissue Preparation

Fresh tissue is preferable.

Procedure

1. Treat the material in potassium hypochlorite (or Eau de Javelle) to destroy tannins and to make lignified walls lose their staining capacity.
2. Stain in cyanin solution (1% cyanin in 50% ethanol + equal volume of glycerin).

Results

Cutin and suberin turn blue.

Procedure 5: Gentian Violet Method[86]

Tissue Preparation

Fresh tissue is preferable.

Procedure

1. Dehydrate the material.
2. Treat in the stain solution (1 gm of gentian violet is dissolved in 100 ml of clove oil).
3. Remove the excess stain by treating it with fresh clove oil.
4. Observe.

Results

Cutin and suberin stain deep violet.

Applications of This Procedure

Faulkner and Kimmins[250] have used this method to study suberin in the tissue bordering lesions induced by wounding, tobacco mosaic virus, and tobacco necrosis virus in bean.

Pearse and Rutherford[266] have used this technique to study the wound associated suberized barrier to the spread of decay in the sap wood of Oak.

Procedure 6: Toluidine Blue O Method[263]

Tissue Preparation

Fresh tissue is preferable.

Procedure

1. Bring the material down to water.
2. Stain in 0.05% (W/V) aqueous solution of the dye for 5 to 30 min.
3. Check staining intensity periodically after rinsing in water.
4. Observe in water itself or if the material is to be made permanent, air dry and mount in synthetic neutral resin and then observe.

Result

Cutin and suberin stain a blue–green color.

Procedure 7: Nile Blue Sulphate Method[267]

Tissue Preparation

Fresh or frozen tissue or tissue fixed in Lewitsky's fluid can be used. The latter can be embedded in paraffin.

Procedure

1. Section the material, deparaffinize (if necessary), and bring down to water.
2. Stain in the nile blue sulphate solution (1% in distilled water) at 37°C for 30 sec.
3. Differentiate in 1% acetic acid (aqueous) at 37°C for 30 sec.
4. Wash in distilled water.
5. Mount in glycerine jelly.

Result

Cutin and suberin stain red.

Applications of This Procedure

This method was used by Brammal and Higgins[265] to study pathogenesis-related cell wall changes.

2.6.6 Minerals

2.6.6.1 Aluminium: Chrome Azural-S Method [268]

Tissue Preparation

Fresh tissue is preferred.

Procedure

1. Section the material.
2. Apply a 0.5% solution of chrome azurol-S (Eastern Kodak company) to the sections. (0.5 g of the dry chrome azuraol-S granules and 5.0 g of sodium acetate are dissolved in 80 ml of distilled water. After the chemicals are completely dissolved, sufficient

distilled water is added to make 100 ml of reagent. There is no deterioration even after years of storage under room temperature.)

3. Observe.

Result

Blue color reveals the presence of aluminium. Pink to orange color shows a negative reaction. In highly aluminium positive materials, the color will appear in a matter of minutes. Others require several hours.

Comments

1. This method is generally used to detect aluminium in woody tissue. The wood has to be sectioned along the end grain surface.
2. According to Eslyn,[269] chrome azurol-S serves as an indicator of some types of wood decaying fungi; care must be taken to avoid areas where fungi are suspected.

2.6.6.2 Calcium

Procedure 1: Silver Method[270,271]

Tissue Preparation

Frozen tissue or tissue chemically fixed and embedded in paraffin.

Procedure

1. Section the material, deparaffinize, and bring down to water.
2. Place sections in 0 to 5% solution of silver nitrate for 10 min.
3. Wash thoroughly in distilled water.
4. Immerse the slides in 0.5% amidol solution for 2 min and then rinse in distilled water.
5. Place the sections in a 2% sodium thiosulphate (hypo) solution for several minutes and then wash in distilled water.
6. Dehydrate, stain lightly with fast green, and mount.

Result

Deposits of metallic silver will indicate the sites of calcium carbonate or calcium phosphate.

Control

The presence of uric acid may pose a problem in the staining reaction. So, instead of silver nitrate, place the sections in a solution made by adding 5 ml of a 5% silver nitrate solution to 100 ml of a 3% methanamine solution and buffer to pH 9. Keep the solution at 37°C for 30 min, then rinse the slide and place it directly into the thiosulphate solution. Metallic silver will be present at sites of uric acid. These control sections should be compared with those prepared for localizing calcium.

Procedure 2: Calcium Sulphate Crystal Method[272]

Tissue Preparation

Tissue should be fixed in neutral 4% formalin. Fresh and frozen tissue can also be used. If chemically fixed, dehydrate, clear, and embed in paraffin.

Procedure

1. Section the material, deparaffinize, and bring down to 40% alcohol.
2. Place a coverslip over the tissue.
3. Now add 3% sulphuric acid through the edges of the cover slip.

Result

If calcium is present, colorless monoclinic needles of calcium sulphate will form.

2.6.6.3 Calcium Carbonate: Von Kossa's Silver Method[271]

Tissue Preparation

Tissue fixed in formalin and embedded in paraffin is preferred.

Procedure

1. Section the material, deparaffinize, and bring down to water.
2. Immerse in 5% aqueous solution of silver nitrate for 10 to 60 min in bright sunlight (but not directly exposed).
3. Wash in distilled water.
4. Treat for 2 to 3 min in 5% aqueous sodium thiosulphate.
5. Wash in distilled water.
6. Counterstain for 1 min in 0.5% aqueous safranin.

7.   Dehydrate, clear, and mount.

Result

The locations of phosphates and carbonates of calcium are indicated as black deposits.

Comments

The reaction of calcium oxalate with this method is variable and uncertain because this method depends on the presence of phosphate and carbonate and not on the presence of calcium.

Rationale of the Color Reaction

Coversion of phosphates and carbonates (of calcium) into its yellow silver equivalent by treatment with silver nitrate and the subsequent reduction of this by sodium thiosulphate, a strong reducing agent, to black metallic silver.

2.6.6.4   Calcium Oxalate[273]

Procedure 1: Silver Nitrate-Rubeanic Acid Method[273]

Tissue Preparation

Fresh or chemically fixed and paraffin embedded tissue is preferred. It is better to avoid fixation in FAA (as it decalcifies the material) or in paraformaldehyde-glutaraldehyde (as it does not retain calcium).[67]

Procedure

1.   Section the material, deparaffinize (if necessary), and bring down to water.
2.   Treat with 5% acetic acid for 30 min. (This treatment removes calcium salts of phosphates and carbonates.)
3.   Treat with 5% aqueous silver nitrate for 15 min.
4.   Treat with saturated rubeanic acid (dithio-oxamide) in 70% ethanol containing 2 drops of strong ammonia for 1 min.
5.   Dehydrate and mount.

Result

Calcium oxalate appears black to dark brown

Comments

This method is superior to all other methods described thus far, especially in light microscopy.[274] It is more consistant and faster than others. This is due to the excessive silvering reaction by rubeanic acid treatment.

Procedure 2: Silver-Hydrogen Peroxide Method[275]

Tissue Preparation

Fresh or formalin fixed and paraffin embedded tissue.

Procedure

1. Section the material, deparaffinize (if necessary), and bring down to water.
2. Treat with 2 N acetic acid for 15 min to remove phosphate and carbonates.
3. Treat with 1% silver nitrate in 15% hydrogen peroxide (equal parts of 30% hydrogen peroxide and 2% silver nitrate) for 15 min at 22°C.
4. Wash in distilled water.
5. Counterstain with 2% safranin for 1 to 3 min.
6. Dehydrate, clear, and mount.

Result

Calcium oxalate deposits and crystals stain black while the background is red.

Rationale of the Reaction

Hydrogen peroxide converts oxalate into carbonate, which reacts with silver nitrate in the presence of light to form black deposits.

2.6.6.5 Calcium Retention in Cell Walls (Especially in Cystoliths)

Procedure 1: Pyroantimonate Precipitation Method[67]

Procedure

1. Fix tissue in 5% glutaraldehyde.
2. Treat in 2% potassium pyroantimonate in 0.1 M potassium phosphate buffer (pH 7.6) for 2 h.

3. Rinse in 2.5% potassium pyroantimonate in 0.05 M potassium phosphate buffer (pH 7.6).

4. Post-fix, if need be, in 1% osmium tetroxide and then in 1.5% potassium pyroantimonate in 0.05 M buffer for 1 h.[279]

Procedure 2: Oxalate Precipitation Method[67]

Procedure

1. Immerse materials for 20 min in 0.01 M potassium oxalate and 0.16 M potassium chloride in 0.1 M cacodylate buffer, pH 7.2.

2. Rinse the material in (1) above but without potassium oxalate.

3. Fix in 2% glutaraldehyde in the rinsing solution for 2 h.

4. Rinse in solution in (2) above.

5. Post-fix, if necessary, in 1% osmium tetroxide in the same buffer for 1 h.[280]

Procedure 3: Phosphate Precipitation Method[67]

Procedure

1. Fix in 5% glutaraldehyde in 0.01 M potassium phosphate buffer, pH 7.6, 2 h.

2. Rinse in 0.05 M phosphate buffer at pH 7.6.

3. Post-fix, if necessary, in 2% osmium tetroxide in the same buffer for 2 h.

Result

All of the above procedures will help in precipitating cell wall calcium.

2.6.6.6   Magnesium: Quinalizarin–Titian Yellow Method[276]

Tissue Preparation

Frozen or chemically fixed and paraffin embedded tissue is preferred.

Procedure

1. Section the material; deparaffinize (if necessary) and bring down to water.

2. Add 1 or 2 drops of quinalizarin reagent (mix 100 mg of quinalizarin and 500 mg of sodium acetate and then dissolve 500 mg of this mixture in 100 ml of 5% sodium hydroxide) or titian yellow solution. Add 1 or 2 drops of 10% sodium hydroxide.

Results

Magnesium will develop a blue color after several hours in quinalizarin treatment and a brick red color in titian yellow treatment.

2.6.6.7 Silica

Procedure 1: Silver-Amine Chromate Method[277]

Tissue Preparation

Fresh tissue is preferable. Sections or peels can be used.

Procedure

1. Rinse tissue in 50% sulphuric acid for 1 to 5 min. In some cases, treatment with 0.1% hydrofluoric acid or ammonium fluoride for less than 1 min is necessary.
2. Thoroughly wash in several changes of water.
3. Mount in the staining solution. [The staining solution is prepared separately by dissolving 34 g silver nitrate and 20 g potassium chromate in 100 ml water each, and then mixing these solutions to precipitate silver chromate. After washing the precipitate in hot (60°C) water to remove soluble potassium nitrate, it is dissolved in 200 ml of 3% ammonia (excess ammonia interferes with the staining reaction). The solution can be stored for many months in a tightly closed bottle.]
4. Observe.
5. If preparations are to be made permanent, dehydrate, clear, and mount.

Result

Silica appears red to red–brown in non-dehydrated preparations but black in dehydrated ones.

Rationale of the Color Reaction

Silver-amine chromate staining is based on the removal of ammonia from the silver-amine complex, leading to red–brown precipitation of silver chromate over the silica surface. Ammonia masks the reddish brown color of silver chromate.

Procedure 2: Methyl Red Method[277]

Tissue Preparation

Fresh tissue is preferable. Sections or peels can be used.

Procedure

1. Immerse tissue in 50% sulphuric acid for 2 to 5 min.
2. Rinse in tap water and dehydrate in benzene.
3. Treat the tissue with the dye solution. (It is prepared by making a saturated and filtered solution of the acid form of methyl red in benzene.)*
4. Observe.

Result

The solution appears yellow–orange, while silica appears bright red in procedures stopped at the Step 2. In prepartations made after the completion of all steps, the background appears green.

Comment

Preparations can be made permanent following customary procedures. Preparation can be counterstained with fast green.

Procedure 3: Crystal Violet Lactone Method[277]

Tissue Preparation

Fresh tissue is preferable. Sections or peels can be used.

Procedure

1. Immerse tissue in 50% sulphuric acid for 2 to 5 min.
2. Rinse in tap water and dehydrate in benzene.

---

* The acid form of methyl red is preferable to the sodium salt or the hydrochloride of methyl red. The latter two compounds do not stain silica as intensely as the acid form. If the hydrochloride of methyl red is to be used, first it should be dissolved in ethanol before the addition of benzene.

3.   Place the tissue from absolute benzene in a 0.1% solution of crystal violet lactone (Hilton Davis Company, Ohio) in benzene. Color develops immediately but longer treatment enhances the color.
4.   Preparations can be made permanent using customary methods.
5.   Preparations can be counterstained in safranin.

Result

The solution is colorless. Silica stains blue to violet. If counterstained in safranin, the background appears red.

References

1.   Jensen, W. A., *Botanical Histochemistry*, W.H. Freeman and Co., San Francisco, CA, 1962, Chap. 4.
2.   Sunderland, N. and Brown, R., Distribution of growth in the apical region of the shoot of *Lupinus albus*, *J. Exptl. Bot.*, 7, 127, 1956.
3.   Jensen, W. A., A morphological and biochemical analysis of the early phases of cellular growth in the root tip of *Vicia faba*, *Exptl. Cell Res.*, 8, 506, 1955.
4.   McLane, S. R., Higher polyethylene glycols as water soluble matrix for sectioning fresh or fixed plant tissue, *Stain Technol.*, 26, 63, 1951.
5.   Bell, I. G. E., The application of freezing and drying techniques in cytology, in *International Review of Cytology*, Vol. I, Bourne, I. G. H. and Danielli, J. F., Eds., Academic Press, New York, 1952, 35.
6.   Chayen, J., Newer methods in cytology, *Bull. Res. Council of Israel*, 8D, 273, 1960.
7.   Chayen, J., Cunningham, G. J., Graham, P. B., and Silcox, A. A., Life-like preservation of cytoplasmic detail in plant cells, *Nature*, 186, 1068, 1960.
8.   Gahan, P. B., McLean, J., Kalina, M., and Sharma, W., Freeze-sectioning of plant tissues: the technique and its use in plant histochemistry, *J. Exptl. Bot.*, 18, 151, 1967.
9.   Gahan, P.B., *Plant Histochemistry and Cytochemistry*, Academic Press, Florida, 1984, Chap. 2.
10. Jensen, W. A., *Botanical Histochemistry*, W.H. Freeman and Co., San Francisco, CA, 1962, Chap. 5.
11. Glick, D. and Bloom, D., Studies in histochemistry XXXIX. The performance of freeze-drying apparatus for the preparation of embedded tissue and an improved design, *Exptl. Cell Res.*, 10, 687, 1956.
12. O'Brien, T. P. and McCully, M. E., *The Study of Plant Structure. Principles and Selected Methods*, Tamarcarphi Ltd., Melbourne, 1981, Chap. 4.
13. Baker, J. R., *Principles of Biological Technique*, Methuen, London, 1958, Chap. 2.
14. Sabatini, D. D., Bensch, K., and Barrnett, R. J., Cytochemistry and electron microscopy. The preservation of cellular ultrastructure and enzymatic activity by aldehyde fixation, *J. Cell Biol.*, 17, 19, 1963.
15. Seligman, A. M., Chauncey, H. H., and Nachlas, M. M., Effects of formalin fixation on the activity of five enzymes of rat liver, *Stain Technol.*, 26, 19, 1951.

16. Christie, K. N. and Stoward, P. J., A quantitative study of the fixation of acid phosphatase by formaldehyde and its relevance to histochemistry, *Proc. Royal Soc. London.*, B186, 137, 1974.

17. Vijayaraghavan, M. R. and Shukla, A. K., *Histochemistry: Theory and Practice,* Bishen Singh Mahendra Pal Singh, Dehra Dun, India, 1990, Chap. 2.

18. Baker, J. R., The histochemical recognition of lipine, *Quart. J. Microsc. Soc.*, 87, 441, 1946.

19. Pearse, A. G. E., *Histochemistry Theoretical and Applied,* Vol. I, 4th ed., Churchill Livingstone, New York, 1980, Chap. 5.

20. Anderson, P. J., Purification and quantification of glutaraldehyde and its effect on several enzyme activities in skeletal muscle, *J. Histochem. Cytochem.*, 15, 652, 1967.

21. Feder, N. and Wolf, M. K., Studies on nucleic acid metachromasy. II. Metachromatic and orthochromatic staining by toluidine blue of nucleic acids in tissue sections, *J. Cell Biol.*, 27, 327, 1965.

22. Klebs, E., Die einschmelzungs—Methode, ein Beitrag zur Mikroskopischen Technik, *Archiv für Microskopische Anatomie und Entwicklungsmechanik,* 5, 164, 1869.

23. Butschli, O., Modifikation der paraffine-inbettung für mikroskopische schnitte, *Biol. Zentrabl*, 1, 591, 1881-1882.

24. Jensen, W.A., *Botanical Histochemistry,* W.H. Freeman and Co., San Francisco, CA, 1962, Chap. 6.

25. Berlyn, G. P. and Miksche, J. P., *Botanical Microtechnique and Cytochemistry,* Iowa State University. Press, Iowa, 1976, Chap. 5

26. Jensen, W. A., *Botanical Histochemistry,* W.H. Freeman and Co., San Francisco, CA, 1962, Chap. 3.

27. Gurr, E., Notes on the theory of staining, in *The Encyclopedia of Microscopy and Microtechnique,* Gray, P., Ed., R. E. Krieger Publishing Co., Malabar, FL, 1982, 547.

28. Pearse, A. G. E., *Histochemistry Theoretical and Applied,* Vol. II, 4th ed., Churchill Livingstone, New York, 1985, Chap. 15.

29. Bergeron, J. A. and Singer, M., Metachromasy: an experimental and theoretical reevaluation, *J. Biophys. Biochem. Cytol.*, 4, 433, 1958.

30. Michaelis, L., The nature of the interaction of nucleic acid and nuclei with basic dye stuffs, Cold Spring Harbour Symp., *Quant. Biol.*, 12, 131, 1947.

31. Sylven, B., Metachromatic dye-substrate interactions, *Quart. J. Micros. Sci.*, 95, 327, 1954.

32. Chayen, J., Bitensky, L., and Butcher, R. G., *Practical Histochemistry,* John Wiley, London, 1973, Chap. 2.

33. Jones, G. R. N., Quantitative histochemistry: design and use of a simple microcell for standardized incubation on the slide, *Stain Technol.*, 39, 155, 1964.

34. Delly, J. G., Mountants, in *The Encyclopedia of Microscopy and Microtechnique,* Gray, P., Ed., R. E. Krieger Publishing Co., Malabao, FL, 1982, 344.

35. Gray, P., Mountant Formulas, in *The Encyclopedia of Microscopy and Microtechnique,* Gray, P., Ed., R. E. Krieger Publishing Co., Malabao, FL, 1982, 337.

36. Vijayaraghavan, M. R. and Shukla, A. K., *Histochemistry: Theory and Practice,* Bishen Singh Mahendra Pal Singh, Dehra Dun, India, 1990, Chap. 6.

37. Jensen, W. A. and Ashton, M., The composition of the developing primary wall in onion root tip cells. 1. Quantitative analyses, *Plant Physiol.*, 35, 313, 1960.

38. Woods, P. S. and Pollister, A. W., An ice-solvent method for drying frozen tissue for plant cytology, *Stain Technol.*, 30, 123, 1955.

39. Glick, D. and Malstrom, B. G., Studies in histochemistry XXIII. Simple and efficient freeze-drying apparatus for the preparation of embedded tissue, *Exptl. Cell Res.*, 3, 125, 1952.

40. Jensen, W. A., The application of freeze-dry methods to plant material, *Stain Technol.*, 29, 143, 1954.

41. Jensen, W. A., A new apparatus to freeze-drying of tissue, *Exptl. Cell Res.*, 7, 572, 1954.

42. Sumner, B. E. H., *Basic Histochemistry*, John Wiley, Chichester, 1988, Chap. 3.

43. Krishnamurthy, K.V., *Methods in Plant Histochemistry*, S. Viswanathan, Madras, 1988.

44. Gahan, P. B., *Plant Histochemistry and Cytochemistry*, Academic Press, Florida, 1984, App. 2.

45. Coetzee, J . and Van der Merwe, C. F., Penetration rate of glutaraldehyde in various buffers into plant tissue and gelatine gels, *J. Microscopy*, 137, 129, 1985.

46. La Cour, L. F., Chayen, J., and Gahan, P. B., Evidence for lipid materials in chromosomes. *Exptl. Cell Res.*, 14, 469, 1958.

47. Berlyn, G. P. and Miksche, J. P., *Botanical Microtechnique and Cytochemistry*, Iowa State University Press, Iowa, 1976, Chap. 4.

48. Feder, N. and O'Brien, T. P., Plant microtechnique: Some principles and new methods, *Am. J. Bot.*, 55, 123, 1968.

49. O'Brien, T. P. and McCully, M. E., *The Study of Plant Structure. Principles and Selected Methods*, Tamarcarphi Ltd., Melbourne, 1981, Chap. 6.

50. Maxwell, M. H., Two rapid and simple methods used for the removal of resins from 1.0 mm thick epoxy resin sections, *J. Microsc.*, 112, 253, 1978.

51. Hotchkiss, R. D., A microchemical reaction resulting in the staining of polysaccharide structures in fixed tissue preparations, *Arch. Biochem.*, 16, 131, 1948.

52. McManus, N. F. A., Histological and histochemical uses of periodic acid, *Stain Technol.*, 23, 99, 1948.

53. Kiernan, J. A., *Histological and Histochemical Methods. Theory and Practice*, Pergamon Press, Oxford, 1981, Chap. 11.

54. Bold, H. C. and Wynne, M. J., *Introduction to the Algae: Structure and Reproduction*, Prentice-Hall of India, New Delhi, 1978, Chap. 6.

55. Nanda, K. and Gupta, S. C., Malfunctioning tapetum and callose wall behaviour in *Allium cepa* microsporangia, *Beitr. Biol. Pflanzen*, 56, 465, 1974.

56. Albertini, L. G., Auberger, H., and Souvre, A., Polysaccharides and lipids in microsporophytes and tapetum of *Rhoeo discolor* Hance. Cytological Study, *Acta Soc. Bot. Polon.*, 50, 21, 1981.

57. Heslop-Harrison, J., Cell walls, cell membranes and protoplasmic connections during meiosis and pollen development, in *Pollen Physiology and Fertilization*, Linskins, H. F., Ed., North-Holland Publishing Co., Amsterdam, 1964, 39.

58. Hale, A. J., The histochemistry of polysaccharides, *Intl. Rev. Cytol.*, 6, 193, 1957.

59. O'Brien, T. P. and Thimann, K. V., Observations on the fine structure of the Oat coleoptile. III. Correlated light and electron microscopy of the vascular tissues, *Protoplasma*, 63, 443, 1967.

60. Jensen, W. A., *Botanical Histochemistry*, W.H. Freeman and Co., San Francisco, CA, 1962, Chap. 9.

61. Vijayaraghavan, M. R. and Bhat, U., Synergids before and after fertilization, *Phytomorphology*, 33, 74, 1983.
62. Vijayaraghavan, M. R., Jensen, W. A., and Ashton, M. E., Synergids of *Aquilegia formosa* - their histochemistry and ultrastructure, *Phytomorphology*, 22, 144, 1972.
63. Chao, C. Y., A periodic acid-Schiff's substance related to the directional growth of pollen tube into embryo sac in *Paspalum* ovules, *Am. J. Bot.*, 58, 649, 1971.
64. Chao, C. Y., Further cytological studies in a periodic acid-Schiff's substance in the ovules of *Paspalum orbiculare* and *P. Longifolium*, *Am. J. Bot.*, 64, 922, 1977.
65. Chao, C. Y., Histochemical study of a PAS-positive substance in the ovules of *Paspalum orbiculare* and *Paspalum longifolium*, *Phytomorphology*, 29, 381, 1979.
66. Sexton, R, Burdon, N., Reid, J. S. G., Durbin, M. L., and Lewis, L. N., Cell wall breakdown and abscission, in *Structure, Function and Biosynthesis of Plant Cell Walls*, Dugger, W. M. and Bartnicki-Garcia, S., Eds., American Society of Plant Physiologists, Maryland, 1984, 195.
67. Watt, W. M., Morell, C. K., Smith, D. L., and Steer, M. W., Cystolith development and structure in *Pilea cadierei*. (Urticaceae), *Ann. Bot.*, 60, 71, 1987.
68. Premamalini, P., Histochemical studies on the foliage leaf and stipules of *Ficus elastica* Roxb. M.Phil. Thesis, Bharathidasan University, Tiruchirappalli, India, 1988.
69. Jensen, W. A., The composition of the developing primary wall in onion root tip. II. Cytochemical localization, *Am. J. Bot.*, 47, 287, 1960.
70. Flemion, F., Cytochemical studies of the developing primary cell wall in the apical shoots of normal and physiologically dwarf peach seedlings, *Plant Physiol.*, 36, Suppl. XXVII, 1961.
71. Doyle, W. L., The morphology and affinities of the liverwort *Geothallus*. Ph.D. thesis, University of California, Berkeley, 1960.
72. Wilson, B. F., Cell wall development of cambial derivatives in *Abies concolor*. Ph.D. thesis, University of California, Berkeley, 1961.
73. Krishnamurthy, K. V., Nature of cell wall of gelatinous fibres of Caesalpiniaceae, in *V Cell Wall Meeting*, Fry, S. C., Brett, C. T., and Reid, J. S. G., Eds., Edinburgh, 1989, 64.
74. Ranjani, K., Wood of some taxa of Caeselpiniaceae. A structural and histochemical study. Ph.D. thesis, Bharathidasan University, Tiruchirappalli, India, 1988.
75. Smith, D. L., Staining and osmotic properties of young gametophytes of *Polypodium vulgare* L. and their bearing on rhizoid function, *Protoplasma*, 74, 465, 1972.
76. Robards, A. W. and Purvis, M. J., Chlorazol black E as a stain for tension wood, *Stain Technol.*, 39, 309, 1964.
77. Wilcox, H. E. and Marsh, L. C., Staining plant tissue with Chlorozol black E and pianese III-B, *Stain Technol.*, 39, 81, 1964.
78. Hickey, E. L. and Coffey, M. D., A cytochemical investigation of the host-parasite interface in *Pisum sativum* infected by the downy mildew fungus, *Peranospora pisi*, *Protoplasma*, 97, 201, 1978.
79. Fisher, J. B., A survey of buttresses and aerial roots of tropical trees for presence of reaction wood, *Biotropica*, 14, 56, 1982.
80. Fisher, J. B., Induction of reaction wood in *Terminalia* (Combretaceae): Roles of gravity and stress, *Ann. Bot.*, 55, 237, 1985.
81. Fisher, J. B. and Stevenson, J. W., Occurrence of reaction wood in branches of dicotyledons and its role in tree architecture, *Bot. Gaz.*, 142, 82, 1981.

82. Kucera, L. J. and Philipson, W. R., Growth eccentricity and reaction anatomy in branch wood of *Drimys winteri* and five native New Zealand trees, *N.Z. J. Bot.*, 15, 517, 1977.

83. Kucera, L. J. and Philipson, W. R., Growth eccentricity and reaction anatomy in branch wood of *Pseudowintera colorata, Am. J. Bot.*, 65, 601, 1978.

84. Kucera. L. J. and Philipson, W. R., Occurrence of reaction wood in some primitive dicotyledonous species, *N.Z. J. Bot.*, 15, 649, 1977.

85. Ranjani, K. and Krishnamurthy, K. V., Gelatinous fibres of Caeselpiniaceae, *Can. J. Bot.*, 66, 394, 1988.

86. Chamberlain, C. J., *Methods in Plant Histology,* University of Chicago Press, Chicago, IL, 1924.

87. Gahan, P. B., *Plant Histochemistry and Cytochemistry,* Academic Press, Florida, 1984, App. 7.

88. Whaley, W. G., Mericle, L. W., and Heimsch, C., The wall of the meristematic cell, *Am. J. Bot.*, 39, 20, 1952.

89. Naylor, G. L. and Russell-Wells, B., On the presence of cellulose and its distribution in the cell walls of brown and red algae, *Ann. Bot.*, 48, 635, 1934.

90. Van Wisselingh, C., Mikrochemische untersuchungen Åber die Zellwande der Fungi, *Jahrb. Wiss. Bot.*, 31, 619, 1898.

91. Aronson, J. M., The cell wall, in *The Fungi*, Vol. I, *The Fungal Cell*, Ainsworth, G. C. and Sussman, A. S., Eds., Academic Press, New York, 1973, 49.

92. Percival, E., The polysaccharides of green, red, and brown sea weeds: their basic structure, biosynthesis and function, *Br. Phycol. J.*, 14, 103, 1979.

93. Matty, P. J. and Johansen, H. W., A histochemical study of *Corallina officinalis* (Rhodophyta, Corallinaceae), *Phycologia*, 20, 46, 1981.

94. Purvis, M. J., Collier, D. C., and Wallis, D., *Laboratory Techniques in Botany,* Butterworths, London, 1966, Chap. 6.

95. Evans, L. V. and Holligan, M. S., Correlated light and electron microscopic studies on brown algae. I. Localization of alginic acid and sulphated polysaccharides in *Dictyota*, *New Phytol.*, 71, 1161, 1972.

96. Pearlmutter, N. L. and Lembi, C. A., Structure and composition of *Pithophora oedogonia* (Chlorophyta) cell walls, *J. Phycol.*, 16, 602, 1980.

97. Vian, B. and Roland, J. -C., Affinodetection of the sites of formation and of the future distribution of polygalactrouranans and native cellulose in growing plant cells, *Biol. Cell.*, 71, 43, 1991.

98. Benhamou, N., Cytochemical localization of ß(1→4) -D glucans in plant and fungal cells using an exoglucanase-gold complex, *Electron. Microsc. Rev.*, 2, 123, 1989.

99. Faulk, W. P. and Taylor, G. M., An immunocolloid method for the electron microscope. *Immunochemistry*, 8, 1081, 1971.

100. Frens, G., Controlled nucleation for the regulation of particle size in monodisperse gold solutions, *Nature Phys. Sc.*, 241, 20, 1973.

101. Horisberger, M. and Rosset, J., Colloidal gold, a useful marker for transmission and scanning electron microscopy, *J. Histochem. Cytochem.*, 25, 295, 1977.

102. Raikhel, N. V., Mishkind, M., and Palevitz, B. A., Immunocytochemistry in plants with colloidal gold conjugates, *Protoplasma*, 121, 25, 1984.

103. Romano, E. L., Stolinski, C., and Hughes-Jones, N. C., An antiglobulin reagent labeled with colloidal gold for use in electron microscopy, *Immunochemistry*, 11, 521, 1974.

104. Slot, J. W. and Geuze, H. J., Sizing of protein A-Colloidal gold probes for immuno-electron microscopy, *J. Cell Biol.*, 90, 533, 1981.

105. Chanzy, H., Henrissat, B., and Vuong, R., Colloidal gold labelling of 1,4-D-glucan cellobiohydrolase adsorbed on cellulose substrates, *FEBS Lett.*, 172, 193, 1984.

106. Berg, R. H., Erdos, G. W., Gritzali, M., and Brown, R. D. Jr., Enzyme-gold affinity labeling of cellulose, *J. Electron Micrsoc. Techn.*, 8, 371, 1988.

107. Heath, M. C. and Perumalla, C. J., Haustorial mother cell development by *Uromyces vignae* on colloidion membranes, *Can. J. Bot.*, 66, 736, 1988.

108. Kaminskyj, S. G. N. and Heath, M. C., Histochemical responses of infection structures and intercellular mycelium of *Uromyces phaseoli* var. *typica* and *V. phaseoli* var. *vignae* to the HNO_3-MBTH-$FeCl_3$ and IKI-H_2SO_4 tests, *Physiol. Plant. Pathol.*, 22, 173, 1983.

109. Roelofsen, P. A. and Huette, I., Chitin in the cell walls of yeasts. Antonie van Leenwenhoek, 17, 297, 1951.

110. Fuller, Biochemical and microchemical study of the cell walls of *Rhizidiomyces sp.*, *Am. J. Bot.*, 47, 838, 1960.

111. Cheadle, V. I., Gifford, E. M., Jr., and Esau, K., A staining combination for phloem and contiguous tissues, *Stain Technol.*, 28, 49, 1953.

112. Reynolds, J. D. and Dashek, W. V., Cytochemical analysis of callose localization in *Lilium longiflorum* pollen tubes, *Ann. Bot.*, 40, 409, 1976.

113. Senthil Kumar, T. and Krishnamurthy, K. V., A Histochemical study of the sieve plate and sieve tube cytoplasm of two Cucurbitaceae taxa, in *Perspectives in Environment*, Agarwal, S. K., Kaushik, J. P., Koul, K. K., and Jain, A. K., Eds., A. P. H. Publishing Corporation, New Delhi, 1998, 303.

114. Shah, J. J. and James, M. R., Observations on companion cells and specialized phloem parenchyma cells of *Nelumbo nucifera* Gaertn., *Ann. Bot.*, 33,185, 1969.

115. Indra, R., Embryology of *Ottelia alismoides* (L.) Pers. A developmental and histochemical study, Ph.D. Thesis, Madras University, India, 1985.

115. Senthil Kumar, S. and Krishnamurthy, K. V., A cytochemical study on the mycorrhizae of *Spathoglottis plicata, Biologia Plant.*, 41, 111, 1998.

117. Beneš, K., On the stainability of plant cell walls with alcian blue, *Biol. Plant.*, 10, 334, 1968.

118. Beneš, K. and Uhlirovà, S., Staining plant tissues with alcian blue (The affinity of alcian blue to polysaccharides), *Comm. Czech. Soc. Histochem. Cytochem.*,12, 858, 1966.

119. Clarke, K. J., McCully, M. E., and Miki, N. K., A developmental study of the epidermis of young roots of *Zea mays* L., *Protoplasma*, 98, 283, 1979.

120. Quintarelli, G., Scott, J. E., and Dellovo, M. C., The chemical and histochemical properties of alcian blue. III. Chemical blocking and unblocking, *Histochemie*, 4, 99, 1964.

121. Scott, J. E., Histochemistry of Alcian blue. I. Metachromasia of Alcian blue, Astra blau, and other cationic phthalocyanin dyes, *Histochemie*, 21, 277, 1970.

122. Scott, J. E. and Dorling, J., Differential staining of acid glucosaminoglycans (mucolpolysaccharides) by alician blue in salt solutions, *Histocheme*, 5, 221, 1965.

123. Cole, K. M., Park, C. M., Reid, P. E., and Sheath, R. G., Comparative studies on the cell walls of sexual and asexual *Bangia atropurpurea* (Rhodophyta). I. Histochemistry of polysaccharides, *J. Phycol.*, 21, 585, 1985.

124. Scott, J. E., Histochemistry of Alcian blue. III. The molecular biological basis of staining by alcian blue 8 GX and analogous phthalocyanins, *Histochemie*, 32, 191, 1972.

125. Scott, J. E., Histochemistry of Alcian Blue. II. The structure of alcian Blue 8GX, *Histochemie*, 30, 215, 1972.

126. O'Colla, P. S., Mucilages, in *Physiology and Biochemistry of Algae*, Lewin, R. A., Ed., Academic Press, New York, 1962, 337.

127. La Claire II, J. W. and Dawes, C. J., An autoradiographic and histochemical localization of sulphated polysaccharides in *Eucheuma nudum* (Rhodophyta), *J. Phycol.*, 12, 368, 1976.

128. Balakrishnan, S., A structural and histochemical study on *Padina boergeseenii* Allender and Kraft. Ph.D. thesis, Bharathidasan University, Tiruchirappalli, India, 1990.

129. Heslop-Harrison, Y., The pollen-stigma interaction: Pollen-tube penetration in *Crocus*, *Ann. Bot.*, 41, 913, 1977.

130. Shivanna, K. R., Ciampolini, F., and Cresti, M., The structure and cytochemistry of the pistil of *Hypercium calycinum*: The stigma, *Ann. Bot.*, 63, 613, 1989.

131. Simpson, M. G., Pollen wall development of *Xiphidium coeruleum* (Haemodoraceae) and its systematic implications, *Ann. Bot.*, 64, 257, 1989.

132. McCully, M. E., Histological studies on the genus *Fucus*. I. Light microscopy of the mature vegetative plant, *Protoplasma*, 62, 287, 1966.

133. O'Brien, T. P., Feder, N., and McCully, M. E., Polychromatic staining of plant cell walls by toluidine blue O, *Protoplasma*, 59, 367, 1964.

134. Rappay, Gy. and Van Duijn, P., Chlorous acid as an agent for blocking tissue aldehydes, *Stain Technol.*, 40, 275, 1965.

135. Mowry, R. W., The special value of methods that color both acidic and vicinal hydroxyl groups in the histochemical study of mucins, with revised directions for the colloidal iron stain, the use of Alcian blue 8GX and their combinations with the periodic acid - Schiff's reaction, *Ann. N.Y. Acad. Sci.*, 106, 402, 1963.

136. Gahan, P. B., *Plant Histochemistry and Cytochemistry*, Academic Press, Florida, 1984, Chap. 7.

137. Dawes, J. M., Ferrier, N. C., and Johnston, C. S., The ultrastructure of the meristoderm cells of the hapteron of *Laminaria*, *J. Mar. Biol. Ass. UK*, 53, 237, 1973.

138. McCully, W. G., The histological localization of the structural polysaccharides of sea weeds, *Ann. N.Y. Acad. Sci.*, 175, 702, 1970.

139. Quatrano, R. S. and Crayton, M. A., Sulfation of fucoidan in *Fucus* embryos. 1. Possible role in localization, *Dev. Biol.*, 30, 29, 1973.

140. Quatrano, R. S., Brawley, S. H., and Hogsett, W. E., The control of the polar deposition of a sulphated polysaccharide in *Fucus* zygote, in *Determinants of Spatial Organiazation*, Subtelny, S. and Konigsberg, I. R., Eds., Academic Press, London, 1979, 77.

141. Quatrano, R. S., Hogsett, W. S., and Roberts, M., Localization of a sulfated polysaccharide in the rhizoid wall of *Fucus distichus* (Phaeophyceae) zygotes, in *Proc. 9th Intl. Seaweed Symp.*, Jensen, A. and Stein, J. R., Eds., Princeton University Press, Princeton, NJ, 1979, 113.

142. Waterkeyn, L. E. and Bienfait, A. V., Localization and the role of the β-1, 3 -glucans in the large sized pennate diatom *Pinnularia*, in *V Cell Wall Meeting*, Fry, S. C., Brett, C. T., and Reid, J. S. G., Eds., Edinburgh, 1989, 186.

143. Lev, R. and Spicer, S. S., Specific staining of sulphate groups with alcian blue at low pH, *J. Histochem. Cytochem.*, 12, 309, 1964.

144. Parker, B. C. and Diboll, A. G., Alcian stains for histochemical localization of acid and sulphated polysaccharides in algae, *Phycologia*, 59, 367, 1966.

145. Ravetto, C., Alcian blue—alcian yellow: A new method for the identification of different acidic groups, *J. Histochem. Cytochem.*, 12, 44, 1964.

146. Spicer, S. S., Diamine methods for differentiating mucosubstances histochemically, *J. Histochem. Cytochem.*, 13, 211, 1965.

147. Höster, H. and Liese, W., Über das vorkommen von reactionsgewebe in Wurzeln und Ästen der dicotyledonen, *Holzforschung*, 20, 80, 1966.

148. Heath, I. D., Staining sulphated mucopolysaccharides, *Nature*, 191, 1370, 1961.

149. Santhakumari, C., A histochemical study of fruit development in banana. Ph.D. thesis., Bharathidasan University, Tiruchirappalli, India, 1990.

150. Sterling, C., Crystal structure of ruthenium red and steriochemistry of its pectin stain, *Am. J. Bot.*, 57, 172, 1970.

151. Luft, J. H., Ruthenium red and violet. I. Chemistry, purification, methods of use for electron microscopy and mechanism of action, *Anat. Rec.*, 171, 347, 1971.

152. Albersheim, P., A cytoplasmic component stained by hydroxylamine and iron, *Protoplasma*, 60, 131, 1965.

153. Reeve, R. M., A specific hydroxylamine-ferric chloride reaction for histochemical localization of pectin, *Stain Technol.*, 34, 209, 1959.

154. Webster, B. D., Anatomical and histochemical changes in leaf abscission, in *Shedding of Plant Parts*, Kozlowski, T. T., Ed., Academic Press, New York, 1973, 45.

155. Gee, M., Reeve, R. M., and McCready, Reaction of hydroxylamine with pectinic acids. Chemical studies of pectic substances in fruit, *Agric. Food Chem.*, 7, 34, 1959.

156. Foster, A. S., The use of tannic acid and iron chloride for staining cell walls of meristematic tissues, *Stain Technol.*, 9, 91, 1934.

157. Swamy, B. G. L. and Krishnamurthy, K. V., *From Flower to Fruit*, Tata McGraw Hill, New Delhi, 1980.

158. Krishnamurthy, K. V., Embryo, in *Growth Patterns in Vascular Plants*, Iqbal, M., Ed., Discorides Press, USA, 1994, 372.

159. Hale, C. W., Histochemical demonstration of acid polysaccharides in animal tissue, *Nature*, 157, 802, 1946.

160. Cawood, A. H., Potter, U., and Dickinson., H. G., An evaluation of Coomassie blue as a stain for quantitative microdensitometry of protein in section, *J. Histochem. Cytochem.*, 26, 645, 1978.

161. Eklavya, C., A technique for making CBB stained sections of Paraffin and resin embedded tissue permanent, *Indian J. Bot.*, 2, 73, 1979.

162. Fisher, D. B., Protein staining of ribboned epon sections for the light microscopy, *Histochemie*, 16, 327, 1968.

163. Vijayaraghavan, M. R. and Shukla, A. K., *Histochemistry: Theory and Practice*, Bishen Singh Mahendra Pal Singh, Dehra Dun, India, 1990, Chap. 5.

164. Gahan, P. B., *Plant Histochemistry and Cytochemistry*, Academic Press, Florida, 1984, App. 4.

165. Lu, J., Mayer, A., and Pickersgill, B., Stigma morphology and pollination in *Arachis* L., *Ann. Bot.*, 66, 73, 1990.

166. Chapman, D. W., Dichromatism of bromophenol blue with an improvement in the mercuric bromophenol technique for protein, *Stain Technol.*, 50, 25, 1975.
167. Mazia, D., Brewer, P. A., and Alfert, M., The cytochemical staining and measurement of proteins with mercuric bromo phenol blue, *Biol. Bull.*, 104, 57, 1953.
168. Deitch, A. D., Microspectrophotometric study of the binding of the anionic dye , naphthol yellow S, by tissue sections and purified proteins, *Lab. Invest.*, 4, 324, 1955.
169. Tas, J., Oud, P., and James, J., The naphthol yellow for proteins tested in a model system of polyacrylamide films and evaluated for the practical use in histochemistry, *Histochemistry*, 40, 231, 1974.
170. Knox, R. B. and Heslop-Harrison, J., Pollen-wall proteins: localization and enzymic activity, *J. Cell. Sci.*, 6, 1, 1970.
171. Hopkin, A. A. and Reid, J., Cytological studies of the M-haustorium of *Endocronortium harknessii*: morphology and ontogeny, *Can. J. Bot.*, 66, 974, 1988.
172. Isaac, W. E. and Winch, N. H., Guiaco-hydrogen peroxide and benzidine-hydrogen peroxide color reactions in beans *Phaseolus vulgaris* L., *J. Pomol. Hort. Sci.*, 23, 23, 1947.
173. Van Fleet, D. S., Histochemical localization of enzymes in vascular plants, *Bot. Rev.*, 18, 354, 1952.
174. Graham, R. C. and Karnovsky, M. J., The early stages of absorption of injected horseradish peroxidase in the proximal tubules of mouse kidney. Ultrastructural Cytochemistry by a new technique, *J. Histochem. Cytochem.*, 14, 291, 1966.
175. Frederick, S. E., The cytochemical study of Diaminobenzidine, in *Handbook of Plant Cytochemistry*, Vol. I., Vaughn, K. C., Ed., CRC Press, Boca Raton, FL, 1987, 3.
176. Essner, E., Hemoproteins, in *Electron Microscopy of Enzymes*, Vol. 2., Hayat, M. A., Ed., Van Nostrand Reinhold, New York, 1974, 1.
177. Seligman, A. M., Karnovsky, M. J., Wasserkrug, H. L., and Hanker, J. S., Nondroplet ultrastructural demonstration of cytochrome oxidase activity with a polymerizing osmiophilic reagent, diaminobenzidine (DAB), *J. Cell Biol.*, 38, 1, 1968.
178. Czaninski, Y. and Catesson, A. -M., Localization ultrastructurale d'activitês peroxidasiques dans les tissus conducteurs végétaux au cours du cycle annuel, *J. Microsc.*, 8, 875, 1969.
179. Czaninski, Y. and Catesson, A. -M., Activites peroxydasiques d' origines diverses dans les cellules d' *Acer pseudoplatanus* (tissus conducteurs et cellules en culture). *J. Microsc.*, 9, 1089, 1970.
180. Goff, C. W., A light and electron microscopic study of peroxidase localization in the onion root tip, *Am. J. Bot.*, 62, 280, 1975.
181. Hall, J. L. and Sexton, R., Cytochemical localization of peroxidase activity in root cells, *Planta*, 108, 103, 1972.
182. Henry, E. W., Peroxidase in tobacco abscission zone tissue. III. Ultrastructural localization in thylakoids and membrane-bound bodies of chloroplasts, *J. Ultrastruct. R.*, 52, 289, 1975.
183. Hepler, P. K., Rice, R. M., and Terranova, W. A., Cytochemical localization of peroxidase activity in wound vessel members of *Coleus*, *Can. J. Bot.*, 50, 977, 1972.
184. Poux, N., Localisation d'activités enzymatiques dans les cellules du méristèmes radiculaire de *Cucumis sativus* L. II. Activité peroxydasique, *J. Microsc.*, 8, 855, 1969.

185. Poux, N., Localisation d'activités enzymatiques dans les cellules du meristémes radiculaire de *Cucumis sativus* L. IV. réactions avec la diaminobenzidine mise en evidence de peroxysomes, *J. Microsc.*, 14, 183, 1972.

186. Nir, I. and Seligman, A. M., Ultrastructural localization of oxidase activities in corn root tip cells with two new osmophilic reagents compared to diaminobenzidine, *J. Histochem. Cytochem.*, 19, 611, 1971.

187. Mesulam, M. M., Tetrametyl benzedine for horse radish peroxidase neurohistochemistry: a non- carcinogenic blue reaction product with superior sensitivity for visualizing neural afferents and efferents, *J. Histochem. Cytochem.*, 26, 106, 1978.

188. Hanker, J. S., Yates, P. E., Metz, C. B., and Rustioni, A., A new specific, sensitive and non- carcinogenic reagent for the demonstration of horseradish peroxidase, *Histochem. J.*, 9, 789, 1977.

189. Imberty, A., Goldberg, R., and Catesson, A. - M., Teteramethyl benzidine and *p*-phenylenediamine-pyrocatechol for peroxidase histochemistry and biochemistry: two new non-carcinogenic chromogens for investigating lignification process, *Plant Sci. Lett.*, 35, 103, 1984.

190. Wachstein, M. and Meisel, E., Demonstration of peroxidase activity in tissue sections, *J. Histochem. Cytochem.*, 12, 538, 1964.

191. Al-Azzawi, M. J. and Hall, J. L., Effects of aldehyde fixation on adenosine triphosphatase and peroxidase activities in maize root tips, *Ann. Bot.*, 41, 431, 1977.

192. Sexton, R. and Hall, J. L., Enzyme Cytochemistry, in *Electron Microscopy and Cytochemistry of Plant Cells*, Hall, J. L., Ed., Elsevier, North Holland, Amsterdam, 1978, 63.

193. Straus, W., Factors affecting the cytochemical reaction of peroxidase with benzidine and the stability of the blue reaction product, *J. Histochem. Cytochem.*, 12, 462, 1964.

194. Gahan, P. B., *Plant Histochemistry and Cytochemistry,* Academic Press, Florida, 1984, App. 5.

195. Poux, N., Localisation des activites phosphatasiques acides et peroxydasiques au niveau des ultrastructures végétatales, *J. Microsc.*, 21, 265, 1974.

196. Fahimi, H. D., Cytochemical localization of peroxidatic activity of catalase in rat hepatic microbodies (peroxisomes), *J. Cell Biol.*, 43, 275, 1969.

197. Papadimitriou, J. M., Van Duijn, P., Brederoo, P., and Streefkerk, J. G., A new method for cytochemical demonstration of peroxidase for light, fluorescence and electron microscopy, *J. Histochem. Cytochem.*, 24, 82, 1976

198. Burstone, M. S., Modification of histochemical techniques for the demonstration of cytochrome oxidase, *J. Histochem. Cytochem.*, 9, 59, 1960.

199. Jensen, W. A., *Botanical Histochemistry,* W.H. Freeman and Co., San Francisco, CA, 1962, Chap. 15.

200. Berlyn, G. P. and Miksche, J. P., *Botanical Microtechnique and Cytochemistry,* Iowa State University Press, Iowa, 1976, Chap. 15.

201. Goldberg, R., Le, T., and Catesson, A. -M., Localization and properties of cell wall enzyme activities related to the final stages of lignin biosynthesis, *J. Exp. Bot.*, 29, 969, 1985.

202. Fielding, J. L. and Hall, J. L., A biochemical and cytochemical study of peroxidase activity in roots of *Pisum sativum*. I. A comparison of DAB-peroxidase and guaiacol-peroxidase with pariticular emphasis on the properties of cell wall activity, *J. Expt. Bot.*, 19, 969, 1978.

203. Kaur-Sawhney, R., Flores, M. E., and Galston, A. W., Polymine oxidase in oat leaves: A cell wall localized enzyme, *Plant Physiol.*, 68, 494, 1981.

204. Olson, P. D. and Varner, J. E., Personal communication, 1993.

205. Novikoff, A. B. and Masek, B., Survival of lactic dehydrogenase and DPNH- diaphorase activities after formal Calcium fixation, *J. Histochem. Cytochem.*, 6, 217a, 1958.

206. Malik, C. P. and Singh, M. B., *Plant Enzymology and Histoenzymology*, Kalyani Publishers, New Delhi, 1980, Chap. 6.

207. Sauter, J. J. and Braun, H. J., Cytochemische Untersuchung der Atmungsaktivitat in den Strasburgerzellen von *Larix* und ihre Bedeutung für den Assimilattransport, *Z. Pflanzenphysiol.*, 66, 440, 1972.

208. Santos, I. and Salema, R., Cytochemical localization of malic dehydrogenase in chloroplasts of *Sedum telephium*, in *Electron Microscopy*, Vol. 2., Proc. 7th. Eur. Congr. Electron Microscopy., Brederoo, P. and de Prister, Eds., European Congress on Electron Microscopy. Leiden Retain, 1980, 246.

209. Wenzel, J. and Behrlisch, D., Elektronenmikroskopischer Nachweis von oxydo-Reduktasen im Herzmuskel der Ratte, *Z. Mikrosk.-Anat. Forsch.*, 84, 372, 1971.

210. Sexton, R., Cronshaw, J., and Hall, J. L., A study of the biochemistry and cytochemistry localization of β-glycerophosphatase activtiy in root tips of maize and pea, *Protoplasma*, 73, 417, 1971.

211. Shnitka, T. K. and Seligman, A. M., Ultrastructural localization of enzymes, *Ann. Rev. Biochem.*, 40, 375, 1971.

212. Gahan, P. B. and McLean , J., Acid phosphatases in root tips of *Vicia faba*, *Biochem. J.*, 102, 47, 1967.

213. Livingston, D. C., Coombs, M. M., Franks, L. M., Maggi, V., and Gahan. P. B., A lead phthalocyanin method for demonstrating acid hydrolases in plant and animal tissues, *Histochemie*, 18, 48, 1969.

214. Gahan, P. B. and McLean, J., Subcellular localization and possible functions of acid β-glycerophosphatases and naphthol esterases in plant cells, *Planta*, 89, 126, 1969.

215. Beadle, D. J., Dawson, A. J., James, D. J., Fisher, S. W., and Livingstone, D. C., The avidity of heavy metal diazotates for animal lysosomes and plant vacuoles during the ultrastructural localization of acid hydrolase, in *Electron Microscopy and Cytochemistry*, Wisse, E., Daems, W. T., Molenar, I., and Van Duijn, P., Eds., Elsevier-North Holland, New York, 1974, 85.

216. Gahan, P. B., Dawson, A. L., and Fielding, J., Paranitrophenyl phosphate as a substrate for some acid phosphatases in roots of *Vicia faba*, *Ann. Bot.*, 42, 1413, 1978.

217. Barka, T. and Anderson, P. J., Histochemical methods for acid phosphatase using hexazonium pararosanalin as coupler, *J. Histochem. Cytochem.*, 10, 741, 1962.

218. Fosket, D. E. and Miksche, J. P., A histochemical study of the seedling shoot apical meristem of *Pinus lambertiana*, *Am. J. Bot.*, 53, 694, 1966.

219. Gomori, G., Microtechnical demonstration of phosphatase in tissue sections, *Pro. Soc. Exp. Biol. Med.*, 42, 23, 1939.

220. Gomori, G., The distribution of phosohatase in the normal organs and tissues, *J. Cell Compar. Physiol.*, 17, 71, 1941.

221. Gomori, G., An improved histochemical technique for acid phosphatase, *Stain Technol.*, 25, 81, 1950.

222. Holt, S. J., Factors governing the validity of staining methods for enzymes and their bearing upon the Gomori acid phosphatase technique, *Exp. Cell Res.*, Suppl. 7, 1, 1959.

223. Jensen, W. A., The cytochemical localization of acid phosphatase in root tip cells, *Am. J. Bot.*, 43, 50, 1956.

224. Moore, R., McClelen, C. E., and Smith, H. S., Phosphatases, in *Handbook of Plant Cytochemistry*, Vol. I, Vaughn, K. C., Ed., CRC Press, Boca Raton, FL, 1987, 37.

225. Takamastsu, H., Histologische und biochemische studien uber die phosphatase (I. Mitteilung). Histochemsche untersuchungsmethodik der phosphatase und deren verteilung in verchiedenen organen und geweben, *Trans. Soc. Pathol. Jap.*, 29, 492, 1939.

226. Tice, L. W. and Barrnett, R. J., The fine structural localization of glucose-6-phosphatase in rat liver, *J. Histochem. Cytochem.*, 10, 754, 1962.

227. Halperin, W., Ultrastructural localization of acid phosphatase in cultured cells of *Daucus carota*, *Planta*, 88, 91, 1969.

228. Poux, N., Localization d'activites enzymatiques dans le meristeme radiculaire de *Cucumis sativus* L. III. Activite phosphatasique acide, *J. Microsc.*, 9, 407, 1970.

229. Marty, F. M., Peroxisomes et compartiment lysosomal dans les cellules du meristeme radiculaire d' *Euphorbia characias* L., *Une etude Cytochimique.*, C.R. Acad. Sci. (Paris), 273, 2504, 1971.

230. Holt, S. J. and Hicks, R. M., The localisation of acid phosphatase in rat liver cells as revealed by combined cytochemical staining and electron microscopy, *J. Biophys. Biochem. Cytol.*, 11, 47, 1961.

231. Maier, K. and Maier, U., Localization of ß-glycerophosphate and Mg^{++} activated adenosine triphosphatase in a moss haustorium and the relation of these enzymes to cell wall labyrinth, *Protoplasma*, 75, 91, 1972.

232. Goldfischer, S., Essner, E., and Novikoff, A. B., The localization of phosphatase activities at the level of ultrastructure, *J. Histochem. Cytochem.*, 12, 72, 1964.

233. Chiquoine, A. D., The distribution of glucose-6-phosphatase in the liver and kidney of the mouse, *J. Histochem. Cytochem.*, 1, 429, 1953.

234. Mahler, H. R. and Cordes, E. H., *Biological Chemistry*, Harper and Row, New York, 1971, 497.

235. Catesson, A. -M. and Roland, J. C., Sequential changes associated with cell wall formation in the vascular cambium. *IAWA Bull. n.s.*, 2, 151, 1981.

236. Knox, R. B. and Heslop- Harrison, J., Cytochemical localization of enzymes in the wall of the pollen grain, *Nature*, 223, 92, 1969.

237. Smith, D. L., Localization of phosphatase of young gametophytes of *Polypodium vulgare* L., *Protoplasma*, 74, 133, 1972.

238. Burstone, M. S., The relationship between fixation and techniques for the histochemical localization of hydrolytic enzymes, *J. Histochem. Cytochem.*, 6, 322, 1958.

239. Kaplow, L. S. and Burstone, M. S., Cytochemical demonstration of acid phosphatase in hematopoietic cells in health and in various hematological disorders using azo dye techniques, *J. Histochem. Cytochem.*, 12, 805, 1964.

240. Smith, R. E. and Fishman, W. H., *p*-(Acetoxymercuric) aniline diazoate: A reagent for the visualizing the napthol A5-BI product of acid hydrolase action at the level of light and electron microscope, *J. Histochem. Cytochem.*, 17, 1, 1969.

241. Gahan, P. B., *Plant Histochemistry and Cytochemistry*, Academic Press, Florida, 1984, Chap. 5.

242. Crevier, M. and Bélanger, L. F., Simple method for histochemical detection of esterase activity, *Science*, 122, 556, 1955.

243. Moloney, W. C., McPherson, K., and Fliegelman, L., Esterase activity in leukocytes demonstrated by use of naphthol AS-D chloroacetate substrate, *J. Histochem. Cytochem.*, 8, 200, 1960.

244. Barnett, R. J. and Seligman, A. M., Histochemical demonstration of esterases by production of Indigo, *Science*, 114, 579, 1951.

245. Holt, S. J., Indigogenic staining methods for esterases, in *General Cytochemical Methods*, Vol. I, Danielli., J. F., Ed., Academic Press, New York, 1958, 375.

246. Deising, H., Nicholson, R. L., Haug, M., Hioward, R. J., and Mendgen, K., Adhesion pad formation and the involvement of cutinase and esterases in the attachment of uredospores to the host cuticle, *Plant Cell*, 4, 1101, 1992.

247. Enright, J. B., Frye, F. L., and Atwal, O. S., Ribonuclease activity of peripheral leukocytes and serum in rabies-susceptible and rabies-refractory mice, *J. Histochem. Cytochem.*, 13, 515, 1965.

248. Smith, R. E. and Van Frank, R.M., The use of aminoacid of 4-methoxy-B-naphthylamine for the assay and subcellular localization of tissue proteinases, *Front. Briol.*, 43, 193, 1975.

249. Campbell, W. G., Bryant, S. A., and Swann, G., The chlorine-sodium sulphite color reaction of woody tissues, *Biochem. J.*, 31, 1285, 1937.

250. Faulkner, G., Kimmins, W. C., and Brown, R. G., The use of fluorochromes for the identification of ß(1→3)-glucans, *Can. J. Bot.*, 51, 1503, 1973.

251. Siegel, S. M., On the biosynthesis of lignins, *Physiol. Plant.*, 6, 134, 1953.

252. Wardrop, A. B., Lignins in plant kingdom. Occurrence and formation in plants, in *Lignins, Occurrence, Formation, Structure and Reactions*, Sarkanen, K. V. and Ludwig., C. H., Eds., Wiley, New York, 1971, 19.

253. Gibbs, R. D., The Maule reaction, lignins and the relationship between woody plants, in *The Physiology of Forest Trees*, Thimann, K. V., Ed., Roland Press, New York, 1958, 269.

254. Iiyama, K. and Pant, R., The mechanism of the Mäule color reaction. Introduction of methylated syringyl nuclei into soft wood lignin, *Wood Sci. Technol.*, 22, 167, 1988.

255. Meshitsuka, G. and Nakano, J., Studies on the mechanism of Lignin reaction (XI). Mäule color reaction (7), *Mokuzai Gakkaishi*, 23, 232, 1977.

256. Meshitsuka, G. and Nakano, J., Studies on the mechanism of lignin colour reaction (XII). Mäule colour reaction (8), *Mokuzai Gakkaishi*, 24, 563, 1978.

257. McLean, R. C. and Cook, W. R. I., *Plant Science Formulae*, Macmillan, London, 1941, Chap. 7.

258. Baayen, R. P., Responses related to lignification and intravascular periderm formation in carnations resistant to Fusarium wilt, *Can. J. Bot.*, 66, 784, 1988.

259. Wardrop, A. B., Lignification and xylogenesis, in *Xylem Cell Development*, Barnett, J. R., Ed., Castle House Publication Ltd., Kent, 1981, 115.

260. Reeve, R. M., Relevance of immature tuber periderm to high commercial losses, *Am. Potato J.*, 81, 254, 1974.
261. Shah, J. J. and Babu, A. M., Vascular occlusions in the stem of *Ailanthus excelsa* Roxb., *Ann. Bot.*, 97, 603, 1986.
262. Southworth, D., Solubility of pollen exines, *Am. J. Bot.*, 61, 36, 1974.
263. Holloway, P. J. and Watterndorf, J., Cutinized and suberized cell walls, in *Handbook of Plant Cytochemistry*, Vol. II, Vaughn, K. C., Ed., CRC Press, Boca Raton, FL, 1987, 1.
264. Jensen, W. A., *Botanical Histochemistry*, W.H. Freeman and Co., San Francisco, CA, 1962, Chap. 12.
265. Brammal, R. A. and Higgins, V. J., A histological comparison of fungal colonization in tomato seedlings susceptible or resistant to *Fusarium* crown and root rot disease, *Can. J. Bot.*, 66, 915, 1988.
266. Pearce, R. B. and Rutherford, J., A wound-associated suberised barrier to the spread of decay in the sap wood of oak (*Quercus robur* L.), *Physiol. Plant Path.*, 19, 359, 1981.
267. Gahan, P. B., *Plant Histochemistry and Cytochemistry*, Academic Press, Florida, 1984, App. 6.
268. Kukachka, B. F. and Miller, R. B., A chemical spot-test for aluminum and its value in wood identification, *IAWA Bull. n.s.*, 1, 104, 1980.
269. Eslyn, W. E., Utility pole decay. Part 3: Detection in pine by color indicators, *Wood Sci. Technol.*, 13, 117, 1979.
270. Jensen, W. A., *Botanical Histochemistry*, W.H. Freeman and Co., San Francisco, CA, 1962, Chap. 13.
271. Sumner, B. E. H., *Basic Histochemistry*, John Wiley, Chichester, 1988, Chap. 13.
272. Marsh, R. P. and Shive, J. W., Boron as a factor in calcium metabolism of the corn plant, *Soil Sci.*, 51, 141, 1941.
273. Yasue, T., Histochemical identification of calcium oxalate, *Acta Histochem. Cytochem.*, 2, 83, 1969.
274. Franceschi, V. R. and Horner, H. T., Calcium oxalate crystals in plants, *Bot. Rev.*, 46, 361, 1980.
275. Pizzolato, P., Inorganic constituents and foreign substances, *J. Histochem. Cytochem.*, 12, 331, 1964.
276. Broda, B., Über die Verwendbarkeit von chinalization Titangelb und Azublau zum mikro-und histo-chemischen Magnesiumnachweis in Pflanzengeweben, *Mikrokosmas*, 32, 184, 1939.
277. Dayanandan, P., Localisation of silica and calcium carbonate in plants, *Scanning Electron Microscopy*, 1983/III, SEM inc., AMF O'Hare (Chicago), 1519, 1983.
278. Krishamurthy, K. V., Personal observations, 1988.
279. Wick, S. M. and Hepler, P. K., Localization of calcium^{++}–containing antimonate precipitates during mitosis, *J. Cell Biol.*, 86, 500, 1980.
280. Jonas, L. and Zelck, V., Subcelluar distribution of calcium in smooth muscle cells of pig coronary artery, *Exptl. Cell Res.*, 89, 352, 1974.

Chapter 3

Fluorescence Microscopic Cytochemistry

Contents

3.1 Introduction

Self-luminous hot bodies such as the sun and stars are called *incandescent*. All the other forms of light emission by objects are *luminescence*. Emission of light as a result of absorbed light radiation is known as *photoluminescence*. Here, the electrons of the object that absorb the light radiation first get excited (i.e., the electrons reach from ground state to excited state) and then revert back to ground state through the loss of vibrational energy which is externally seen as light emission. The electron takes less than 10^{-3} sec for light absorption. Photoluminescence is of two types: (1) *Phosphorescence*, in which the excited electrons take a long time to return to ground state; therefore, phosphorescent objects emit light for a longer time, and (2) *fluorescence*, in which the excited electrons reach the ground state very quickly, emitting electromagnetic radiation in that process. The emission ceases within about 10^{-9} sec of the removal of the excited radiations.[1]

Fluorescent substances get excited when irradiated with high energy radiations of shorter wavelengths, such as UV, blue–violet, and blue, and emit low energy light of longer wavelengths such as green, yellow, and red, thus following *Stoke's Law*. This difference in absorbed and emitted wavelengths is the basis for observation of fluorescence in fluorescence microscopy. Substances exhibiting fluorescence are called *fluorophores* or *fluorochromes*. Different fluorochromes have different spectral characteristics, as a result of differing electronic configurations. Peak absorption and emission of light take place at different regions of the light spectrum for each fluorochrome. These two peaks are respectively called *excitation peak* and *emission peak*. To obtain intense fluorescence, irradiation with a light of wavelengths close to the peak of excitation spectrum is desirable. In addition, the different fluoro-chromes differ in the fluorescence intensity obtained by excitation at the optimum wavelength. *Quantum efficency* (Q) is the ratio between energy emitted and energy absorbed and, theoretically, the Q-value should be between 0 to 1. Although Q-values give an indication of the relative brightness of fluorescence for various fluorochromes, fading can result due to bleaching owing to photochemical reactions which cause decomposition of fluorescing molecules; reduction in intensity may also be due to quenching caused by the presence of other fluorochrome-oxidizing agents such as oxygen, halogens, salts of heavy metals, etc.[1]

Fluorescence may be *autofluorescence* (*inherent* or *primary fluorescence*) or *induced fluorescence* (*secondary fluorescence*). In the latter, non-fluorescent objects are combined with fluorchromes to induce fluorescence in them. Examples of cell wall substances that exhibit autofluorescence include lignin, cutin, suberin, pheno-lics, sporopollenin, etc. Any cell wall substance, other than autofluorescing ones, can be induced to fluoresce by tagging suitable flurochromes to them.

The fluorescence microscope can easily be fabricated from a bright field light microscope. The three main requirements are flurochromes, light source, and fluo-rescence detecting unit; besides these are the appropriate filters: the *exciter filter* and the *barrier* (also called *absorption* or *secondary*) filter. The former filter, also called the *primary filter*, is placed in the optical axis of the microscope between the light source and the object, while the latter filter is placed in the observation path, between the object and the eye. The exciter filter helps in the selection of the desired exciter radiations by filtering away the unwanted wavelengths from the light source. Excitation wavelengths are only slightly weakened as they are transmitted by the exciter filters. The barrier filter absorbs the residual ultraviolet radiations and thus protects the vision of the observer.

There is a lot of confusion in filter terminology. Exciter filters are categorized depending upon their range of transmission and can be used singly or in combination for eliminating unwanted low-energy radiations. *UG* and *BG* are the two commonly used series. The former transmit maximally in the UV range and are used mostly for the UV fluorescence studies alone, while the latter transmit maximally in the blue–violet and blue and are used mostly for the blue and, in addition, sometimes for the UV-fluorescence observations. Besides UG and BG, *Corning 7 to 37* is another series of filters that is commonly available. There are various numbers (or types) of UG and BG filters, each coding for the narrow specific range of wavelengths which the filter transmits and each of particular thickness. The barrier filters are also labeled with standard markings designating their range of absorption. All of them absorb in the UV range and the difference between various filters lies mainly on their transmission of visible light wavelengths. For example, OG1 does not transmit waves other than about 500 nm whereas GG9 permits the transmission of waves through the entire range of the visible light spectrum and beyond up to 302 nm, although this transmission is negligibly small for radiation with wavelengths shorter than 436 nm. Depending on requirement, any UG and BG filters can be combined for use.[2]

Normal optical glass has sufficient transmittance for the most widely used exciter wavelengths. Thus, routine fluorescence microscopy can be done using the regular optics in the microscope. In certain cases, quartz or crown glass lenses may be put in the optical path between the light source and the object. Use of quartz condenser is beneficial in the detection of very weak primary fluorescence. Since the exciter radiations are weakened as they pass through the glass, the number of lenses and also of the exciter filters should be the minimum necessary between the object and the light source. The intensity of exciter radiations at the object level increases in proportion to the square of the numerical aperture of the condenser. Thus, the diaphragm should be fully open. Once the excitation light has reached the specimen, its further transmission in the optical axis is normally undesirable. It should become extinct by the barrier filter.

Brightness of the fluorescence image is direcly proportional to the numerical aparture of the objective and, inversely, to the magnification. Apochromatic objectives are ideal for fluorescence microscopy. The eye piece should be of low power so that the objective of greater magnification (and thus of higher numerical aperture) can be used to attain the desired total magnification.

Light source is chosen depending on the excitation spectrum of the fluorochrome and its Q-value. Weak fluorochromes (=low Q) require more exciting light for viewing. Tungsten halogen lamps are suitable and inexpensive light sources for routine studies; these are unsuitable for illumination with UV light. They are less suitable for green excitation of red fluorescence. High pressure mercury lamps are recommended if UV light and high-energy excitation is required. Xenon lamps emit a spectrum of constant intensity from UV to red. Although relatively expensive, these are recommended.

To prevent the glaring due to superposition of fluorescent details, the specimen should be as thin as possible.

Fluorescence microscopy is ideally done in a dark room so that the eyes become accustomed to detecting even low intensities and the weakly fluorescent details are not missed from observation. Preparations to be studied in fluorescence microscopy are best stored at 4°C in the dark since fluorescence intensity fades with time, light exposure, and increase in temperature.

The other details to be taken care of in fluorescence cytochemistry are as follows:

1.   If immersion oil is used during observation, it should be of the non-fluorescent type.
2.   Glass of the slide (also coverslip) holding the specimen should be as thin as possible and non-fluorescent. Quartz slides may be used to advantage when weakly fluorescent objects are studied.
3.   Although fluorescence of objects remains unaffected by the RI of the mountant, an ideal mountant should be primarily non-fluorescent and should lack UV-absorption. Plain distilled water is the best mountant. Glycerol can be used. Euparal is known to preserve aniline blue-induced fluorescence of callose.[3] Harleco mountant, an isobutyl methacrylate (synthetic), is often recommended. DPX is also most commonly used. Piccolytes are fluorescent and should not be used.

Fluorescent dyes used as stains are known as fluorochromes. While some dyes are exclusively fluorochromes, many serve both as diachromes (dyes intended for transmitted light) and fluorochromes. Examples of both categories, as well as their other characteristics, are detailed in Rost.[4]

3.2   Fluorescence Cytochemical Methods

3.2.1 Cell Wall Carbohydrates

a. Total Insoluble Polysaccharides (Fluorescence Schiff's or Pseudo-Schiff's Reaction)[5,6,7-10]

Tissue Preparation

Fresh, chemically fixed, dewaxed, and glycol methacrylate or epoxy-embedded tissue can be used. It is better to avoid aldehyde fixatives.

Procedure

1.   Bring the tissue down to water.
2.   Block the tissue aldehydes in a saturated solution of 2,4-DPNH in 15% acetic acid for 30 min.
3.   Wash thoroughly for 30 min in tap water.
4.   Oxidize for 10 min in 1% periodic acid.
5.   Stain in any one of the following:
     (a) Stain for 5 to 10 min in a pseudo-Schiff reagent such as freshly prepared 0.1% w/v acridine yellow, acriflavine (in HCl), coriphosphine O, or benzoflavine in distilled water saturated with SO_2 just before use. Pearse[5] recommends a 0.01% solution in aqueous or 30% ethanolic medium saturated with SO_2; 2 to 3 drops of thionyl

chloride are added to 100 ml of the dye solution; the pH is to be adjusted to 2.5 to 3.5 with dilute NaOH.

(b) Stain for 5 to 10 min in a 0.01% v/v solution of pararosaniline-Schiff's reagent. (The reagent sold by Fisher Scientific Company is adequate.)

6. Rinse for 5 min in running tap water.

7. Mount in water; if necessary, dehydrate in suitable solvents or air-dry and mount in non-fluorescing immersion oil.

8. If stained in pseudo-Schiff's reagent, observe in UV or short-wavelength blue, in a fluorescence microscope. If pararosaniline-Schiff's reagent has been used, observe in a fluorescence microscope using green light as an exciter.

Results

If stained in pseudo-Schiff's reagent, insoluble polysaccharides fluoresce a yellow to yellowish orange color. If the other stain has been used, a red color would be obtained with insoluble polysaccharides.

Rationale of the Color Reaction

Treatment with periodic acid would result in aldehydes formed from polysaccharides. These aldehydes combine with pseudo-Schiff's reagent such as acriflavine and the combined product would emit the characteristic wavelength when excited. The mechanism of combination of periodate-induced aldehydes with pseudo-Schiff's reagents such as acriflavine is different from the one described under PAS Techniques. Pararosaniline, by itself, is not fluorescent at any pH from 1 to 7. But after linking Schiff's reagent to the aldehydes (formed by periodate oxidation from polysaccharides) in the presence of SO_2, the bound pararosaniline is fluorescent, emitting a red light when excited with green light.[11]

Controls

Any one of the following controls can be used :

1. Acetylation
2. Benzoylation
3. Omission of periodate treatment

Comments

1. If done with proper controls, this technique is one of the most reliable in cytochemistry (see also Ref. 6).
2. It is a very specific test.
3. Overstaining may result in reduced or even no fluorescence.

Application of This Procedure

Although very potential, reliable, and specific, the pseudo-Schiff reaction has not found a wide application in cell wall cytochemistry. Scott and Peterson[12] used this technique to study the endodermal wall thickenings in the roots of *Ranunculus acris*. Heath and Perumalla[13] employed this procedure to study the response of host cells to fungal attack.

b. Total Cell Wall Polysaccharides (Calcofluor White* or Optical Brightener Method)[6,14,15]

Tissue Preparation

Use fresh, freeze-dried, freeze-substituted, or chemically fixed tissue. If fixed tissues are used, embed them in paraffin or glycol-methacrylate. If epoxy resins are used for embedding, care should be taken to remove the resin completely from the material before staining.[16]

Procedure

1. Wash the material thoroughly in water; if embedded, bring it down to water.
2. Stain for 30 sec to 2 min in 0.1% aqueous solution of calcofluor white M2 R New or calcofluor white ST,[18] if the material is fairly thick; if thin, stain in 0.01% aqueous solution for about 20 sec. Dumas and Knox[19] and DeMason[18] used a 0.7% aqueous solution. Wood et al.[20] used a 0.01% solution in 50 mM phosphate buffer, pH 7.8.
3. Wash briefly in water.
4. Mount in water or other aqueous mountants. Embedded materials can be dehydrated or air-dried and mounted. The mountant should be of nil or low autofluorescence.
5. Examine under fluorescence microscope using UV light (400 to 410 nm is ideal).

Result

Cell wall carbohydrates with (1-4)-β-D-glucans (cellulose) and (1-3)-β-D glucans (callose) fluoresce to a pale blue color.[6]

Rationale of the Color Reaction

Although the basis of the color reaction is not clear, calcofluor is said to temporarily bind to the 1,3 (glucose or mannose) and 1,4 linked glucans by forming complexes.[20,21] Presumably specific sites are present on these polysaccharide chains

---

* This is now called Cellufluor.[17]

which immobilize the fluorochrome so that the energy received on irradiation is dissipated as a fluorescence emission rather than as enhanced molecular vibration.

Controls

1. Acetylation
2. Benzoylation

Use either of the above two control procedures to prevent the wall polysaccharides from getting complexed to calcofluor. However, if specific cell wall carbohydrates such as cellulose or callose are to be localized with this technique, then follow control methods specific to those carbohydrates. It is advisable to follow differential extraction procedures, before calcofluor staining.

Comments

1. Initially calcoflour was suggested to specifically bind to cellulose[22] and, therefore, in many research papers calcofluor-fluorescing walls were taken for granted as cellulosic. The following may be cited: Watt et al.[23] on the cystoliths of *Pilea cadierei*, Simpson[24] on the developing pollen wall of *Xiphidium coeruleum*, Shivanna et al.[25] on the stigmatic cells of *Hypericum calycinum*, and Darken[26] on algae and fungi. In all of these instances, what has been stained by calcofluor might be cellulose, but chances for the presence of other carbohydrates are also present. The non-specificity of this procedure was pointed out by many workers in recent years, and substances such as callose, cellulose, chitin, pectin, alginic acid, etc. were shown to be stained by calcofluor.[6,13,15,27,28,30,129] Therefore, caution must be exercised in interpreting the results obtained. However, with the right type of specific control measures and differential extraction procedures, it is possible to judiciously localize cellulose, callose, chitin, (1→3, 1→4) linked carbohydrates of grasses, or any other polysaccharide to which it binds. Our experience[28] shows, however, that sulphated polysaccharides like fucoidan are not localized by this technique.

2. To localize the cell walls sharply and in a better way, it is ideal to remove cell contents prior to staining. For removing cell contents, the materials may be treated in 6% sodium hypochlorite for 30 sec to 2 min followed by a wash in water for a minimum period of 2 min.[6]

3. Calcofluor, at least in low concentrations, is not toxic to the plants. Plants can be grown in the weak calcofluor solution for some time and the dye can be located even after fixation and embedding procedures in any part of the plant at a subsequent time.[15]

4. Photine HV is another optical brightener which can be used instead of calcofluor.[15,22,31]

Application of This Procedure

The major application of this procedure, in the past, has been to localize cellulose in a variety of materials ranging from algae to angiosperms (see representative literature cited earlier). Waaland and Waaland[32] have exploited this technique to find out the population of newly formed cells in some red algae. They have "fluorescent labeled" the existing cells by growing the algae on the dye-solution and then shifted them to dye-free solution. Subsequent analysis would reveal the newly labeled cells free of fluorescent walls.

Another potential application of this procedure is to use calcofluor along with aniline blue, the latter serving as a counter-fluorochrome.[15] Jefferies and Belcher[33] have used this stain combination to study pollen tubes *in vivo*. If calcofluor staining is followed by aniline blue, the latter will replace calcofluor wherever it has bound itsef to 1,3-β-glucans; calacofluor would be retained only in places rich in 1,4-β-glucans. Consequently, the two regions would appear yellow and blue, respectively, giving strong contrast.

Calcofluor has been used advantageously to study the sites of cellulose synthesis in the root hairs of *Peperomia* species, thus proving that cellulose can be localized by calcofluor even in the formative stages.[31] It has also been used to detect the time and site of cellulose wall regeneration in cultivated protoplasts.[34] Stone[27] and Wood et al.[20] have employed this technique to localize the differential distribution of (1→3, 1→4)-β-glucans in the aleurone layers of wheat grains. The cells of aleurone layer have a two-layered cell wall of which the inner layer is very strongly reactive to calcofluor in comparison to the outer layer.

This technique was used by Quatrano and Stevens[35] to prove that when the zygote of *Fucus* germinates, the rhizoidal wall does not represent a local stretching of the preexisting zygote wall but that it is formed *de novo*. They also showed that alginates (within 15 min) and cellulose (within 30 min) were the first components to be detected in the rhizoidal cell wall, white fucoidin, which is calcofluor negative, appears after 1 h (see also Nakazawa et al.[36]).

Herth and Hausser[37] used calcofluor to study chitin and cellulose microfibrillar assembly and its experimental alteration. Since calcofluor has a dipole character and is incorporated parallel to cellulose or chitin chains, it then prevents the hydrogen bond formation between adjacent chains and thus prevents crystallization. The experimental organisms included centric diatoms, the pollen tubes of *Lilium longiflorum* and the root hairs of *Zea mays* and *Trianea bogetensis*. The systems were grown on media containing 0.1% calcofluor white. Chitin fibril crystallization was arrested in calcofluor white treated diatoms and tip growth altered in pollen tubes and root hairs.

Calcofluor has been used to investigate the nature of cell walls in the developing sporangium of *Padina*, a brown alga.[28] Since calcofluor does not show any affinity for sulphated fucans, but shows for alginic acid, it was possible to demonstrate that the young sporangial wall in this alga was totally free from alginic acid, relatively poor in cellulose and consisted mainly of sulphated glucans. Subsequently, however, the sporangium develops a thick wall rich in cellulose and alginic acid.

c. Cellulose, Callose, and 1→3, 1→4,-β-glucans (FITC-Bacillus 1-3, 1-4-β-Glucan Hydrolase Conjugate Method)[27,38]

Tissue Preparation

Fresh or glutaraldehyde fixed tissues can be used.

Procedure

1. Purify the bacillus 1-3, 1-4-β-glucan hydrolase and couple it to fluorescein isothiocyanate (FITC) as a specific carbohydrate-binding probe.
2. Pretreat the material to be studied with bromine (1.6% v/v) in chloroform to eliminate autofluorescence.
3. Treat the material in the dark with FITC-β-glucan hydrolase in tris buffer (0.05 M, pH 8.5) for 45 min.
4. Rinse in distilled water.
5. Mount in fluorescence-free mountant.
6. Observe in the filter combination BP 485 and FT 510.

Results

Cellulose, callose, and 1→3,1→4-β-glucan-rich regions fluoresce.

Control

Prior treatment of material with 1→3,1→4-β-glucan hydrolase.

Application of This Procedure

1. Seibert et al.[38] have used this probe for locating cellulose in *Dictyostelium* stalks.
2. This procedure was employed by Stone[27] to study wheat aleurone cells. He found that the inner cell wall layer of these cells was rich in 1→3,1→4-β-glucans.

d. Callose

Procedure 1: Aniline Blue Fluorescence Method[6,16,39-44]

Tissue Preparation

Fresh tissue is preferable. Materials fixed in any aldehyde can be used. Embedment can be in wax or glycol methacrylate. If epoxy resins are used as embedding media, staining should be attempted only after the resin has been completely removed from the material.

Procedure

1. Bring the material down to water.
2. Place the material in the staining solution for about 1 h. The stain can be prepared according to any one of the following procedures:

   (a) Linskens and Esser:[45] Decolorized aniline blue 0.1% in 0.15 ML^{-1} Na$_2$ HPO$_4$.

   (b) Smith and McCully[41] and O'Brien and McCully:[6] 0.005% w/v decolorized aniline blue in 0.067 M phosphate buffer, pH 8.5.

   (c) Hopkin and Reid:[42] 0.05% aniline blue in 0.01 M phosphate buffer, pH 8.5.
3. Mount in gelatin or in immersion oil of low fluorescence. Stained sections of embedded materials can be permanently mounted by gently air-drying materials until moisture just disappears. Do not wash out the stain before drying. The gentle air-drying ensures that some water molecules remain, as these appear to be essential for the formation of the fluorescence complex, at least in sieve plates and new cell walls.
4. Examine in a fluorescence microscope using UV excitation filter BG 12 (330 to 380 nm), dichroic mirror DM 400 (365 nm) and a barrier filter which will minimize chlorophyll autofluorescence.

Result

Callose-containing walls fluoresce in a yellow color. Fluorescence fades away during observation but can be restored by thoroughly washing the material in tap water for 12 h followed by restaining. Ramanna[3] reported retention of fluorescence in aniline blue stained pollen tubes which were solvent dehydrated and mounted in Euparal.

Rationale of the Color Reaction

The exact nature of the fluorochrome and its reaction with cell walls was not established for a long time. The first problem has been solved. The fluorochrome is *sirofluor*[46] which forms a minor component of the commercial aniline blue, as was predicted by Smith and McCully.[44] This binds to β-1,3 glucans (see details in *Sirofluor Method*).

Control

See lacmoid blue procedure (Page 67).

Comments

1. The staining solution preferably should be freshly prepared every time. But solution stored in dark bottles in a refrigerator can do well for weeks. Discard when the blue color changes to green.

2. The validity of the aniline blue fluorescence technique to detect callose has been questioned on the basis of experimental studies in pollen tubes by Reynolds and Dashek.[47] These authors consider the lacmoid blue technique as a more reliable one. Faulkner et al.[48] have demonstrated that aniline blue lacks specificity for linkage configuration (α or β) or for linkage groups (-1,3-,- 1,4-,1,6-). Smith and McCully,[44] in a more detailed study on aniline blue specificity, found similar results. These authors concluded that the strong fluorescence found in areas such as sieve plates is probably due to localized structural packing of wall polymers, rather than to specific linkage.

3. Crude aniline blue also binds, although weakly, to all cell walls and, therefore, care must be taken to interpret the fluorescence obtained.

4. In materials subjected to PAS reagent before staining with aniline blue, the "non-specific" binding of aniline blue to wall areas not containing callose can be completely eliminated. It suggests that the binding of Schiff's reagent to wall areas rich in 1, 2 glycols prevents access of the stain to binding sites on these compounds. In other words, dye binding to callose is unaffected by PAS treatment. Similarly, non-specific binding of aniline blue to lignified walls is eliminated by prior staining with toluidine blue O.[41]

Application of This Procedure

This is one of the most widely used techniques thus far to detect callose. This has been used by several investigators to study sieve plates, pit fields, new cell walls and cell plates, microsporogenesis and megasporogenesis, pollen tubes, etc.[40,43,44,49]

Vithanage and Knox[50] have applied this technique with good results for studying the induction of callose deposition in stigmas following pollination by foreign pollen. Knox and Heslop-Harrison[51] employed this technique and established the absence of callose in the walls of pollen grains of several angiosperms. The root hair tip is shown to possess callose essentially due to the application of aniline blue fluorescence technique (See Cooper[52] and the literature cited therein).

Procedure 2: Sirofluor Fluorescence Method[27,46,53,54]

Tissue Preparation

Fresh or chemically fixed tissue can be used. Fixation in 2% formaldehyde and 0.5% glutaraldehyde in 0.1 M phosphate buffer, pH 7.4, for 12 h gives good results. After fixation in the above, the material should be rinsed in 0.1 M phosphate buffered saline, dehydrated in ethanol, and embedded.

Procedure

1.   Bring the material down to water.
2.   Treat the material for 30 min at 20°C with 20 ml of a solution of sirofluor (0.03 mg/ml in distilled water).
3.   Rinse the material in water and air dry.
4.   Observe under fluorescence microscope under conditions mentioned for aniline blue fluorescence method.

Result

Callose (1,3 β-glucans) gives a brilliant yellow fluorescence with very little background fluorescence. (1→3, 1→4)-β-glucan lichenin is positive to this but not (1-3, 1-4) grass β-glucans.[27]

Rationale of the Color Reaction

The ordered and open helical conformation of (1→3)-β-D-glucans is very essential for complexing with sirofluor. Model building studies have shown that sirofluor, which is a flexible molecule with a length equivalent to a glycosyl tetrasaccharide, could interact specifically with the surface of a (1→3)-β-D-glucan triple helix possibly through hydrophobic regions.[46,55]

Controls

As in lacmoid blue procedure (Page 67).

Comments

It appears to be a very specific procedure and even laminarin which is closely related to callose is not showing good fluorescence with sirofluor.

Application of This Procedure

This procedure has not been used much because of its recent introduction. Bonhoff et al.[53] have used this procedure to detect callose in roots of resistant and susceptible soybean infected with *Phytophthora megasperma* f. sp. *glycinea*. Callose deposition (only in resistant cases) was found only at cell walls which were in contact with fungal hyphae. Pollen tube wall callose has been demonstrated in *Nicotiana* (see Stone et al.;[54] Meikle et al.[56]) using this procedure. Waterkeyn and Bienfait[57] used this procedure for studying callose in the pennate diatom, *Pinnularia*. McConchie and Knox[59] employed this technique to study callose deposition during the pollen-stigma interaction process in *Posidonia australis*, a seagrass.

e. Sulphated Polysaccharides

Acriflavine Method[5]

Tissue Preparation

Formalin-fixed, freeze-dried, or freeze-substituted tissue can be used.

Procedure

1. Bring the material down to water.
2. Stain in 1:20,000 acriflavine in 0.1 M citrate buffer, pH 2.5. (The stock solution is prepared by dissolving 500 mg of purified acriflavine in 100 ml distilled water at 80°C. It is cooled and stored in the dark until use. The staining solution is prepared by adding 1 ml of stock solution to 99 ml of buffer.)
3. Differentiate in 70% isopropanol for 1 min.
4. Dehydrate in isopropanol, clear and mount in a medium that is not fluorescent.
5. Examine in a fluorescence microscope under UV light.

Result

Clear yellow fluorescence marks the sites of sulphated polysaccharides.

Rationale of the Color Reaction

Acriflavine gives insoluble salts with sulphated polysaccharides. Further details are not known.

Controls

See Page 71.

f. Alginic Acid[58]

Tisse Preparation

Fresh or fixed tissue can be used.

Procedure

1. Purify alginate from any brown seaweed. Commercially available purified alginate can also be used.
2. Conjugate it with fluorescein. Fluorescein gets conjugated to the unesterified polygalacturonate blocks of alginate.
3. Bring the material to be studied down to water.
4. Treat it with alginate fluorescein conjugate.
5. Wash and observe.

Results

Sites positive to alginate will fluoresce.

Application of This Method

This procedure has been used to study alginate distribution in brown algae, especially *Fucus* by Vreeland and her colloborators.[58]

3.2.2 Enzymes

a. Peroxidase[60,61]

Incubation Medium

Same as in Procedure 6 under light microscopic localization of peroxidase (Page 95); add to this medium 1 mg of rhodamine B or rhodamine 6G.

Procedure

Repeat Steps 1 through 4 from the procedure mentioned on Page 95.

5. Wash for 10 min in three changes of 50% ethanol to remove uncoupled rhodamine.
6. Hydrate and mount in Farrants medium or dehydrate and mount in Euporal.
7. Examine under fluorescence microscope (excitation with green light).

Result

Sites showing reddish white fluorescence indicate the places of peroxidase activity.

Rationale of the Color Reaction

As in the procedure on Page 95. However, here the dimeric homovanillic acid is precipitated by lead ions in the presence of rhodamine to make it fluorescent. The reaction involved is shown in Figure. 3.1.

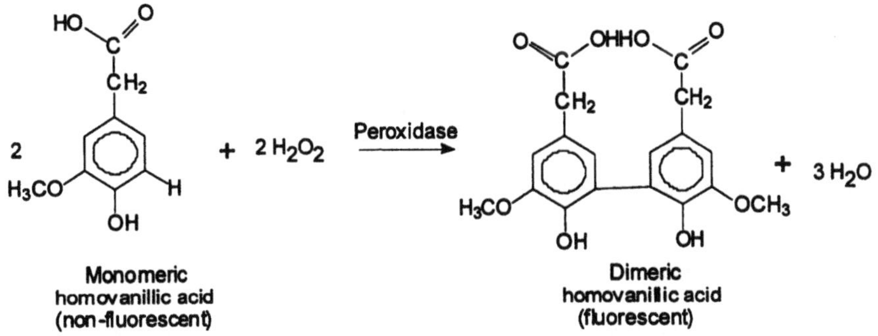

FIGURE 3.1
Monomeric non-flourescent homovanillic acid is converted into flourescent dimeric form by peroxidase in the presence of rhodamine (From Papadimitriou, J. M., et al., A new method for cytochemical demonstration of peroxidase for light, fluorescence and electron microscopy, *J. Histochem. Cytochem.*, 24, 82, 1976. With permission.)

Comment

The enzyme can be quantified by using fluorospectrophotometry.

Control

As in the procedure cited above.

3.2.3 Lignin

Procedure 1. Autofluorescence Method[62,63]

Tissue Preparation

Practically any method of tissue preparation which does not degrade lignin can be used.

Procedure

Observe under UV light

Result

Lignin emits a bluish green autofluorescence with an emission maximum at 400 nm.

Comment

This test alone cannot be taken as a test to identify lignin.

Procedure 2: Ethidium Bromide Method [64]

Tissue Preparation

Fresh material is prepared. The material should be fairly thin and free of strong cuticularization, as ethidium bromide is a poor penetrant.

Procedure

1. Stain the material for 30 sec to 10 min in 0.01% w/v ethidium bromide in tap water.
2. Wash briefly and mount in water.
3. Observe under excited light (above 400 nm).

Result

Lignified walls fluoresce to a brilliant orange.

3.2.4 Wall Bound Ferulic Acid

Autofluorescence Method Coupled with Ammonia Treatment [27,65]

Tissue Preparation

Fresh tissue is preferable.

Procedure

1. Observe the material under UV light.
2. While observing, slowly add more and more liquid ammonia to the material.

Result

Ferulic acid and lignin will autofluoresce to a blue color initially. But with the increased addition of ammonia, the fluorescence of ferulic acid will change from blue to green with great intensity while that of lignin will remain with the same blue color, but with greater intensity.

Application of the Procedure

Harris and Hartley[65] used this method to study the distribution of ferulic acid in the cell walls of grasses. Stone[27] has used this technique to study the distribution of esterified ferulic acid in the cereal grains and found its striking presence in the walls of aleurone cells.

3.2.5 Sporopollenin

Autofluorescence Method [66]

Tissue Preparation

Fresh pollen from any source can be used.

Procedure

1. Acetolyze the pollen.* (Acetolysis will eliminate all substances except sporopollenin, cutin, suberin, lignin and lipids.**)
2. Observe the pollen under UV light.

Result

Sporopollenin emits a yellowish green autofluorescence and can be distinguished from cutin and suberin which emit a whitish green fluorescence and from lignin which emits a bluish green fluorescence.

3.2.6 Cutin and Suberin

Procedure 1: Auramine O Method[68,69]

Tissue Preparation

Fresh, freeze-dried, or freeze-substituted tissue can be used. Chemically fixed and embedded materials are normally not suitable.

Procedure

1. Incubate tissues in chloroform for 5 min to remove lipids.
2. Stain in 0.01 to 0.02% auramine O in water or 0.05 M tris-HCl buffer, pH 7.2.
3. Mount in glycerine and observe under fluorescence microscope.

Result

Cutin and suberin fluoresce to a yellow–green color.

Control

It is difficult to formulate a control for suberin. For cutin, the control procedure is as follows: Pretreat the material repeatedly in ether-ethanol (1:1, v/v) before proceeding with the staining schedule. This treatment removes cuticle and cutin. To

---

* Acetolysis can be carried out as per the procedure of Erdtman:[67] Treat the pollen in 1:9 v/v concentrated H_2SO_4 and acetic anhydride at 70 to 80°C for 15 to 30 min. If desired, bleach the charred acetolyzed pollen by treating them in a mixture of 20 ml glacial acetic acid, 2 ml of concentrated HCl and 1 g of potassium chlorate.
** Lipids do not autofluoresce.

remove lignin autofluorescence, prior treatment in sodium hypochlorite solution is to be used.

Application of the Procedure

This procedure was used by a few investigators to study the cuticle in the stigmatic papillae (see Lu et al.,[70] Malti and Shivanna,[71] and Shivanna et al.[25]).

Procedure 2: 3,4-Benzpyrene-Caffeine Method [72]

Tissue Preparation

Use fresh tissue; if sections are to be used, they should be as thin as possible. Tissues fixed in Lewitsky's fluid can be used, but not those fixed in formal calcium.

Procedure

1. Stain for 20 min in the benzpyrene-caffeine solution. Prepare a saturated solution of caffeine in water (1.5% caffeine) at 20°C. Add 2 mg of benzpyrene to 100 ml of the filtered solution of caffeine and keep the same at 37°C for 2 d. Filter the solution, add 100 ml of distilled, allow it to stand for 2 h and then refilter. The solution is now ready for use. It will keep well for several months.
2. Wash in water, and mount in glycerine.
3. Examine under UV light in a fluorescence microscope.

Result

Cutin and suberin will fluoresce to a blue or bluish-white fluorescence.

Comments

1. Lipids and sterols also will fluoresce to the same colors, but they can be removed by treatment with pyridine extraction, before proceeding with the staining schedule.
2. The preparations are not stable and will fade very rapidly.
3. Caffeine has no role in color production; it is used merely to get the dye into solution.

Procedure 3: Chlorophyll Treatment Method [73,74]

Tissue Preparation

Fresh tissue is preferred.

Procedure

1. Place the tissue in a strong and fresh alcoholic solution of chlorophyll for 15 to 30 min in dark.
2. Wash out the staining solution.*
3. Observe in fluorescence microscope using blue light.**

Result

Cutin and suberin stain to a red fluorescence.

Comment

Although the preparations are not permanent, they are comparable to those made in the sudan dyes.

Application of the Procedure

This procedure has been used by Wattendorf[75] to identify and study the suberin layer in the crystal idioblasts of *Larix decidua*. Weerdenburg and Peterson[76] have used this procedure to study suberin in apple roots.

Procedure 4: Phosphate Buffer Method[77]

Tissue Preparation

Fresh, freeze-dried, or freeze substituted tissue can be used. Chemically fixed and embedded material is not recommended.

Procedure

1. Mount materials in 0.02 M phosphate buffer, pH 9.1.
2. Observe under UV radiation.

Result

Cutin and suberin show a yellow fluorescence.

---

* This makes the loss of chlorophyll from the specimen slow but continuous.
** UV light destroys chlorophyll quickly with the consequent loss of fluorescence. Moreover, the blue autofluorescence (consequent on UV employment) of cutin and suberin could interfere with the chlorophyll fluorescence. With blue light excitation, any autofluorescence cannot pass through barrier filter.

Rationale of the Reaction

The phenol component of these substances is responsible for the optical property exhibited, although the mechanism is not very clear.

Control

Mount material in concentrated H_2SO_4, wash, and then follow the staining procedure. Resistance to this treatment is indicative of cutin and suberin.

Procedure 5: Phosphine 3R Method [72,78-80]

Tissue Preparation

The tissue must be as thin as possible. Use only fresh tissue.

Procedure

1. Stain the material for 3 to 5 min in an 0.1% aqueous solution of phosphine 3R at room temperature.
2. Briefly rinse in water.
3. Mount in 90% glycerin.
4. Observe under UV light.

Result

Cutin and suberin will show a clear, silvery-white fluorescence.

Control

Pyridine extraction.

Procedure 6: Scarlet H Method [81]

Tissue Preparation

Fresh tissue is preferable.

Procedure

1. Treat the material with a 0.1% aqueous solution of scarlet H at room temperature.
2. Wash in water briefly.
3. Mount in water and observe under UV light.

Result

Cutin and suberin are stained to red or orange–red color.

Comment

Lignin should be removed by treating the material in aqueous sodium hypochlorite solution for effective staining of cutin and suberin.

3.2.7 Unsaturated Acidic Waxes

Auramine O Method[77]

Tissue Preparation

Fresh, freeze-dried, or freeze-substituted materials should be used.

Procedure

1. Mount materials in auramine O solution (auramine O, 0.01% solution in 0.05 M tris-HCl buffer, pH 7.2).
2. Observe in fluorescence microscope under blue light.

Result

Unsaturated acidic waxes fluoresce to a bright greenish–yellow color.

References

1. Ploem, J. S. and Tanke, H. J., *Introduction to Fluorescence Microscopy*, Oxford University Press, Oxford, 1987, Chap. 1.
2. Ploem, J. S. and Tanke, H. J., *Introduction to Fluorescence Microscopy*, Oxford University Press, Oxford, 1987, Chap. 2.

3. Ramanna, M. S., Euparal as a mounting medium for preserving fluorescence of aniline blue in plant material, *Stain Techol.*, 48, 103, 1973.

4. Rost, F. W. D., Fluorescence Microscopy, in *Histochemistry Theoretical and Applied*, Vol. I, 4th ed., Pearse, A. G. E., Ed., Churchill Livingstone, Edinburgh, 1980, 346.

5. Pearse, A. G. E., *Histochemistry Theoretical and Applied*, Vol. II 4th ed., Churchill Livingstone, New York, 1985, Chap. 15.

6. O'Brien, T. P. and McCully, M. E., *The Study of Plant Structure. Principles and Selected Methods*, Tamarcarphi Ltd., Melbourne, 1981, Chap. 6.

7. Kasten, F. H., Schiff-type reagents in Cytochemistry I. Theoretical and Practical considerations, *Histochemie*, 1, 466, 1959.

8. Kasten, F. H., Additional Schiff-type reagents for the use in Cytochemistry. I., *Stain Technol.*, 33, 39, 1958.

9. Ornstein, L., Mautner, W., Davis, B. J., and Tamura, R., New horizons in fluorescence microscopy, *Mt. Sinai J. Med.*, 24, 1066, 1957.

10. Stoward, P. J., Studies in fluorescence histochemistry, *J.R. Microsc. Soc.*, 87, 237, 1967.

11. Böhm, N. and Sprenger, E., Fluorescence cytophotometry: A valuable method for the quantitative determination of the nuclear Feulgen-DNA, *Histochemie*, 16, 100, 1968.

12. Scott, M. G. and Peterson, R. L., The root endodermis in *Ranunculus acris*. I. Structure and ontogeny, *Can. J. Bot.*, 57, 1040, 1979.

13. Heath, M. C. and Perumalla, C. J., Haustorial mother cell development by *Uromyces vignae* on colloidion membranes, *Can. J. Bot.*, 66, 736, 1988.

14. Gahan, P.B., *Plant Histochemistry and Cytochemistry*, Academic Press, Florida, 1984, App. 8.

15. Hughes, J. and McCully, M. E., The use of an optical brightener in the study of plant structure, *Stain Technol.*, 50, 319, 1975.

16. Sutherland, J. and McCully, M. E., A note on the structural changes in the walls of pericycle cells initiating lateral roots, *Can. J. Bot.*, 54, 2083, 1976.

17. Peterson, R. L. and Currah, R. S., Synthesis of mycorrhizae between protocorms of *Goodyera repens* (Orchidaceae) and *Ceratobasidium cereale*, *Can. J. Bot.*, 68, 1117, 1990.

18. DeMason , D. A., Endosperm structure and storage reserve histochemistry in the palm, *Washigtonia filifera*, *Am. J. Bot.*, 73, 1332, 1986.

19. Dumas, C. and Knox, R. B., Callose and determination of pistil viability and incompatibility, *Theor. Appl. Genet.*, 67, 1, 1983.

20. Wood, P. J., Fulcher, R. G., and Stone, B. A., Studies on the specificity of interaction of cereal cell wall components with Congo Red and Calcofluor. Specific detection and histochemistry of (1→3), (1→4)- β-D- glucan, *J. Cereal Sci.*, 1, 95, 1983.

21. Maeda, H. and Ishida, N., Specificity of binding of hexapyranosyl polysaccharides with fluorescent brightener, *J. Biochem.*, 62, 276, 1967.

22. Gahan, P. B., *Plant Histochemistry and Cytochemistry*, Academic Press, Florida, 1984, Chap. 7.

23. Watt, W. M., Morell, C. K., Smith, D. L., and Steer, M. W., Cystolith development and structure in *Pilea cadierei* (Urticaceae), *Ann. Bot.*, 60, 71, 1987.

24. Simpson, M. G., Pollen wall development of *Xiphidium coeruleum* (Haemodoraceae) and its systematic implications, *Ann. Bot.*, 64, 257, 1989.

25. Shivanna, K. R., Ciampolini, F., and Cresti, M., The structure and cytochemistry of the pistil of *Hypercium calycinum*: The stigma, *Ann. Bot.*, 63, 613, 1989.
26. Darken, M. A., Natural and induced fluroscence in microscopic organisms,*Appl. Microbiol.*, 9, 354, 1961.
27. Stone, B. A., Non-cellulosic β-glucans in cell walls, in *Structure, Function and Biosynthesis of Plant Cell Walls*, Dugger, W. M. and Bartniki-Garcia, S., Eds., American Society of Plant Physiologists, Rockville, MD, 1984, 52.
28. Balakrishnan, S., A structural and histochemical study on *Padina boergeseenii* Allender and Kraft, Ph.D. thesis, Bharathidasan University, Tiruchirappalli, India, 1990.
29. Brammal, R. A. and Higgins, V. J., A histological comparison of fungal colonization in tomato seedlings susceptible or resistant to *Fusarium* crown and root rot disease, *Can. J. Bot.*, 66, 915, 1988.
30. Krishnamurthy, K. V., Unpublished data, 1998.
31. Tampion, J., McKendrick, M. E., and Holt, G., Use of fluorecent brighteners to visualize the sites of cellulose synthesis in root hairs of species of *Peperomia, Physiol. Plant.*, 29, 440, 1973.
32. Waalland, S. D. and Waaland, J. R., Analysis of cell elongation in red algae by fluorescent labeling, *Planta*, 126, 127, 1975.
33. Jefferies, C. J. and Belcher, A. R., A fluorescent brightner used for pollen tube identification *in vivo*, *Stain Technol.*, 50, 199, 1974.
34. Cooper, J. B., Hydroxyproline synthesis is required for cell wall regeneration, in *Structure, Function and Biosynthesis of Plant Cell Walls*, Dugger, W. A. and Bartnicki-Garcia, S., Eds., American Society of Plant Physiologists, Rockville, MD, 1984, 397.
35. Quatrano, R. S. and Stevens, P. T., Cell wall assembly in *Fucus* zygotes. 1. Characterisation of the polysaccharide components, *Plant Physiol.*, 58, 224, 1976.
36. Nakazawa, S., Takamura, K., and Abe, M., Rhizoid differentiation in *Fucus* eggs labeled with calcofluor white and birefringence of cell wall, *Bot. Mag.*, 82, 41, 1969.
37. Herth, W. and Hausser, I., Chitin and Cellulose fibrillogenesis in vivo and experimental alteration, in *Structure, Function and Biosynthesis of Plant Cell Walls*, Dugger, W. M. and Bartnicki-Garcia, S., Eds., American Society of Plant Physiologists, Rockville, MD, 1984, 89.
38. Seibert, G. R., Benjamison, M., and Hoffman, M., A conjugate of cellulase with fluorescein isothiocyanate- a specific stain for cellulose, *Stain Technol.*, 53, 103, 1978.
39. Currier, H. B., Callose substances in plant cells,*Am. J. Bot.*, 44, 478, 1957.
40. Eschrich, W. and Currier, H. B., Identification of callose by its diachrome and fluorochrome reactions, *Stain Technol.*, 39, 303, 1964.
41. Smith, M. M. and McCully, M. E., Enhancing aniline blue fluorescent staining of cell wall structures, *Stain Technol.*, 53, 79, 1978.
42. Hopkin, A. A. and Reid, J., Cytological studies of the M-haustorium of *Endocronortium harknessii*: morphology and ontogeny, *Can. J. Bot.*, 66, 974, 1988.
43. Currier, H. B. and Strugger, S., Aniline blue and fluorescence microscopy of callose in bulb scales of *Allium cepa* L., *Protoplasma*, 45, 552, 1956.
44. Smith, M. M. and McCully, M. E., A critical evaluation of the specificity of aniline blue fluorescence, *Protoplasma*, 95, 229, 1978.

45. Linskens, H. F. and Esser, K., Über eine spezifische Anfärbung der Pollenschlauche im Griffel und die Zahl der kallosepropfen nach selfstung und Fremdung, *Naturwissenschaften*, 44, 16, 1957.

46. Evans, N. A., Hoyne, P. A., and Stone, B. A., Characteristics and specificity of the interaction of a fluorochrome from aniline blue (sirofluor) with polysaccharides, *Carbohy. Polymers*, 4, 215, 1984.

47. Reynolds, J. D. and Dashek, W. V., Cytochemical analysis of callose localization in *Lilium longiflorum* pollen tubes, *Ann. Bot.*, 40, 409, 1976.

48. Faulkner, G. and Kimmins, W. C., Staining reactions of the tissue bordering lesions induced by wounding tobacco mosaic virus and tobacco necrosis virus in bean, *Phytopathology*, 65, 1396, 1975.

49. Martin, F. W., Staining and observing pollen tubes in the style by means of fluoresence, *Stain Technol.*, 34, 125, 1959.

50. Vithanage, H. I. M. V. and Knox, R. B., Development and cytochemistry of stigma surface and response to self and foreign pollination in *Helianthus annuus*, *Phytomorphology*, 27, 168, 1977.

51. Knox, R. B. and Heslop-Harrison, J., Pollen-wall proteins: localization and enzymic activity, *J. Cell. Sci.*, 6, 1, 1970.

52. Cooper, K. M., Callose-deposit formation in radish root hairs, in *Cellulose and Other Natural Polymer Systems: Biogenesis, Structure and Degradation*, Brown, R. M. Jr., Ed., Plenum Press, New York, 1982, 167.

53. Bonhof, A., Rieth, B., Golecki, J., and Griesebach, H., Race cultivar-specific differences in callose deposition in soybean roots following infection with *Phytophthora megasperma* f. sp. *Glycinea*, *Planta*, 172, 101, 1987.

54. Stone, B. A., Evans, N. A., Bönig, I., and Clarke, A. E., The application of Sirofluor, a chemically defined flurochrome from aniline blue for the histochemical detection of callose, *Protoplasma*, 112, 191, 1984.

55. Evans, N. A. and Hoyne, P. A., A fluorochrome from aniline blue: Structure, synthesis and fluorescence properties, *Aust. J. Chem.*, 35, 2571, 1982.

56. Meikle, P. J., Bonig, I., Hoogenrad, N. J., Clarke, A. E., and Stone, B. A., The localization of $(1\rightarrow3)$- β-glucans in the walls of pollen tubes of *Nicotiana alata* using a $(1\rightarrow3)$- β-glucan specific monoclonal antibody, *Planta*, 185, 1, 1991.

57. Waterkeyn, L. E. and Bienfait, A. V., Localization and the role of the β-1, 3 -glucans in the large sized pennate diatom *Pinnularia*, in *V Cell Wall Meeting*, Fry, S. C., Brett, C. T., and Reid, J. S. G., Eds., Edinburgh, 1989, 186.

58. Vreeland, V. and Laetsch, W. M., Identification of associating carbohydrate sequences with labeled oligosaccharides: Localization of alginate gelling subunits in cell walls of a brown alga, *Planta*, 177, 423, 1989.

59. McConchie, C. A. and Knox, R. B. K., Pollen-stigma interaction in the sea grass *Posidonia australis*, *Ann. Bot.*, 63, 235, 1989.

60. Gahan, P. B., *Plant Histochemistry and Cytochemistry*, Academic Press, Florida, 1984, App. 5.

61. Papadimitriou, J. M., Van Duijn, P., Brederoo, P., and Streefkerk, J. G., A new method for cytochemical demonstration of peroxidase for light, fluorescence and electron microscopy, *J. Histochem. Cytochem.*, 24, 82, 1976.

62. Goldschmidt, O., Ultraviolet spectra, in *Lignin: Occurrence, Formation, Structure and Reactions*, Sarkanen, K. V. and Ludwig, C. H., Eds., Wiley InterScience, New York, 1971, 241.
63. Wardrop, A. B. and Bland, D. E., The process of lignification in woody plants, *4th Inter. Congr. Biochem.*, 2, 92, 1959.
64. McCully, M. E., The use of fluorescence microscopy in the study of fresh plant tissues, *Proc. Microc. Soc. Canada*, 3, 172, 1976.
65. Harris, P. J. and Hartley, R. D., Detection of bound ferulic acid in cell walls of the Graminae by ultraviolet fluorescence microscopy, *Nature*, 259, 508, 1976.
66. Vijayaraghavan, M. R. and Shukla, A. K., *Histochemistry: Theory and Practice*, Bishen Singh Mahendra Pal Singh, Dehra Dun, India, 1990, Chap. 5.
67. Erdtmann, G., *Handbook of Palynology*, Hafner, New York, 1969, 213.
68. Heslop-Harrison, Y., The pollen-stigma interaction: Pollen-tube penetration in *Crocus*, *Ann. Bot.*, 41, 913, 1977.
69. Berg, R. H., Preliminary evidence for the involvement of suberization in infection of *Casuarina*, *Can. J. Bot.*, 61, 2910, 1983.
70. Lu, J., Mayer, A., and Pickersgill, B., Stigma morphology and pollination in *Arachis* L., *Ann. Bot.*, 66, 73, 1990.
71. Malti, and Shivanna, K. R., Structure and cytochemistry of the pistil of *Crotolaria retusa* L., *Proc. Indian Natl. Sci. Acad.*, B50, 92, 1984.
72. Jensen, W. A., *Botanical Histochemistry*, W.H. Freeman and Co., San Francisco, CA, 1962, Chap. 12.
73. Holloway, P. J. and Watterndorf, J., Cutinized and suberized cell walls, in *Handbook of Plant Cytochemistry*, Vol. II, Vaughn, K. C., Ed., CRC Press, Boca Raton, FL, 1987, 1.
74. Chamberlain, C. J., *Methods in Plant Histology*, University of Chicago Press, Chicago, IL, 1924.
75. Wattendorf, J., Feinbau und Entwicklung der verkorkten calciumoxalate-kristallzellen inder Rinder von *Larix decidua* Mill. Z, *Pflanzenphysiol.*, 60, 307, 1969.
76. Weerdenburg, C. A. and Peterson, C. A., Structural changes in phi thickenings during primary and secondary growth in roots. 1. Apple (*Pyrus malus*) Rosaceae, *Can. J. Bot.*, 61, 2570, 1983.
77. Gahan, P. B., *Plant Histochemistry and Cytochemistry*, Academic Press, Florida, 1984, App. 6.
78. Biggs, A. R., Intercellular suberin: Occurrence and detection in tree bark, *IAWA Bull. N.S.*, 5, 243, 1984.
79. Biggs, A. R., Detection of impervious tissue in tree bark with selective histochemistry and fluorescence microscopy, *Stain Technol.*, 60, 299, 1985.
80. Popper, H., Distribution of vitamin A in tissue as visualised by fluorescence microscopy, *Physiol. Rev.*, 24, 205, 1944.
81. Sumner, B. E. H., *Basic Histochemistry*, John Wiley, Chichester, 1988, Chap. 3.

Chapter **4**

Transmission Electron Microscopic Cytochemistry

Contents

4.1   Introduction

Transmission electron microscopy (TEM) has made great strides in the last three decades in the elucidation of cell wall structure and complexity. A number of techniques have been developed in TEM cytochemistry, essentially adapted from light microscopic cytochemistry. However, five basic problems have cropped up during such adaptations: (1) Electromagnetic radiation in the visible range cannot be used and so colored reactions cannot be observed under TEM; (2) the observations have to be made in vacuum; (3) sufficiently thin sections, through which electrons can pass through, are required; (4) there is a need to enhance the contrast of subcellular structures prepared for cytochemical observations; and (5) advancement in tissue preparation procedures to obtain optimal preservation of biological specimens have not kept equal pace with the increasing precision and sophistication in electron microscopic instrumentation.

4.2   Specimen Preparation

4.2.1 Fixation

For TEM studies, the use of fresh tissue will not normally provide adequate ultrastructural preservation and so some form of chemical fixation is necessary. Use of fixatives is the most widely employed method in TEM tissue preparation.

Among the reasons for its use are: (1) Adequate preservation of many cellular components including some enzymes, (2) clarity of structural details, (3) ease of application, and (4) cost-effectiveness. Several aspects relating to the need for fixation and problems encountered are already discussed in detail in Chapter 2 and will not be repeated here.

Size of the tissue is a critical factor in TEM fixation. Fixation by immersion of tissue in the fixative is limited to 2 to 3 cell layers from the surface (about 90 μm). So, the smaller the specimens, the better and more uniform the fixation. Selection of size of tissue also depends on the fixatives used, as the latter differ in their penetration rate. Immersion of tissue into the fixative usually produces non-uniform fixation.

Details have already been provided on glutaraldehyde and formaldehyde, which are the most commonly used fixatives in TEM cytochemical studies. However, since details were not adequately given on osmium tetroxide and potassium permanganate, fixatives also used in TEM studies, they are provided here.

Osmium tetroxide is also known as osmic acid. It is an additive and non-coagulant fixative. It increases the density of tissue by 2.5 times upon fixation. It is a strong oxidant, a property that enables it to crosslink and stabilize molecules, particularly lipids and proteins. Therefore, it fixes very well unsaturated lipids (but not saturated ones) and proteins. The chief reactive groups in proteins are guanidine group of arginine, E-amino group of lysine, imidazole group of histidine, and pyrolidone group of proline. Tryptophan, cystein, and cystine are also highly reactive.[1] Nucleic acids and carbohydrates are chemically completely inert towards OsO_4. Enzyme activity is affected by this fixative. Osmium tetroxide has a very slow penetration rate and so the material should be as small as possible. Even then there is an unavoidable fixation gradient from the surface to the center of the tissue and damage can occur before the OsO_4 has reacted completely with all structures. Accordingly, a pretreatment or *prefixative* is necessary and this is done in aldehydes, such as glutaraldehyde, glyoxal, hydroxy-adipaldehyde, crotonaldehyde, pyruvic aldehyde, acetaldehyde, and acrolein. A 0.5 to 2% solution in veronal/acetate buffer (pH 7.3 to 7.5) should be used. Since OsO_4 dissolves very slowly, the fixative should be prepared a day before use. Fixation is usually done at 0 to 4°C for 1 to 16 h. Since it is photosensitive, the fixative should be kept in a brown bottle covered with aluminium foil. The fixative is prohibitively costly.

Potassium permanganate is a coagulant fixative. A 0.6% solution in veronal-acetate buffer pH 7.3 to 7.4 is often used for fixation. It is a strong oxidizing agent. Mucopolysaccharides and proteins are fixed. Very little is known on the reaction of permanganate with cell wall polysaccharides, but it is known to affect the enzymes.

4.2.2 Rinsing, Dehydration, and Embedding

4.2.2.1 *Rinsing*

Fixed specimens are to be rinsed before dehydration. Rinsing removes/minimizes the excess fixative in the tissue, which may otherwise react with the dehydrant that would follow. In double fixation, excess aldehyde in the tissue has to be removed before being post-fixed in OsO_4. Washing the tissue after first and second (if any) fixations with the same buffer in which the fixative was prepared is usually done to remove the excess fixative. This will also avoid a different fluid environment for the tissues if some other fluid is used to remove the excess fixative. Either electrolytes or non-electrolytes may be added to the rinsing solution to maintain the same osmolarity as that of the fixative mixture. In general, 2 or 3 quick rinses in the buffer for a total period of 10 to 20 min is adequate for most specimens. But, since some tissues require a longer duration (like in enzyme cytochemistry), it is better to fix the optimum rinsing required for each and every new tissue used. As a matter of principle, rinsing should not be longer than necessary. If fixation is done in cold, rinsing should also be between 0 and 4°C.

4.2.2.2   Dehydration

If water-immiscible resins are to be used for embedding, all the free water from the fixed and rinsed specimens must be replaced with a suitable organic solvent before infiltration and embedding. If water-miscible resins are to be used, dehydration is unnecessary.

Ethanol is one of the commonly used dehydrants. Epoxyresins are soluble in ethanol and so can be used for specimens to be embedded in that resin. However, in such cases the use of propylene oxide at the last stage of dehydration is practiced as the resin is fully miscible in this solvent but not in ethanol. But propylene oxide is very reactive, even at very low temperatures, and thus may combine with reactive groups in the cells which may affect certain cytochemical staining reactions.

Acetone is a relatively non-reactive solvent and causes less specimen-shrinkage and less extraction of cellular chemicals; it does not react with residual OsO_4 in the tissue. Dehydration is faster in acetone than in ethanol.

4.2.2.3   Infiltration and Polymerization

A complete and uniform penetration of tissues by any embedding medium is a prerequisite for satisfactory sectioning. This is accomplished through infiltration followed by embedding. The former process essentially involves a gradual and continuous replacement of the dehydration agent with an embedding substance. The latter process consists of a complete impregnation of the interstices of a tissue with the medium. If the dehydrant is miscible with the embedding medium, then the tissue can be placed directly in a mixture of the agent and the medium. In the event that the dehydrant is immiscible with the embedding medium, a transitional solvent that is miscible with the embedding medium must replace the dehydration agent prior to infiltration.[2]

The specimen is finally embedded most conveniently in pre-dried gelatin or polyethylene capsules or flat embedding moulds. The capsules or moulds containing the embedding mixture and the specimens are then polymerized. Polymerization of the resin is done either by heat or UV radiation.[1] Epon and its substitutes are usually polymerized by heat; LR white, GMA, Lowicryls, etc. can be polymerized by heat or UV. Heat polymerization consists of heating at high temperature while UV polymerization is done at room temperature or in the cold.

The UV method avoids the heat effects, including specimen shrinkage and bubble formation. Volume shrinkage of Epon, Araldite, Vestopal W, butyl methacrylate, and methyl methacrylate, by heat polymerization is 3%, 3.6%, 6.5%, 15%, and 20%, respectively. However, the actual shrinkage of the specimen is less because most of the shrinkage occurs in the liquid phase of the resin before gelatin embedding. UV method is desirable for cytochemical studies, where high temperature must be avoided. The least structural distortion occurs when polymerization is accompanied at controlled low temperatures. However, available data are insufficient to discourage polymerization by heat. Moreover, this process does not require elaborate apparatus. Any standard oven is adequate.

4.2.2.4 Embedding

Embedding facilitates the production of thin and uniform sections even from insufficiently firm and extremely small tissues. Embedding media should have the following features: Preserve the fine structure of cells, no extraction of cellular constituents, soluble in solvents, adequate viscosity (neither so low as to induce specimen shrinkage during polymerization nor so high as to impede rapid and uniform penetration of the specimen by the embedding mixture), uniform polymerization, little change in volume during polymerization and hardening, thermostability during electron bombardment, sufficient homogeneity, hardness, plasticity and elasticity to counter the cutting impact, sufficient cohesion and compactness, low to nil electron density (otherwise, imaging of material is affected), adequate specimen stainability, no barring the penetration of solutions of heavy metal stains or probes, and easy availability, among others. Since granular or crystalline materials lack the necessary cohesion and compactness, they are unsuited for thin sectioning; therefore, paraffin is not suitable for EM work. Most modern resins fulfill most of the above requirements. Choice of a resin is determined by the type of the specimen and objective of the study. The materials most commonly employed are epoxy resins, polyester resins or acrylic methacrylates.

Water Immiscible Embedding Media.

Epoxy resins are transparent yellow substances and are far superior to others as embedding media. They range from viscous liquids to fusible solids, depending on molecular weight. Chemically, these are polyaryl esters of glycerol having terminal epoxy groups and hydroxyl groups spaced along the length of the chain. They are made by condensing epichloroxydrin with polyhydroxy compounds. These resins degrade on exposure to light.

Epoxy resins require the addition of curing (crosslinking) agents to convert them into tough, extremely adhesive, and highly inert solids. This is done by various bifunctional setting agents. The hardening is accomplished without significant shrinkage or uneven polymerization. Embedding has to be done at RH below 50%.

Epoxy resins are relatively stable to heat (especially by electron bombardment).[1] However, there should be absolutely no moisture on the tissue, because these resins are strongly hygroscopic.

Epon 812 (or Epicote) and its substitutes are the most widely used epoxy resins for EM studies. Materials embedded in these resins can easily be sectioned and stained. Since Epon production was discontinued in the 1970s, several substitutes such as EmBed 812, LX-112, Pelco Medcast, and Poly/Bed 812 came into the market. All of these are light-colored, glycerol-based aliphatic epoxy resins of relatively low viscosity. The resins are not very good for high magnification and high resolution TEM. These resins are hardened uniformly at relatively low temperatures with the addition of acid anhydrides and an amine accelerator. The most commonly used acid anhydrides are DDSA (dodecenyl succinic anhydride) and NMA (nadic methyl anhydride), and the amine accelerators are 2,4,6-tri (dimethylaminomethyl) phenol (DMP-30) and benzyldimethylamine (BDMA).[1]

Araldite is a yellowish, transparent, and glycerol-based aromatic epoxy resin. It is a very viscous substance with very little volume shrinkage after polymerization. It has a rather low softening temperature. Its hardness can be varied by plasticizers such as dibutyl phthalate. The tissue can be dehydrated with any commonly used dehydrant (ethanol or methanol is not encouraged unless a transitional solvent such as propylene oxide is used). Araldite resin is mixed with suitable proportions of a liquid anhydride hardener and an amine accelerator, and polymerized usually by heating in an oven at 60°C for 24 h or longer. Araldite sections can be mounted on unsupported grids and can be easily stained with heavy metals. The final hardness of the block can be varied by changing the proportion of hardener or plasticizer. The formulae for araldite embedment are given on Page 187.

Vinylcyclohexane dioxide (ERL 4206 or Spurr resin*) is a cycloaliphatic diepoxide. It has the lowest viscosity (7.8 cP) of all known resins. It penetrates very rapidly into plant tissues, is miscible with ethanol and acetone, and is relatively easy to section with glass knives. The recommended formulation is given on Page 187.

Methacrylates are colorless, transparent, and compact embedding media, completely miscible with ethanol and acetone and with most other organic solvents. These are relatively less viscous and, therefore, can penetrate easily into dehydrated tissues. These also polymerize spontaneously if left for a sufficient length of time, but polymerization can be accelerated in bright light and warm surroundings. Therefore, these resins are available in a form mixed with a hydroquinone inhibitor which prevents polymerization during shipping and storage. Although, it is not necessary to remove the hydroquinone before blocking, it is better to add a catalyst like benzoyl peroxide to about 2% of the total embedding mixture.

Methacrylates as embedding media pose serious problems such as (1) uneven polymerization, (2) too much shrinkage (~20%), (3) gas bubbles, and (4) lack of stability under electron bombardment (as much as 50% of its mass may be lost under normal operating conditions of EM); this loss is followed by a flow of the

---

* Carcinogenic.

remaining resin, which results in the distortion of the macromolecular structure of the tissue.

More uniform polymerization of methacrylates can be obtained by adding 0.01% uranyl nitrate or by OsO_4 postfixation or polymerization with UV than by heat or by adding uranyl nitrate followed by UV polymerization.

Polyethylene glycol (PEG) or di-butylphthalate, as a plasticiser, and benzoyl peroxide or divinyl benze, as a catalyst in the proportion of 10 to 12% and 1%, respectively, of the total embedding mixture have been reported to eliminate gas bubbles, promote more even polymerization, and improve sectioning quality.[1]

Shrinkage damage can be reduced by using partially prepolymerized resin. Elimination of oxygen from the gelatin capsule during the final polymerization results in a block with little shrinkage.

Water-Miscible Embedding Media.

The ideal preparation of specimens for cytochemical studies requires a minimal or no loss of biochemical activity, especially of enzymes and antigenicity. The answer for this is water-miscible embedding media (e.g., Lowicryl K4M) of low solvent power. There is no need for dehydration or use of organic solvents. These resins can be polymerized, in the presence of a higher percentage of water, by UV radiation at low temperature. Therefore, high temperature treatment of tissues (and thereby altering cellular activities) can be avoided.

Most water-miscible resins are not cross-linked. So they are easily permeated by cytochemical reagents (such as stains and enzymes), which, in turn, find easy access to the embedded specimen. The relatively low viscosity permits easy infiltration into the specimen at low temperature. Also, an easy correlation between LM and EM is possible owing to the ease in cutting thick sections. These resins are also chemically inert.

Acrylamide–gelatin–jung resin can be used for cryosectioning. It preserves ultrastructure and is good for LM and EM. The details on the preparation of this resin and embedding are provided on Page 187.

*Durcupan** is an epoxy resin, colorless and a relatively less viscous resin with very slow penetration. It is not in common use.

Glycol methacrylate (GMA) is a colorless liquid with a viscosity of 50.8 cP at 25°C. It is a polar chemical with complete miscibility in water, ethanol (alcohols), ether, etc. This was introduced by Rosenberg et al.[3] for EM; it is now widely used even for LM. Since 0.5 to 4.0 μm thick sections of materials embedded in it can be obtained even with a steel knife, this is particularly useful in LM for enzyme localization and in EM and LM for immunocytochemistry.

Polymerization of GMA can be done at low temparatures with UV irradiation. The latter may affect the activity of some enzymes but not all. The embedding medium is also permeable to antibodies and many enzymes. Dehydration is not necessary. GMA sections can be stained without removing the embedding matrix, thus avoiding clearing agents such as xylene. Superior preservation of tissue details is guaranteed with GMA.

---

* Durcupan is a strong skin irritant.

Commercial GMA contains hydroquinone (in order to prevent spontaneous polymerization) and thus needs to be removed from GMA before polymerization is initiated. Another impurity to be removed is methacrylic acid (MA). NaOH removes both these and the procedure is given on Page 188.

GMA has the following limitations for use in TEM:

1. GMA tends to swell in many solvents, particularly in water. This can be reduced by using a partially prepolymerized embedding mixture.

2. GMA is not stable under electron beam. The specimen sections break under electron beam. This can be reduced by use of support films on the grids or by adding a cross-linker to GMA such as ethylene glycol dimethacrylate.

3. GMA block is often excessively brittle and thus is difficult to trim and section. This is solved for LM by adding a plasticizer such as PEG 400 and for TEM by acrylates.

The GMA embedding formulations are given on Page 188.

Lowicryls are acrylate-methacrylate mixtures that polymerize to form a saturated carbon chain structure. These are low viscosity embedding media, used at low temperatures. There is an improved preservation of molecules of the tissue including antigenecity when lowicryls are used. A rapid and uniform penetration by the resin into the tissues, especially at low temperatures, was always noticed.

Lowicryls are very good for immunocytochemistry, because protein structure preservation is markedly improved, increasing the efficiency and accuracy of immunolocalization. There is also a reduction in the non-specific background immunostaining.

Lowicryls can be polymerized by long-wave (360 nm) indirect UV light at temperatures ranging from −50 to 0°C with the initiator benzoylmethyl ether for 1 d and from 0 to 30°C with benzoyl ethyl ether for days or weeks. The addition of the cross-linker triethyleneglycerol dimethacrylate improves the sectioning properties of these resins. During embedding, oxygen is excluded, because free radical polymerization is strongly inhibited by oxygen. The procedures involved in lowicryl embedding are given on Page 189.

Some of the limitations of lowicryls include: Shrinkage during polymerization, uneven polymerization, bubble formation in the block, evolution of heat during polymerization, etc.

LR Gold stands for London Resin Company's product. LR Gold is an acrylic resin often used for embedding large tissue specimens (3 × 3 × 3 mm). It is very useful in enzyme cytochemistry and immunocytochemistry. Polyvinyl pyrollidone (PVP) is used in combination with methanol for dehydration and infiltration to protect the specimens from osmotic changes. Oxygen inhibits polymerization of LR Gold, which is usually done at room or low temperatures. The embedding procedure for LR Gold is given on Page 190.

LR White is a polar monomer of polyhydroxylated aromatic acrylic resin. Polymerization can be done by thermal or cold treatment, when an accelerator is used. Anaerobic polymerization is usually done. The monomeric LR White is hydrophobic, while the polymeric form is hydrophilic. LR White has the following advantageous features: Aqueous solutions of cytochemical reagents such as immunoglobulins, colloidal gold, and lectins can easily penetrate LR White and, therefore, etching of sections for immunocytochemistry is not necessary. It also has the minimal nonspecific staining. LR White rapidly infiltrates tissue because of its low viscosity (8 cP). It also has low toxicity and so a fume hood is not necessary. LR-embedded thin sections of tissues generally demonstrate a positive immuno-cytochemical reaction after very short incubation in the primary antiserum. Gradual polymerization, combined with partial dehydration (in 70% ethanol) facilitates in-depth staining with the immunoperoxidase method. LR White also seems to be superior to epoxy resins for post-embedding immunogold labeling. LR White can be used in both LM and TEM cytochemistry. LR White should be preferably used fresh, but storage until 6 months is possible.

4.3 Procedures / Formulae for Tissue Preparation in TEM Cytochemistry

4.3.1 Dehydration Protocols

4.3.1.1 *Ethanol Dehydration*
5 to 100% (5, 10, 20% and so on) ethanol—5 min treatment in each grade, followed by two changes in 100% propylene oxide, 5 min each, with continuous stirring in each step.

4.3.1.2 *Acetone Dehydration*
5 to 100% (5, 10, 20% and so on) acetone—5 min treatment in each grade, followed by two changes in 100% acetone, with continuous stirring in each step.

4.3.2 Recommended Routine Method for Infiltration

1. Bring the tissue from the 100% acetone or propylene oxide to 100% acetone or propylene oxide:resin mixture sequentially in the ratios 3:1, 2:1, 1:1, 1:2, 1:3, each 20 min,* with constant stirring.**
2. Transfer to resin mixture, 20 min,* with constant stirring.**
3. Polymerize.

---

* Timing to be adjusted for each specimen.
** Stirring is done to achieve a uniform and rapid infiltration. Various types of mechanical shakers are available for stirring purposes. Rotational shaking is more efficient than back-and-forth shaking. Rotation is normally done at an angle (4 to 15°) and should be slow enough (65 to 130 rev/min) for gravity to be able to act effectively against the forces of adhesion, but fast enough to achieve maximum streaming movement.

4.3.3 Heat Polymerization of Resins

1. Do the polymerization in an oven for 12 h at 30 to 45°C.

2. Follow this by polymerization at 60°C in the oven until polymerization is completed. The time for complete polymerization varies with the resin. On the average, 2 d at 60°C is sufficient to complete polymerization.

4.3.4 UV Polymerization of Resins

Use a UV lamp with peak emission around 300 nm and hold it about an inch above the capsule containing the resin for 12 to 72 h depending on requirement at 1 to 3°C. For methacrylates, use a fluorescent source of UV in order to minimize the effect of heat.

4.3.5 Recommended Formulae and Procedures for Embedding Resins

4.3.5.1 Epon 812

Formula 1:[1]

Epon 812 (or a substitute)	30 ml
DDSA (hardener)	54 ml
Hexahydropthalic acid anhydride	3 ml
Retail (accelerator)	
(To be added just before use)	0.87 to 1.74 ml (to make a final concentration of 1 to 2% in the medium)

Formula 2:[1]

Stock Solution A:

Epon 812	62 ml
DDSA (hardener)	100 ml

Stock Solution B :

Epon 812	100 ml
NMA (hardener)	89 ml

Mix Solution A and B in the ratio of 7:3. Just before use, add DMP30 (catalyst) to make a final concentration of 1.5 to 2%.

4.3.5.2   *Aradite*

Formula 1

Aradite 502	10.00 g
DDSA	7.89 g
DMP-30	1.5 to 2%

Formula 2[1]

Aradite 502 (Resin)	50 ml
DDSA (Hardener)	1 ml
DMP-30 (Catalyst)	2 ml
Dibutylphthalate (Plasticizer)	1,000 g

4.3.5.3   *Vinylcyclohexane Dioxide (ERL 4206 or Spurr Resin)[1]*

ERL 4206 (epoxide)	50 ml
DER (Dow Epoxy Resins) 736 (epoxide, flexibilizer)	40 ml
NSA (hardener)	130 ml
DMAE (accelerator) (add before use)	1 ml

4.3.5.4   *Acrylamide–Gelatin–Jung Resin*

Acrylamide (3.938 M) containing 0.048M bis	20 ml
Sorensen phosphate buffer (0.2M, pH 7.1)	10 ml
Gelatin (16%)	35 ml
Jung embedding resin	20 ml
N,N,N',N'-Tetramethylenediamine (TEMED)	0.2 ml

1. Mix this solution continuously with a magnetic stirrer in a water bath at 30°C before use to prevent phase separation. Store this solution in the refrigerator for at least 8 weeks. Use DDH$_2$O for all aqueous solutions.

2. Place specimens in 8.5 ml of well-mixed stock solution which is overlayered in sequence with 0.5 ml of 2% ammonium peroxodisulphate and 0.1 ml of fixative (a mixture of 10% formaldehyde and 6.25% glutaraldehyde in 0.1 M phosphate buffer at pH 7.1). Rapidly mix all components and pour into a paraffin oil-coated plastic box.

3. Orient the specimen and keep it in position until the mixture solidifies (within 30 to 60 sec). The block with embedded specimen has a rubber–like consistency.

4. Rapidly freeze the block in isoterpene cooled by liquid nitrogen.

For light microscopy, Steps 5 and 6 can be followed.

5. Using a cryotome at -18 to -22°C, cut 5 to 20 μm sections.

6. Mount the sections on a glue-coated glass slide. The glue contains 0.8% agarose, 1.8% glycerol, and 0.5% glutaraldehyde or 0.75% formaldehyde. The slides are coated by vertical dropping in the glue at 80°C and stored in a dust-free box until use.

For TEM studies, the block has to be cut in an ultramicrotome.

4.3.5.5 Removal of Hydroquinone and Methacrylic Acid Impurities from Glycol Methacrylate

Procedure

1. Dilute 1000 ml of crude GMA with 1000 ml of $CHCl_3$ in a separating funnel.
2. Add 250 ml of 0.5 N NaOH.
3. Shake the mixture vigorously.
4. Allow the mixture to separate into two phases and discard the brownish aqueous phase.
5. Repeat Steps 2 to 4 five more times.
6. Remove the $CHCl_3$ with a rotary evaporator.
7. Purified GMA is obtained. This has a light tint. Store the GMA at -20ºC to prevent autolysis.

4.3.5.6 Glycol Methacrylate (GMA) Formulations[4]

Formulation 1:[4]

GMA	66.5 ml
DH_2O	3.5 ml
n-butyl methacrylate	28.5 ml
Ethylene dimethacrylate	5.0 ml
Benzoyl peroxidase	1.5 g

Formulation 2:[1]

GMA (monomer)	92.5 ml
PEG (softener)	7.5 ml
Azo-bis-iso-butyronitrile (cross-lining agent)	0.3 g

4.3.5.7 Procedure Using Formulation 2 of GMA[1]

1. Dehydrate the tissue in ethanol.
2. Give three changes of mixture containing 92.5 ml GMA and 7.5 ml of PEG, every 24 h at 3 to 4ºC.
3. Give one change of the embedding mixture shown in Formulation 2 after 24 h at 3 to 4ºC.
4. Transfer the tissue to polythene-based molds ($12 \times 16 \times 15$ mm) fitted with aluminium block holders and containing 2.5 ml of a 30:1 mixture of Solutions A and B.
5. Place the molds in an atmosphere of nitrogen at 20ºC to prevent inhibition of the polymerization by atmosphere oxygen. Polymerization is complete in 2 h.
6. Cut sections under a constant humidity of 60% within a week after polymerization.

7. Stretch the sections on water at room temperature.

8. Mount the sections on glass slides or grids and dry at 20°C in a hot plate.

4.3.5.8 Procedure for Prepolymerization of GMA

Prepolymerization of GMA increases the speed of final polymerization and reduces swelling artefacts.

1. Heat the embedding mixture to 40 to 45°C on a magnetic stirrer hot plate with continuous stirring until the benzoyl peroxidase is dissolved.

2. Remove the mixture from heat and transfer to an Erlenmayer flask with a Teflon-coated magnet.

3. Heat it with continuous stirring to 98°C on the hot plate.

4. Remove the flask and plunge it in a dry ice–ethanol bath with rapid swirling to cool the mixture rapidly to ~2°C. Complete the whole process in ~5 min.

The prepolymer has the viscosity of a thick syrup at 0 to 4°C and can be stored almost indefinitely.

4.3.5.9 Procedure for Lowicryl Embedding

1. Fix the tissue in glutaraldehyde; do not use osmium tetroxide.

2. Dehydrate in graded ethanol series. Start dehydration at 0°C and then lower the temperature in conjunction with increasing concentrations of ethanol until it has reached -35 to -50°C. Avoid freezing the tissue. Follow the following time schedule:

30% ethanol:	0°C—30 min
50% ethanol:	-20°C—1 h
70% ethanol:	-20°C—1 h
100% ethanol:	-20°C—2 h

3. Add resin gradually as per the specifications given below.

100% ethanol +resin (1:1):	-35°C—1 h
100% ethanol +resin (1:2):	-35°C—1 h
Pure resin:	-35°C—1 h

4. Infiltrate overnight at the same temperature or at -48°C. Accomplish infiltration by the basically non-polar and hydrophobic resin HM 20 or the more polar and hydrophilic K4M at the same temperature.

5. Embed at -48°C with the following resin mixture which has been polymerized:

Hydroxypropyl methacrylate	48.4 g
Hydroxyethyl acrylate	23.7 g
n-Hexyl methacrylate	9.0 g
Triethyleneglycol dimethacrylate	13.9 g

6. Store the blocks under partial vacuum and over dessicant because polymerized resin is hygroscopic.

7. Use a diamond knife for sectioning at a speed of 2 to 5 mm/sec.

4.3.5.10 *Procedure for LR Gold Embedding*

1. Treat the specimens in the following conditions in sequence:

Methanol 50% + PVP 20%	4°C	15 min
Methanol 70% + PVP 20%	-20°C	45 min
Methanol 90% + PVP 20%	-20°C	45 min
Methanol 90% + PVP 20% + LR Gold (5:1:5)	-20°C	30 min
Methanol 90% + PVP 20% + LR Gold (3:1:7)	-20°C	60 min
LR Gold 90%	-20°C	60 min
LR Gold + 5% benzoin-methyl ether	-20°C	60 min
LR Gold + 5% benzoin-methyl ether	-20°C	overnight

2. Embed the tissue in gelatin capsules covered and polymerized under UV light at -20°C for 1 week.
3. Cut sections with glass or diamond knives.

4.4 EM Staining

Colors cannot be distinguished in electron microscopy. Therefore, EM staining is essentially a precipitation procedure where a biological substance is localized by the formation of insoluble precipitates using inorganic reagents. Regions of electron density and opacity can be recognized in a specimen when viewed under EM and precipitated areas have greater electron density. Any substance that can bring about electron density of certain sites in the specimen can act as a "stain" and substances with a higher atomoic number can do this, e.g., heavy metals. The cellular components that have been "stained" are less prone to sublimation under the electron beam than those that remain unstained. This selective stabilization also contibutes substantially to image contrast or differential staining in TEM. Thus, the necessary requisites for an electron stain are: (1) it should increase the local electron-scattering power of the specimens sufficiently to result in an appreciable increase in image contrast to result, (2) it should be localized selectively to result in positive identification of a given chemical, and (3) its density should be higher than that of the embedding medium.

Gold (atomic no. 79) is sufficiently heavy to impart electron density to biological macro-molecules. Even at a relatively low magnification, the gold particles have sufficient electron density. Moreover, the binding constants of gold markers to cell surface are very high and the specimens can be processed for EM without significant loss of bound gold particles. The major objective in using colloidal gold is the *in situ* localization of cellular macromolecules. Methods for multiple labeling of the same specimen with colloidal gold particles of different sizes (2 to 150 nm diameter) are also available. The enhancement of gold staining with silver is also possible (see

Page 265).[5] Colloidal gold can be adsorbed to a wide variety of molecules such as lectins, glycoproteins, enzymes, carbohydrates, etc. Nonspecific adsorption is extremely low. Smaller gold particles are better for EM. 3 nm particles are best suited for high resolution TEM.

Enzyme-colloidal gold method can be used to localize various cellular or tissue substrates in thin sections.[6] It is a direct, one-step post-embedding method. Since this method is based on the affinity existing between an enzyme and its substrate, the results are specific for a given substrate. Also, the activity of the enzyme molecules adsorbed onto gold particles is retained. The exact mechanism responsible for the attachment of an enzyme-gold complex to the substrate is not known, but several types of interactions can occur (like electrostatic, hydrogen bonding, etc.). The enzymes are pH-sensitive and so labeling must be carried out at pH values compatible with the activity of the enzyme used. Cellulose, xylan, chitin, etc. are some of the cell wall substrates that have been localized by this procedure,[6-11] but progress is limited by the scarcity of cell wall enzymes of sufficient purity and having a sufficiently restricted sensitivity.[12]

TEM staining can be performed during fixation and/or dehydration (*en block staining*) or after embedding. In general, the staining carried out after fixation and embedding is decidedly superior to that accomplished before embedding for the following reasons: (1) there is minimal extraction and displacement of the stain as the tissue is already embedded, (2) there is faster and greater contrast, (3) a more uniform staining is obtained, (4) staining is usually simple and can be easily controlled at this time, and (5) staining before embedding may make the tissue hard so that sectioning becomes difficult.

The most effective and widely used method of staining is by floating the grid-mounted sections, section-side downward on the surface of a drop of the staining solution. The section may also be stained and then mounted onto grids.

Duration of staining is a complex function of the rate of adsorption of metal ions, type of tissue and embedding medium, the coated and uncoated surface of the grid, pH, the type and concentration of the staining solution, the type of fixatives, section thickness, etc. It is advisable to keep the staining time as short as possible, for, prolonged exposure will cause extraction of cellular materials or even destaining. 2 to 15 min with heavy metals is adequate. Thicker sections require a longer duration.

The formation of large stain aggregates in the specimen is undesirable as it will obscure cellular fine details. Smaller stain aggregates are particularly useful in high resolution EM. Very little is known on the size of aggregates formed by different electron stains. Silver forms relatively large aggregates (4 to 6 nm in diameter). The size of the stain aggregates is again a function of several factors mentioned earlier, although the duration of staining and the concentration of the staining solution are the most important.

TEM has a great depth of field, as for instance, a TEM with a resolution of 0.6 nm has a depth of focus of ~ 400 nm. A thin section for EM is around 80 nm and, therefore, its entire depth can be focussed by an EM. Hence, image accuracy depends only on the penetration and distribution ability of the stain. Many factors control these abilities of the stain (like embedding medium, type of stain, the matrix in which the stain is prepared, its concentration, pH, temperature duration of staining, section thickness, nature and location of cellular chemical, etc.). Heavy metals generally penetrate incompletely and non-uniformly. The embedding medium is a major obstacle because most embedding resins are non-polar while most stains are ionic or polar.

Stain specificity is the most crucial requirement for EM cytochemistry. Most of the EM stains are general purpose stains. The most effective ways of imparting specificity to EM stains are to control the pH of the stain solution and to employ specific blocking agents. The most common EM cytochemical methods are: (1) direct attachment of heavy metal salts, (2) attachment of reaction products of heavy metals, (3) attachment of organic reagents, and (4) Attachment of organic reagents substituted by heavy metals, etc.

4.5   TEM Cytochemical Protocols

4.5.1 Cell Wall Carbohydrates

4.5.1.1   Total Insoluble Polysaccharides

Periodic Acid-Silver Methenamine Reaction or PASM Reaction[13-17]

Tissue Preparation

Fix materials in 3% glutaraldehyde in 0.025 M cacodylate buffer, pH 6.8, for 1 h under partial vacuum at room temperature, followed by an additional 48 h at 4°C. Post-fix for 4 h at 4°C in 2% OsO_4. Dehydrate in a gradual ethanol series. Follow this by solvent exchange in 100% propylene oxide and EPON 812-Araldite 6005 embedment.

Procedure

1. Treat the material in any aldehyde blocking reagent to block not only the tissue aldehydes but also the remnant aldehydes of the fixatives used.
2. Wash in water.
3. Place the material in 1% aqueous periodic acid 20 to 30 min. This bleaches the material by removing Os.
4. Wash briefly in distilled water by rinsing three times.*
5. Treat the material in silver methenamine solution for 45 to 90 min at 60°C in darkness Silver methenamine solution is made by mixing 0.3 g hexamethylenetetramine, 12

---

* Ramboug[16] and Pearse[18] suggested leaving the specimen in a distilled water bath for 18 to 24 h.

ml distilled water, 10 ml silver nitrate solution, and 8 ml of 5% sodium borate. Alternatively it can be prepared as follows: 3% aqueous methenamine (hexamethylenetriamine) 50 ml, 1% silver nitrate (stir to dissolve the milky oplescence) 5 ml, 3% aqueous sodium borate (borax) enough to adjust the pH to the range of 8-9 (pH should be measured by indicator papers, since the pH-meter electrodes introduce chloride ions which will precipitate silver). Mix all of the above before use, as the mixture has a shelf-life of only 24 h. Rambourg[16] and Pearse[18] suggested treating the materials in a 25 ml of silver methenamine filtrate poured into a covered petri dish heated to 60°C. The silver methenamine filtrate is prepared as follows: 5 ml of a freshly prepared 5% aqueous silver nitrate solution is added to 45 ml of a freshly prepared 3% aqueous solution of methenamine in a 100 ml pyrex cylinder. A white precipitate would appear and this is dissolved by shaking. Five milliliters of a 2% aqueous solution of sodium borate is added to this. The final solution must be completely clear. After a thorough mixing, the solution is filtered through two sheets of Whatman No. 42 paper in a funnel inserted into an Erlenmeyer Flask. Subsequently the specimens are floated in a freshly prepared 5% aqueous silver nitrate solution at 60°C, for 30 min with the cover in place. The specimens should appear goldish. The specimens then are quickly rinsed in distilled water and transferred to a second bath of silver methenamine filtrate and reaction is allowed to proceed for 30 min at 60°C. The specimens would now appear brownish.

6. Wash in distilled water.
7. Treat for 5 min 0.5% sodium thiosulphate to remove the unreduced silver ions.
8. Wash in distilled water and allow the materials to dry.
9. Observe under TEM.

Results

Electron-dense silver deposits are formed at the sites of insoluble polysaccharides.

4.5.1.2   Cellulose

Cellobiohydrolase-Colloidal Gold Technique[11,19-26]

Tissue Preparation

Fix the material in 2 to 2.5% glutaraldehyde in 0.1 M Na-cacodylate buffer, pH 7.2 for 1.5 to 3 h at room temperature. For delicate materials, do fixation in 1% glutaraldehyde/4% paraformaldehyde/50 mM lysine. Rinse in the same buffer and optionally post-fix 1.5 h in 1 to 2% OsO_4 in sodium cacodylate buffer. Wash in distilled water and dehydrate in ethanol series for 2 h at room temperature. Embed in Spurr's resin, Epon-araldite, Durapan ACM, LR White, Lowricryl K4M, or Epon 812. Take ultrathin sections and mount them on coated or uncoated nickel or gold grids. See the procedure on Pages 61 to 63, where the method, as followed for LM, is given. For EM, "labeling" should be done as follows:

1.   Treat with 0.05 M citrate phosphate buffer (added with 0.5% gelatin)* pH 4.5 for 5
      to 10 min.

2.   Treat for 30 min in a 1/30 dilution of the enzyme cellobiohydrolase-gold stock
      solution with incubation buffer. The procedure for preparation of the stock solution
      is as follows:

      Cellobiohydrolase (EC 3.2.1.91)** was isolated from *Trichoderma reesei.* The com-
      plex cellobiohydrolase-colloidal gold was prepared as follows:

      A 100 µl aliquot of a stock solution of cellobiohydrolase (2mg ml[-1]) is added
      under stirring to colloidal gold solution. The pH of 10 ml of 15 nm colloidal gold
      solution is to be adjusted to 4.5 before this addition. After 5 min, 500 µl of 1%
      polyethylene glycol (mol.wt. 20,000) is added to the enzyme-gold mixture. The
      latter is centrifuged at 14,000 to 18,000 g using a Beckman Rotor T for 1 h at 4°C.
      The mobile pellet is collected in 5 ml of 0.05 M citrate-phosphate buffer, pH
      4.9, to which 0.02% PEG and 3 mM sodium azide are added. The complex is
      centrifuged again under the same conditions and pellets collected a second time
      in the same buffer. The complex is stored at 4°C. For a long-term storage, aliquotes
      with 20% glycerol added were frozen at -80°C.

3.   Float the grids on the buffer for 5 min, rinse twice in deionized water for 5 min each
      and then dry.

4.   Stain the sections with uranyl acetate and/or lead salts.

Result

The 15 nm gold particles bound to cellobiohydrolase 1 were directly visible.
Labeling was not significantly affected by fixative or embedding resin.

Controls

1.   Treat the sections with an enzyme-gold complex solution to which carboxymethyl
      cellulose (CMC) (sodium salt of CMC with medium viscosity) had been added, at
      1.0 mg ml[-1] final concentration, 1 h before use.[11]

2.   Predigest the sections by free cellulase enzyme (1.0 mg m[-1]) in an incubation buffer
      for 18 h prior to incubation in the cellulase-gold probe.

Comments

Neither plastic deembedding before enzyme-gold treatment nor silver intensification done for immunocytochemistry (see Page 265) are necessary for EM
localization.

---

* Gelatin is an effective blocking agent and prevents nonspecific labeling of embedding plastic.
** The enzyme can be obtained from any source. Chanzy et al.[21] used the enzyme purified from the
fungus, *Trichoderma reesei.*

Application of the Procedure

In cultured cells of mung bean, the cell walls showed light labeling here and there. In organized tissues of hypocotyl, the labeling was intense throughout the whole cell walls, and was particularly thickest in the outer wall of epidermis and over parenchyma corners. Here the gold particles are evenly distributed.[254] This technique was used by Bonfante-Fasolo et al.[24] to specifically localize cellulose in the cell walls of hosts colonized by VAM fungi and to compare the distribution of cellulose in the host wall and in the material occurring in the contact zone (interface) between the host and the fungus. The labeling was heavy in the thick epidermal cell walls and moderate in the primary walls of hypodermal and cortical cells. In the interface as well, labeling was observed, thus enabling the authors to conclude that the interface around the VAM fungus showed cellulose molecules as in the host cell walls and that it (interface) was of host origin. Similar studies have been carried out by Berg[23] to localize cellulose and xylan in the interface capsule in symbiotic cells of actinorrhizae.

Berg et al.[11] worked on *Casuarina cunninghamiana* branchlets, spores of *Dictyostelium discoideum* and marine algae such as *Udotea* and *Codium* using this technique. In *Casuarina*, labeling was good in the cell walls and poor in the middle lamella and cell corners.

4.5.1.3   Acidic Cell Wall Poysaccharides (Carboxylated, Sulphated, and Phosphated Polysaccharides)

Colloidal Iron Method[17]

Tissue Preparation

The specimens must be as small as possible because of the poor penetration of the stain. Fix the specimens in any aldehyde. Embed in GMA or butyl methyl methacrylate. Either the whole block or sections out of it can be subjected to staining.

Procedure

1. Place the blocks or sections therefrom in water.
2. Pick up the sections with platinum loops or plastic rings and float successively in either 12% acetic acid, 3 to 10 min, or 1% colloidal iron in 12% acetic acid, 3 to 10 min.
3. Rinse with 12% and 6% acetic acid 2 to 3 min in each.
4. Wash with water slowly to avoid convection currents.
5. Mount on grids and observe.

Result

Sites containing acid polysaccharides are evident by dense precipitates of colloidal iron.

Rationale of the Reaction

The charged colloidal micelles interact with strong acidic groups such as carboxyl, sulphate, and phosphate groups of acidic polysaccharides.

Comments

Colloidal thorium, introduced by Revel[27] can be substituted for colloidal iron, as they produce deposits of high intensity, but the problem with them is that they are no longer commercially available.

4.5.1.4   Pectic Acids

Endopolygalacturonate Hydrolase (endo PGH)—Colloidal Gold Method[9,28]

Tissue Preparation

Specimens can be chemically fixed, dehydrated in ethanol series, and embedded in Spurr's resin. Ultrathin sections (about 100 nm) are to be taken.

Procedure

1.   Bathe the sections in the endo PGH*-gold** complex solution with an absorbance of 1 at 522 to 523 nm. The enzyme adsorption onto colloidal gold particles is dependent on pH and on the quantity of the enzymic protein. The pH of the solution should be above the isoelectric point of the enzyme, i.e., 5-9 and no contaminating ions should be present. Under such conditions, 90% of the gold particles could be stabilized and the zsigmondy number is 7.10^3. This number is defined as the number of milligrams of protective substances to prevent flocculation of 10 ml of solution by the addition of 1 ml of 10% sodium chloride solution. In parallel with the increase of the amount of protein introduced into the solution, the number of

---

* This enzyme can be from any source. Cabin-Flaman et al.[28] obtained this as a crude preparation of *Aspergillus* sp. possesing a molecular weight of 39 kDa (by gel filtration) and 33 kDa (by electrophoresis), and an isoelectric point between 4.40 and 4.50. The optimum pH for its activity was close to 5.0. The enzyme was in acetate buffer.
** Colloidal gold can be prepared as per the procedure of Frens.[29] The mean diameter of particle was 9.6 nm and statistically monodisperse.

proteins adsorbed on the gold particles increased. For example, with 6 µg of proteins per milliliter of gold, 14 molecules of proteins were adsorbed per gold particle solution with an absorbance of 1 at 522 to 523 nm.

2. Observe under TEM.

Results

Loci with pectic acids in the cell walls could be identified by the presence of colloidal gold particles.

Application of This Procedure

Cabin-Flaman et al.[28] located pectic acids in the cell walls of *Linum usitatissimum* (flax) using this method. They found that pectic acids were mainly located on the tangential wall of epidermal cells, on the intercellular junctions of cortical parenchyma, and on the middle lamellae of the phloem tissue.

4.5.1.5   Pectins

Ruthenium Red method[30,31]

Ruthenium Red with an atomic number of 44 is a moderately heavy element. Although pectin–Ruthenium Red complex does not produce much electron density, the latter can be increased by treatment of OsO_4. The stain is applied by adding it to the fixative or as aqeous or buffer (avoid PO_4 buffers) solutions.

Procedure

Solution A:

Aqueous glutaraldehyde	5 ml
0.2 M Cacodylate buffer, pH 7.3	5 ml
Ruthenium Red stock solution (1500 ppm in water)	5 ml

Solution B:

Aqueous OsO_4	5 ml
0.2 M Cacodylate buffer, pH 7.3	5 ml
Ruthenium Red stock solution (1500 ppm in water)	5 ml

1. Fix and stain tissue in Solution A for 1 h at room temperature.

2. Rinse briefly in cacodylate buffer, pH 7.3.

3. Post-fix and stain in Solution B, which is prepared just before use, for 3 h at room temperature.

4. Briefly rinse in cacodylate buffer, dehydrate, and embed as usual.
5. Section the block and observe.

Results

The pectin containing regions of cell wall are evident by electron-dense areas.

Rationale of the Reaction

See LM procedure on Page 79. The staining intensity depends on the number of ionizable carboxylic acid groups available. The stain binds to them electrostatically (salt linkage).

4.5.1.6   Esterified Pectins

Alkaline Hydroxylamine   Hydrochloride Method[32,17,33]

Tissue Preparation

Tissue is fixed in any aldehyde, although glutaraldehyde is better. Post-fixation in OsO_4 is recommended.

Procedure

1. Place the material in 25% ethanol for 15 min.
2. Transfer it into 60% ethanol for 15% min.
3. Immerse the materials for 20 min to 1 h in a freshly prepared solution of alkaline hydroxylamine (equal volumes of 14% NaOH in 60% ethanol and 14% $NH_2OH.HCl$ in 60% ethanol).
4. Wash in 0.1 N HCl in 60% ethanol for 15 min.
5. Immerse in 60% ethanol followed by dehydration and embedding in resin by standard procedures.
6. Cut ultrathin sections, remove resin, and observe in TEM.

Result

Regions containing esterified pectin in the cell walls will be indicated by electron dense areas.

Rationale of the Reaction

Esterified pectins are made selectively dense to electrons through treatment with basic hydroxylamine. This treatment produces pectic hydroxamic acids, which, in turn, are treated with ferric ions to form insoluble electron-dense complexes. Albersheim and Killias[34] have indicated that hydroxylamine produces pectic hydroxyamic acids via nucleophilic substitution at the carboxyl carbon. The amount of iron deposited is dependent upon the concentration of reactive pectin substances, since pectic hydroxyamic acids are formed by substituting hydroxylamine for the methoxyl groups of pectin.

Application of the Procedure

Albersheim et al.[33] and Albersheim and Killias[34] have used this method to localize the site of pectin in the cell wall of onions. The reaction was concentrated in the middle lamellae and also in young primary walls.

Catesson and Roland[35] used this method for studying the vascular cambial cells of a few temperate hard wood trees.

4.5.1.7 Hemicelluloses

Procedure 1: Periodic Acid-Thiocarbohydrazide*-Silver Proteinate Method (PA-TCH-SP Method or PATAg Method)]** [17,36-42]

Tissue Preparation

Tissue is fixed in 3% glutaraldehyde in 0.025 M cacodylate buffer, pH 6.8, for 1 h under partial vacuum at room temperature, followed by an additional 48 h at 4°C. After post-fixation for 4 h at 4°C in 2% OsO_4, specimens are dehydrated in graded ethanol series. This is followed by solvent exchange in 100% propylene oxide and Epon 812-Araldite 6005 embedment.[43] However, the percentage of glutaraldehyde to be used and duration of fixation can be varied depending on the material. For example, Burns et al.[13] used 1.6% v/v glutaraldehyde in 0.05 M sodium cacodylate buffer containing 1% w/v caffeine, followed by a wash in the same buffer and post-fixation in 1% v/v OsO_4 in caffeine-free buffer; for embedding they used Epon. Fixation in only OsO_4 should be avoided. The sections are to be mounted on gold grids.

---

* Thiosemicarbazide (TSC) can be used instead of thiocarbohydrazide.
** According to Vian and Roland,[20] Sandoz was the first to apply to plant cells this technique, which was developed by Thiery[42] for animal cells.[44]

Procedure

1. Float the sections on grids on a drop of 1% v/v aqueous periodic acid for 30 min. It bleaches the material and removes OsO_4.

2. Rinse in distilled water.

3. Float on a drop of 0.2% thiocarbohydrazide (TCH) in 20% aqueous acetic acid for 48 h. To prevent evaporation, place drops of TCH in a petri dish on parafilm and invert the top of the dish over the grids. Burns et al.[13] used a 1% TSC w/v solution in 10% v/v aqueous acetic acid for 4 to 5 h to overnight.

4. Place the grids on drops of 10% aqueous acetic acid followed by three washes of distilled water.

5. Incubate the grids in darkness for 24 h in a 1% aqueous silver proteinate* solution. A 45 min duration was advocated by Burns et al.[13] while 20 to 30 min was suggested by Roland.[17]

6. Remove silver proteinate by washing with distilled water.

7. Examine under TEM.

Result

Deposits of electron-opaque silver granules, which correspond to those areas of the material where hemicelluloses are present, are formed.

Rationale of the Reaction

TCH and TSC are osmophilic reagents for demonstrating aldehydes (i.e., the resultant aldehydes are coupled to these) produced by periodate oxidation of the vicinal groups of hemicelluloses. Silver proteinate (or silver vitellinate) precipitated the TSC or TCH-aldehyde complex.

Controls

1. Treatment of material with hemicellulase enzyme before proceeding further. It is found as a mixture in commercial enzymes such as Driselase and Macerozyme.

2. Omission of periodic acid treatment.[13]

3. Using sodium borohydride to block aldehydes produced by periodic acid oxidation.[45]

4. Eliminating TCH or TSC from the procedure.

5. Eliminating silver proteinate.

---

* Silver proteinate was replaced by 1% w/v silver vitellinate (pH 12 adjusted by 1 M NaOH) by Silva and Macedo[46] and Owen and Thompson[47]). This change was used to overcome the problems in the Thiery procedure of the instability of the silver proteinate solutions and the granularity of the silver grains that appear over time which limits the high resolution examinations. Also, a more finely grained silver deposition than that which occurs with the original procedure was noticed.

6.   Staining of material processed without OsO_4 post-fixation.

Comments

1.   Joseleau and Ruel[39] used enzymes such as xylanase to partially hydrolyze hemicellulose and followed this by coupling newly formed reducing ends of hemicellulose to TCH and silver proteinate. This enzyme pretreatment resulted in greater labeling of hemicellulose.

2.   This technique gives good contrast; the reaction specificity is also good. Very fine details are visible when applied to plant cells.

3.   Polysaccharides having 1→6, 1→4, and 1→2 linked sugar residues in their chain react to this method, although 1→4 linked carbohydrate backbones (like those of hemicelluloses) are the most potential substances for this staining reagent. Some reaction of pectic materials and amorphous cellulose within the cell wall may also occur[39] as long as they have, in their structure, sugar units possessing two neighboring unsubstituted hydroxyls. Although the specificity of this method for hemicellulose has been questioned by these authors, it does not positively react with crystalline forms of cellulose in the cell walls. However, enough evidences are available to show that the primary cell wall component that reacts positively is hemicellulose.[48]

4.   1→3 linked carbohydrates are not stained as they are immune to periodate oxidation.

Application of the Procedure

Like PAS technique in LM, this has been employed by several investigators and a few representative cases are given below.

1.   Burns et al.[13] used this procedure to detect under EM, 1-4 linkages of polyglucosides and glycoproteins containing hexose, fucose, etc. in the cell walls of the brown alga, *Sphacelaria furgicera*.

2.   This method has been shown to be more specific to localize hemicelluloses than other cell wall components and has been used for demonstrating hemicelluloses in woody cell walls.[8,48-50]

3.   Catesson and Roland[35] used this technique to study in detail the chemical nature of the radial and tangential cell walls of fusiform and ray initials of the vascular cambium.

4.   Els Bakker and Gerritsen[51] used this method to study the mucilage cells of *Cinnamomum*.

5.   The cell wall changes preceding abcission in the separation zone of senescing leaves were studied by Sexton et al.[52] using this technique.

6.   McConchie and Knox[54] employed this technique to detect polysaccharides with vicinal glycol groups in the stigma surface cell walls of the sea grass, *Posidonia australis*.

7.   This procedure was employed by Owen and Thompson[47] to study the structure and function of the specialized cell wall (with enormous wall ingrowths) in the trichomes of the carnivorous bromeliad, *Brocchinia reducta*.

8.   Vian et al.[6] used this technique to detect vicinalglycol groups of polysaccharides like xyloglucans in the cell walls of *Nasturtium* seeds.

Procedure 2: Enzyme-Colloidal Gold Complex Method[7,8,53,55]

Preparation of Enzyme-Gold Complex

Enzymes that can be used are endo-β-D-glucanase and β-D-galactosidase, which are specific to xyloglucans; low molecular weight β-galactosidases can also be used. Dialyze the purified enzymes against distilled water.

Endo-β-D-glucanase-gold complex (Au-Glu) can be formed by using colloidal gold (nominal particle diameter 5 nm) prepared by the phophorus procedure (see details in Roth[56]). The native enzyme (0.25 ml; 20.5 μg protein) inactivated by heat treatment is to be mixed with colloidal gold (5 ml, pH 7.3). Heat inactivation of native enzyme can be done at 100°C for 2 to 5 min. This treament brings about a mild denaturation of the enzyme molecule, sufficient to abolish enzymatic activity but not substrate recognition. The importance in enzyme-gold cytochemistry of conditions for the probes, which retain the specificity of enzyme-substrate binding and yet eliminate the possibility of modification of the target molecule by enzyme action, need to be stressed at this point.

The enzyme-gold complex can be recovered by centrifugation at 40,000 g and the "mobile pool" collected.[57] The complex has to be diluted to 1 ml in distilled water to which 0.02% polyethylene glycol (PEG 20000) has been added, and to be stored at 4°C. Just before use, the pH of the complex has to be adjusted by mixing with an equal volume of McIlvaine phosphate-citrate buffer diluted 1 to 10, at the desired pH of 5.0.

β-D-galactosidase-gold complex (Au-Gal) has to be formed using colloidal gold (nominal particle diameter 15 nm) prepared by the citrate method (see details in Frens[29] and Roth[56]). The enzyme (70 μg protein) is to be mixed with colloidal gold (5 ml, pre-adjusted to pH 7.2). The complex has to be recovered by centrifugation (20,000 g, 30 min), diluted to 1 ml with distilled water, to which 0.02% of PEG 20000 was added, and stored at 4°C.

Tissue Preparation

Fix the tissue in 2.5% glutaraldehyde in 0.1 M sodium cacodylate buffer, pH 7.5, for 2 h. It may be post-fixed in OsO_4 for 1 h. Dehydrate the tissue and embed in LR white or epon-araldite. Collect the sections on nickel or gold grids.

Procedure

1. Float the grids with sections on Au-Glu or Au-Gal complexes for 30 to 40 min at room temperature.
2. Wash several times in distilled water.
3. Stain with 1% uranyl acetate, followed, if necessary, by lead citrate.

Result

The regions in the cell wall containing hemicelluloses will show the deposits of gold particles.

Controls

1. Pretreat the tissue for 4 h with native endo-β-D-glucanase.
2. Pretreat the tissue with boiling water prior to fixation to extract the xyloglucans and other hemicelluloses.

Application of the Procedure

This method was especially used for studying the distribution of hemicelluloses like xyloglucans in the woody cell walls. Ruel et al[50] used this technique to study the cell walls of spruce wood decayed by fungi. Joseleau et al[53] employed this technique to study the distribution of hemicelluloses in the suspension-cultured cells of *Rubus fruticosus*.

4.5.1.8 Xyloglucans

ETAg or Enzyme-Thiocarbohydrazide-Silver Proteinate Method[8,58]

Tissue Preparation

Fix the specimens in 6% glutaraldehyde in 0.1 M phosphate buffer, pH 7.0, for 3 h at room temperature. Wash in 0.1 M phosphate buffer for 1 min. Dehydrate in ethanol and embed in methyl-butyl methacrylate with 1% benzoyl peroxidase as catalyst. Take thin sections of about 500 Å thickness and collect them on gold grids or plastic rings.

Procedure

1. Float sections on the enzyme solution for 10 to 30 min at room temperature. (Xylanase enzyme is very ideal. Optimum pH for activity of this enzyme is 4.5 to 5.0. It can be obtained from any source, such as the fungus *Aspergillus niger*.)
2. Thoroughly rinse the sections in distilled water.
3. Treat in thiocarbohydrazide (0.2% in 20% acetic acid) for 24 h.
4. Wash successively in 20% and 10% acetic acid, several times, each time for about 30 min.

5. Treat the sections in a series of acetic acid baths of decreasing concentration from 20% downwards to end up with distilled water.
6. Contrast the sections in aqueous solution of silver proteinate (1% in the dark).
7. Rinse twice in distilled water.
8. Mount on carbon-coated grids and observe.

Result

Xyloglucan-containing regions of cell wall become evident as electron-dense areas.

Controls

1. Enzyme was preheated for 10 min to suppress its activity.
2. Omission of incubation in enzyme.
3. Omission of TCH.

Rationale of the Reaction

Xylanase hydrolyzes xyloglucans creating new reducing end groups along the polysaccharide chain. The disclosed reducing sugars then exhibit the reactions of aldehyde groups and are, in particular, potential sites for coupling with thiocarbohydrazide or thiosemicarbozide. Then silver proteinate is added to precipitate the complex. This is a direct labeling of hemicellulosic polysaccharides.

Comments

In ETAg method, there is no ambiguity in the position of the silver granules and there are no risks of unspecific adsorption of the enzyme on the network of cell wall polymers. On the other hand, the critical point of this method is in the finding of appropriate time of contact between enzyme and substrate such that enough new reducing end groups are created without concomitant solubilization of the substrate. Hence, different incubation times are to be experimented.

4.5.1.9   Chitin

Enzyme–Gold Method[59,60]

It is very similar to the methods 4.5.1.2 and 4.5.1.4. Instead of cellobiohydrolase or Endo PGH respectively of these methods, here β-chitinase is used. This enzyme is conjucated to colloidal gold particles. The further procedures are almost similar.

4.5.2 Cell Wall Proteins

4.5.2.1 *Enzymes*
Peroxidase.

Procedure 1:[13,61]

Tissue Preparation

Preferably tissue is to be fresh, although, in general, peroxidase is relatively unaffected by fixatives.* If fixatives are to be used, 1.6% (v/v) glutaraldehyde in 0.05 M sodium cacodylate buffer, pH 7.0 containing 1% w/v caffeine**[62] or a mixture of the same concentration (1.6%) of glutaraldehyde and formaldehyde are effective. Wash for 1.5 to 4 h in the same buffer after fixation. Transfer to 0.05 M trismaleate buffer, pH 7.0 and store overnight at 4°C. The tissue should be as small as possible, as otherwise the substrates for peroxidase (like DAB, H_2O_2, etc.) do not easily penetrate the tissue.

Incubation Medium

> 50 mM tris-maleate buffer, pH 7.0
> 1 mM $MgSO_4$
> 0.3 mM $Pb(NO_3)_2$
> 0.3% H_2O_2 (substrate)

Procedure

1. Pre-incubate the material in the incubation medium mentioned above minus the substrate for 30 min at room temperature, in darkness.
2. Incubate in the incubation medium for 1 h at room temperature.
3. Wash the material in 0.05 M sodium cacodylate buffer, pH 7.0.
4. Refix the material in 1.6% (v/v) glutaraldehyde in the same buffer for 2 h at room temperature.
5. Post-fix the material overnight at 4°C in 1% v/v OsO_4 in sodium cacodylate buffer.***

---

* The effect of fixatives on peroxidase activity has been investigated by Al-Azzawi and Hall.[63] 1% and 5% glutaraldehyde and 1% and 5% formaldehyde for 2 h at 4°C had an inhibition (of enzyme activity) percentage of 14, 12, 7, and 8, respectively. The use of fixation seems to help in minimizing the effects of diffusion of enzyme and in improving the rate of penetration of DAB into tissues, which is normally low.[64]
** Caffeine treatment greatly reduces the loss of peroxidase from the tissue.
*** Attempts to increase the electron opacity for EM studies have been made by the following approaches (see details in Frederick[65]): (1) Incorporation of other heavy metals such as gold or cobalt plus nickel in place of osmium, and (2) use of a potassium-osmium-cyanide complex (POCC) in place of osmium tetroxide.

6.   Dehydrate in methanol/propylene oxide and embed in Epon.
7.   Section the material, remove resin, and examine under EM.

Result

Sites of peroxidase activity are osmophilic, electron-dense, and colored black.

Controls

1.   Omit the substrate from the incubation medium, or
2.   Boil the material for 5 min prior to incubation. This inactivates the enzyme.

Application of the Procedure

This method was used by Burns et al.[13] to study the peroxidase activity in the cell walls of *Sphacelaria* (brown alga). Negrel and Lherminier[66] employed this technique to investigate the integration of tyrosine into xylem cell walls of tobacco cells. Hall and Sexton[61] used this technique to localize peroxidase in pea roots, where it was particularly concentrated in the cell walls and intercellular spaces.

Procedure 2:[67]
See LM procedure on Page 95.

Procedure 3:[68]
See LM procedure on Page 95.

Procedure 4:[69]

Tissue Preparation

Tissue is fixed in 4% depolymerized paraformaldehyde (dissolved in 0.05 M phosphate buffer, pH 7.0) at 0 to 4°C for 30 min. Rinse for 30 min in phosphate buffer (0.5 M, pH 7.0) which contains 5% sucrose. The materials were dehydrated in ethanol and embedded in ethanol. Thin sections were taken.

Incubation Medium

0.05 M Phosphate buffer, pH 7.0
BAXD or BED or DAB 1 mg/ml
Incubation time 50 to 60 min at 37°C

Result

Regions of cell wall containing peroxidase are electron dense.

Control

Tissue heated to 80°C in 4% formaldehyde for 20 min. Rinse tissue for 30 min in cold phosphate buffer (0.05 M, pH 7.0) containing 5% sucrose.

Procedure 5:[70,71]

Tissue Preparation

Tissue is fixed for 1 to 4 h in cold 3 to 5% glutaraldehyde in 0.1 M phosphate buffer, pH 7.2 to 7.6 at 4°C. It is then washed at least overnight in 0.1 M phosphate buffer containing 5% sucrose. The tissue can otherwise be fixed for 5 h at room temperature in a fixative containing 4% formaldehyde and 5% glutaraldehyde in 0.1 M cacodylate buffer, pH 7.2 containing 25 mg of calcium chloride; it is then washed at least overnight in 0.1 M cacodylate buffer, pH 7.2.

Procedure

1. Incubate sections for 3 to 10 min at room temperature in a saturated solution of 3,3'diaminobenzidine (free base) (DAB) in 0.05 M Tris-HCl buffer, pH 7.6, containing 0.01% H_2O_2. Prepare the saturated solution by shaking 2 to 3 mg of 3,3'-DAB with 10 ml buffer and then by filtering.
2. Wash sections in three changes of distilled water or in 0.05 M propandiol buffer, pH 9.0.
3. Post-fix for 90 min in 1 to 3% OsO_4 in S-collidine buffer (other buffers can be used), pH 7.2 containing 5% sucrose.
4. Dehydrate in graded ethanol and embed in Epon 812.
5. Cut ultrathin sections, stain with lead, and examine under TEM.

Result

Cell wall regions containing peroxidase activity can be identified by electron dense precipitates.

Glucose-6-Phosphatase[72].

Tissue Preparation

Fix the tissue in hydroxyadipaldehyde (12.5% in 0.05 M trismaleate buffer to which 0.4 M sucrose was added) at 4°C for 1 to 2 h. Wash in the same buffer.

Procedure

Incubation medium*

0.4 M sucrose

4 mM Glucose-6-phosphate

6 mM $Pb(NO_3)_2$

1. Incubate for 15 to 20 min at room temperature.
2. Transfer to 1% OsO_4 containing sucrose for 20 min.
3. Dehydrate in ethanol.
4. Embed in a mixture of n-butyl and methyl methacrylates (7:3).
5. Take thin sections and observe under TEM.

Result

Regions of cell wall containing the enzyme appear as electron dense precipitates.

Control

Omit the substrate in the incubation medium.

Acid Phosphatase .

Procedure 1: Lead Salt Method[73]

Tissue Preparation

Fix the tissue for 5 min at room temperature in 1.6 to 3% glutaraldehyde in 0.05 M cacodylate buffer, pH 6.8 with or without 1% caffeine[62] overnight at 4°C.** The tissue can also be washed in a series of acetate buffers of decreasing pH, 7.2, 6.5, 6.0, 5.5, and 5.0;[74,75] Poux[76,77] recommends that the washing medium be kept isotonic with the cell sap by adding 10% sucrose. Another modification introduced by Poux[78,79] is a citrate buffer wash (0.2 M, pH 4.8), which is considered to

---

* Prepared as follows: 2.6 g sucrose in 10 ml H_2O, 4.0 ml of 0.02 M glucose-6-phosphate in enough water to make 18.8 ml, 1.2 ml of 0.1 M $Pb(NO_3)_2$. The reagents should be added in the same order with agitation of the mixture, which was then brought to pH 6.8 with NaOH (0.1 and 0.01 N). A moderate amount of flocculent white precipitate will appear. After infiltration, 6.63 mg dry $Pb(NO_3)_2$ (1 mM in the 20 ml medium) was added to the filtrate and the pH adjusted to 6.85.

** The post-fixation wash is extremely important in order to remove both traces of fixative, which may interfere with the staining and the fixative buffer, so that, when the tissue is placed in the incubation medium, the pH will rapidly adjust to 5.0.

remove insoluble salts of calcium, including calcium phosphate, which may make identification of the reaction product more difficult.

Procedure

Incubation Medium

See Page 104.

1. Put the tissue in the incubation medium for 30 min at 37°C. Periods longer than 45 min are disadvantageous.
2. Wash in the same buffer.
3. Fix for 1 h in 3% glutaraldehyde in 0.05 M cacodylate buffer.
4. Post-fix for 2 to 4 h in unbuffered OsO_4.
5. Dehydrate in acetone, embed in Epon, and take ultrathin sections.
6. Observe under TEM.

Results

Regions of cell wall rich in acid phosphatase will be indicated by black precipitates.

Controls

1. Tissue heated to 100°C for 5 min prior to incubation.
2. Fix tissue preincubated in 10 to 2 M sodium fluoride.
3. Omit substrate in the incubation medium.

Application of the procedure

Halperin[73] used this technique to localize the activity of acid phosphatase in the cultured cells of *Daucus carota*. The enzyme activity was seen in the cell walls, especially in the central portion of the wall.

Procedure 2: Azodye Method[80]

Tissue Preparation

Fix the tissues at 0 to 4°C for 1 h in 3% glutaraldehyde in 0.1 M cacodylate buffer, pH 7.4. Wash overnight in the same buffer and then rinse briefly in 0.2 M acetate buffer, pH 5.2 at 0 to 4°C. Post-fix in OsO_4 for 2 h. Dehydrate in ethanol and embed in araldite.

Procedure

Incubation Medium

Naphthol AS-TR or AS-BI Phospahate (substrate)	5 g
Dimethylformamide or dimethylsulphoxide	0.25 ml

Dilute the obtained solution with 25 ml distilled water. Then, add the following in sequence:

0.2 M acetate buffer, pH 5.2	25 ml
p-nitrobenzene diazonium-tetrafluro-borate	30 mg

After the above two are dissolved by constant stirring, add the following:
10% $MnCl_2$ (as activator)   0.1 ml

1. Incubate the tissue for 1.5 to 3 h at room temperature in the above medium.
2. Observe under TEM.

Result

Distinctly crystalline azo dye is formed at the sites of acid phosphatase, and these areas appear electron dense.

Controls

1. Add sodium fluoride (0.01M) to the medium
2. Omit substrate in the medium.

Applications of This Procedure

This procedure was employed by Charvat and Esau[81] to study the enzyme activity in the phloem cells of the petioles of *Phaseolus vulgaris*.

Procedure 3: Azodye Method[82]

Tissue Preparation

Fix the tissue in formal-calcium, pH 7.4 for 18 to 20 h at 4°C or in cold phosphate-buffered (pH 7.4) 10% formaldehyde containing 0.1% chloral hydrate for 18 to 20 h or in cold 1.5% glutaraldehyde containing 1% sucrose and buffered at pH 7.4 with phosphate or 0.067 M sodium cacodylate buffer. After fixation wash the tissue and store at 4°C in 0.1 M cacodylate or veronal acetate buffer, pH 7.4 (containing 7% sucrose) for varying periods of time up to 1 week. Dehydrate the tissue in ethanol series and embed in araldite. Take ultrathin sections.

Procedure

Incubation medium:
Stock solution

Naphthol AS-BI phosphate (substrate) 100 mg dissolved in N,N'-Dimethylacetamide, 10 ml.

Veronal acetate buffer (5 ml stock veronal acetate buffer, 7 ml of 0.1 N HCl and 13 ml of distilled water).

Working solution

15 ml buffer, 35 ml distilled water, and 0.6 ml stock substrate solution, pH adjusted to 5.0 with 0.1 N HCl.

1. Float the tissue from one to several hours at 37°C in the incubation medium under constant agitation.

2. Wash tissue in two changes of ice-cold distilled water containing 7% sucrose, or in 0.1 M veronal-acetate wash buffer, pH 7.4, containing 7% sucrose for less than 15 min.

3. Post-couple in diazotized acetoxymercuric aniline (AMA-4) for 2 to 3 min, pH 7.25, at room temperature. Prepare the postcoupler as follows: *p*-acetoxymercuric aniline (AMA) is the post coupling agent. Make a 4% AMA in 50% acetic acid at room temperature, with constant shaking. Plunge the container into ice. An ice-cold solution is stable for some hours, but always prepare fresh. Make a 4% aqueous sodium nitrite solution by dissolving 2 g of the salt in 50 ml distilled water. Keep the container in an ice-box. The solution is stable for 2 d in a refrigerator but it is better to prepare it fresh before use. After the first and second solutions mentioned above are in ice-boxes at least for 45 min, mix equal parts of these two. In 15 min diazotization takes place. Diazotization is indicated by a change of color from orange to red–brown with a brown precipitate. Filter through Whatman 1 and keep the filtered solution in the ice bath. The solution should be a clear yellow orange, and should be used within 12 h after preparation. Dilute 0.7 ml of this solution to 50 ml with 1 N phosphate buffer, pH 7.6. Adjust the pH of the final solution to from 7.25 to 7.3. Do not refrigerate this. It should be used within 15 min and at room temperature not lower than 15°C.

4. Rapidly pipette off the coupler.

5. Wash the tissue with ice-cold veronal-acetate buffer, three changes, 5 min each.

6. React tissue 30 to 45 min with a 1% solution of thiocarbohydrazide containing 2% sucrose, at 25°C to room temperature. Constantly agitate the tissue.

7. Wash tissue in two changes of 0.1 M veronal acetate buffer as in Step 2.

8. Following the second washing, transfer tissue to clean vials.

9. Wash once again in veronal-acetate buffer.

10. React tissue in 1.5 to 2% cold aqueous OsO_4, pH 7.2 to 7.6, with constant agitation at room temperature for 90 min or for 45 to 60 min at 35°C; or in 2% lead citrate, pH 7.6, for 10 to 15 min. Wash well with buffer and post-fix, if desired, in 1% OsO_4 buffered with acetate for 15 min.

11. Pipette off OsO_4, dehydrate, transfer to ethanol-araldite mixtures and finally embed in araldite overnight.

Results

Sites of acid phosphatase are indicated by electron dense precipitates.

Controls

1. Heated tissue.
2. Substrate mixed with sacharolactone 20 mg/ml H_2O; 0.01 ml saccharolactone to 0.99 ml substrate.
3. Omission of substrate in the medium.

Comments

The coupling of mercury-substituted diazotate with naphthol AS-BI is to increase the intensity of the precipitate for better EM vision.

Procedure 4: [Barka and Anderson's[83] method modified by Burns et al.[13] for EM]

Tissue Preparation

Any convenient EM fixation procedure.

Procedure

Incubation medium:

  50 mM tris-maleate, pH 7.0

  3 mM $Pb(NO_3)_2$

  8 mM sodium metaglycerophosphate (substrate)

1. Transfer materials from fixative to 0.05 M tris-maleate buffer, pH 7.0, and store overnight at 4°C.
2. In the next day, pre-incubate the material for 30 min at room temperature in darkness in the incubation medium except the substrate.
3. Incubate in dark for 60 min, at room temperature.
4. Wash in 0.05 M sodium cacodylate buffer, pH 7.0
5. Refix for 2 h at room temperature in 1.6% v/v glutaraldehyde in the same buffer.
6. Post-fix for overnight at 4°C in 1% v/v OsO_4 in sodium cacodylate buffer.
7. Dehydrate in methanol/propylene oxide series and embed in Epon.
8. Section and observe.

Result

Electron dense deposits at the sites of enzyme activity.

Controls

1. Omit the substrate from the incubation medium.
2. Boil the material for 5 min prior to incubation.
3. Add 10 mM sodium fluoride to the incubation medium to specifically inhibit the enzyme activity.

Malate Dehydrogenase[84,85].

Tissue Preparation

Fix the tissue in paraformaldehyde 3% for 2 h at 3°C. The other steps can be that of any conventional EM protocol.

Procedure

Incubation medium:

Stock solution	Final concentration (mM)
0.1 M Sorensen's phosphate buffer, pH 7.6, 1 ml	80
0.5 M K-Na tartarate 3.0 ml	300
0.3 M CuSO$_4$ 0.35 ml	21
0.1 M D,L-maleate 0.5 ml	10
NAD or NADP	1 to 5 mg/ml of medium
0.05 M K-ferricyanide 0.15 ml	1.5
Adjust pH to 7.0	

   Treat the material in the incubation medium. Time and conditions to be adjusted by trial and error.

Results

Regions of cell wall containing malate dehydrogenase are evident as electron-dense precipitations.

Controls

1. Heat the sample at 70 to 80°C during fixation.
2. Omit the substrate in the medium.

Non-Specific Esterases[86,87].

Tissue Preparation

Standard EM processing of tissues can be followed from fixation to sectioning.

Procedure

Incubation medium:

>   Thioacetic acid (substrate)
>   Lead nitrate

1. Subject the sections to the incubation medium.
2. Observe under EM.

Results

Regions containing non-specific esterase activity are evident as electron-dense precipitated areas.

Rationale of the Reaction

Thioacetic acid is hydrolyzed to give H_2S and acetic acid as products by the enzyme esterase. Lead nitrate will combine with H_2S and form electron-dense lead sulphide precipitates.

>   *Cellulase*[88,89] .

Tissue Preparation

Fresh tissue is to be used.

Procedure

Incubation medium*

---

* Ekes[90] used the following medium and procedure:
0.7 ml 1M sodium succinate
0.15 ml 50 mM potassium ferricyanide
0.35 ml 0.3M $CuSO_4$
0.8 ml 0.1 M Svensen's phosphate buffer, pH 7.6
3 ml 0.5 M sodium potassium tartarate in phosphate buffer, pH 7.6.
Tissue incubated 20 to 30 min at room temperature in the above medium. Further steps are from 3 above.

0.2% carboxymethylcellulose (substrate for cellulase)

0.1 M phosphate buffer (pH 6.0)

1. Incubate tissue slices in the above medium (preferably 0.5 mm slices) for 10 min at 25°C.
2. Transfer to hot Benedict's solution at 80°C for 10 min.
3. Wash thoroughly with distilled water.
4. Fix the tissue slice in glutaraldehyde.
5. Post-fix in OsO$_4$

Result

Sites of cellulase activity are evident through electron dense precipitates.

Control

Add 140 mM sodium malonate, as inhibitor, to the incubation medium.

Rationale of the Reaction

The cupric salts are reduced to an electron-dense precipitation of cupric oxide by the glucose produced from cellulase activity[88,89] on reacting with carboxymethyl cellulose, a soluble cellulase substrate. Cupric oxide crystals are deposited at the site of reducing sugar production.

Application of the Procedure

Bal[88] and Bal et al.[89] have employed this technique to study the distribution of cellulase activity in pea epicotyl. The activity was found to be concentrated in the region between the plasma membrane and the cell wall.

Lieberman et al.[91] studied the distribution of cellulase in the abcission zone of tobacco flower pedicels, using this procedure.

Pectinase[92].

Tissue Preparation

Specimens (0.5 to 1.0 mm^3) are fixed 1 to 2 h on ice by immersion in a modified Karnovsky fixative composed of 3% paraformaldehyde and 2% glutaraldehyde in 0.05 M phosphate buffer, pH 7.2. Wash specimens thoroughly with at least 20 changes of ice-cold 0.05 M phosphate buffer (pH 7.2) to prevent the reaction of Benedict's solution with the aldehyde groups in the fixative. After washing, equilibrate the tissues further with the same buffer by storing them overnight at 0°C.

Procedure

1. Incubate specimens in 0.5% pectin solution in 0.1 M sodium acetate buffer, pH 5.0. Vary the incubation time slightly depending on the material, but it should be relatively short so as to minimize the diffusion of galacturonic acid residues away from the sites of enzyme activity.

2. Transfer the specimen immediately to Benedict's solution and boil for 10 min.

3. Allow tissue to cool.

4. Rinse with several changes of 0.05 M sodium cacodylate buffer, pH 7.2.

5. Post-fix for 2 h in 1% OsO_4 in the same buffer.

6. Rinse in water.

7. Dehydrate through a graded ethanol series into acetone or propylene oxide.

8. Embed in Spurr or Epon-araldite.

9. (optional) Stain with uranyl acetate for 10 min at 40°C followed by 10 min in lead citrate at room temperature.

10. Section and observe under TEM.

Result

Regions of cell wall rich in pectinase are evident by dense precipitates of cuprous oxide crystals.

Control

1. Avoid substrate treatment.
2. Heat the specimen before processing further.
3. Avoid treatment in Benedict's solution.

Rationale of the Reaction

When heated in the presence of reducing sugars, Benedict's solution changes from clear blue to cloudy red with the formation of an insoluble precipitate of cuprous oxide crystals by the reduction of cuprous salts by sugar aldehydes from galacturonic acid residues by the action of pectinase.

Application of This Procedure

Using this procedure, Nessler and Allen[92] have localized pectinase in the growing tips of laticiferous initials of *Nerium oleander.*

4.5.2.2 Cell Wall Structural Proteins

Procedure 1: Periodic Acid-Phosphotungstic Acid or PA-PTA Method[13,93]

Preparation of Tissue

Same as PATAg procedure (see Page 199), except that glycolmethacrylate is used as the embedding medium after ethanol dehydration.

Procedure

1. Treat in 1% periodic acid, 30 min.
2. Wash in distilled water.
3. Treat in 1% PA-PTA in 1 N HCl (pH 3.0), 30 min.
4. Wash in distilled water.

Result

Regions of cell wall containing glycoproteins are indicated by electron dense precipitates.

Controls

1. Protease 1 mg/ml, 37°C, 60 min treatment.
2. Treatment in NaOH (0.1 N, 100°C, 15 min) prior to Step 3 above.[13]

Procedure 2: Swift's Method[94]

Tissue Preparation

As in PATAg method (see Page 199). Epon is the embedding medium.

Procedure

1. Immerse grids in Swift's reagent, 30 sec to 2 h, at 45°C, in darkness. Prepare Swift's reagent by adding 25 ml of Solution A (5 ml of 5% silver nitrate, 100 ml of 3% methenamine) to 5 ml of Solution B (10 ml of 1.44% of borate, 100 ml of 1.9%

borax) and 25 ml of distilled water. The final staining solution should have a pH of 9.2. Store the solution in the dark at 0°C for up to one week, before use.

2.   Wash in distilled water
3.   Treat in 10% sodium thiosulphate, 10 min.
4.   Wash in distilled water.

Result

Electron-dense silver deposits are formed at the sites of proteins.

Control

Protease (1 mg/ml, 37°C, 30 min) treatment of material prior to Step 1.

Rationale of the Reaction

Silver methenamine interacts with cystein-containing proteins under alkaline conditions and silver deposits are formed.

Application of This Procedure

This procedure has been used for visualizing proteinaceous wound healing material in *Bryopsis hypnoides*, a marine alga.

4.5.2.3   Cutin and Suberin

Procedure 1:[36]

Tissue Preparation

Tissue is to be fixed in buffered glutaraldehyde followed by post-fixation in aqueous OsO_4. Simultaneous glutaraldehyde-OsO_4 fixation can also be employed. Low viscosity resins, like epoxy resin or araldite, have to be used for embedding, as cutin/suberin is highly impervious.

Procedure

Cut ultrathin sections and examine them directly under TEM.

Result

Regions of cell wall rich in cutin and suberin are evident by strongly osmophilic areas.

Rationale of the Reaction

The precise nature of the reaction groups in this procedure is not clear. It is, however, believed that OsO_4 is first concentrated within the lipid phases of the cell during fixation and then after the oxidation of the lipids the colored reduction products of OsO_4 become localized at the lipid/water (apolar/polar) interface (see Reference 36). The most documented reaction of the reagent is with the double bonds of unsaturated lipids which proceeds via the osmate ester. A color-complexing reaction is also known to occur with particular phenolic compounds. In cutins and suberins, unsaturation is unlikely to be the main reason for reactivity because both unsaturated and saturated types respond similarly.

Procedure 2: KMnO$_4$ Method[95]

Protocol A: En Bloc Staining

1. Immerse small pieces of fresh or glutaraldehyde fixed tissue in a freshly prepared 2% (w/v) aqueous $KMnO_4$ solution for 30 min to 2 h at room temperature.
2. Rinse well in several changes of distilled water.
3. Dehydrate in acetone or alcohol series, infiltrate, and embed in a suitable resin.
4. Cut ultrathin sections and examine them under TEM.

Protocol B: Section Staining

1. Prepare a 1% (w/v) solution of $KMnO_4$ in distilled water. Use a clean flask fitted with a clean glass stopper. The reagent is stable for several months, if kept in the dark.
2. Clean a glass petri plate in chromic acid solution, rinse in water, and finally clean in concentrated HCl.
3. Remove a small quantity of the $KMnO_4$ solution with a clean pasteur pipette from a position in the flask midway from the surface and bottom of the bulk solution.
4. Discard the first one or two drops of solution, then transfer the remainder to the cleaned petri dish to form a shallow pool a few millimeters deep.
5. Place grid, section side down, on the solution and cover the dish.
6. Stain for 10 to 30 min.
7. Remove grids and rinse under a jet of distilled water, followed by an immersion and a final rinse in distilled water. Dry grids on filter paper.
8. If there is excess stain, treat in a 0.1 to 0.5% aqueous solution of citric acid.
9. Observe in TEM.

Result for Both Procedures

Cutinized/suberized walls are strongly contrasted. The staining is more granular.

Rationale of the Reaction

Permanganate ions penetrate into lipid phases very slowly, proceeding by the oxidative destruction of superficially located lipid molecules. This results in the deposition of MnO_2, which remains located wherever it has been formed.

Comment

Section staining is preferred as the en bloc method often results in artefacts in staining and also in unequal staining.

References

1. Vijayaraghavan, M. R. and Shukla, A. K., *Histochemistry: Theory and Practice*, Bishen Singh Mahendra Pal Singh, Dehra Dun, India, 1990, Chap. 3.
2. O'Brien, T. P. and McCully, M. E., *The Study of Plant Structure. Principles and Selected Methods*, Tamarcarphi Ltd., Melbourne, 1981, Chap. 4.
3. Rosenberg, M., Bartl, P., and Lesko, J., Water soluble methacrylate as an embedding medium for the preparation of ultrathin sections, *J. Ultrastr. Research.*, 4, 298, 1960.
4. Spaur, R. C. and Moriarty, G. C., Improvements of glycol methacrylate. I. Its use as an embedding medium for electron microscopic studies, *J. Histochem. Cytochem.*, 25, 163, 1967.
5. Danscher, G. and Nörgaard, J. O. R., Light microscopic visualization of colloidal gold on resin embedded tissue, *J. Histochem. Cytochem.*, 31, 1394, 1983.
6. Bendayan, M., Enzyme-gold electron microscopic cytochemistry: a new affinity approach for the ultrastructural localization of macromolecules, *J. Electron Microsc. Techn.*, 1, 349, 1984.
7. Vian, B., Brillouet, J. M., and Satiat-Jaunemaitre, B., Ultrastructural visualization of xylans in the cell walls of hardwood by means of xylanase-gold complex, *Biol. Cell*, 49, 179, 1983.
8. Ruel, K. and Joseleau, J. -P., Use of enzyme-gold complexes for the ultrastructural localization of hemicelluloses in the plant cell wall, *Histochemistry*, 81, 573, 1984.
9. Benhamou, N. and Ouellette, G. B., Use of pectinases complexed to colloidal gold for the ultrastrucural localization of galacturonic acids in the cell walls of the fungus *Ascocalyx abietina*, *Histochem. J.*, 18, 95, 1986.
10. Ruel, K., Joseleu, J. -P., and Franz, G., Aspects cytologiques de la formation des xyloglucanes dans les cotyledons des graines de *Tropaeolum majus* L. (Capucine), *C.R. Acad. Sci.*, 310, 89, 1990.
11. Berg, R. H., Erdos, G. W., Gritzali, M., and Brown, R. D., Jr., Enzyme-gold affinity labeling of cellulose, *J. Electron Micrsoc. Techn.*, 8, 371, 1988.
12. Vian, B., Nairn, J., and Reid, J. S. G., Enzyme-gold cytochemistry of seed xyloglucans using two xyloglucan-specific hydrolases. Importance of prior heat - deactivation of the enzymes, *Histochemical J.*, 23, 116, 1991.

13. Burns, A. R., Oliveira, L., and Bisalputra, T., A cytochemical study of cell wall differentiation during bud initiation in the brown alga, *Sphacelaria furgicera, Bot. Marina,* 27, 45, 1984.

14. Pickett-Heaps, J. D., Preliminary attempts at ultrastructural polysaccharide localization in root tip cells, *J. Histochem. Cytochem.,* 15, 442, 1967.

15. Pickett-Heaps, J. D., Further ultrastructural observations on polysaccharide localization in root tip cells, *J. Histochem. Cytochem.,* 3, 55, 1968.

16. Rambourg, A., An improved silver methenamine technique for the detection of periodic acid-reactive complex carbohydrates with the electron microscope, *J. Histochem. Cytochem.,* 15, 409, 1967.

17. Roland, J. -C., General preparation and staining of thin sections, in *Electron Microscopy and Cytochemistry of Plant Cells* Hall, J. L., Ed., Elsevier, Amsterdam, 1978, 1.

18. Pearse, A. G. E., *Histochemistry Theoretical and Applied,* Vol. II. 4th ed., Churchill Livingstone, New York, 1985, Chap. 15.

19. Benhamou, N., Cytochemical localization of β-(1→4) -D glucans in plant and fungal cells using an exoglucanase-gold complex, *Electron Microsc. Rev.,* 2, 123, 1989.

20. Vian, B. and Roland, J. -C., Affinodetection of the sites of formation and of the future distribution of polygalactrouranans and native cellulose in growing plant cells, *Biol. Cell.,* 71, 43, 1991.

21. Chanzy, H., Henrissat, B., and Vuong, R., Colloidal gold labeling of 1,4-D-glucan cellobiohydrolase adsorbed on cellulose substrates, *FEBS Lett.,* 172, 193, 1984.

22. Benhamou, N., Chamberland, H., Ouellette, G. B., and Pauze, F. J., Ultrastructural localization of β-1,4-D-glucans in two pathogenic fungi and in their host tissues by means of an exoglucanase-gold complex, *Can. J. Microbiol.,* 33, 405, 1987.

23. Berg, R. H., Cellulose and xylans in the interface capsule in symbiotic cells of actinorrhizae, *Protoplasma,* 159, 35, 1990.

24. Bonfante-Fasolo, P., Vian, B., Perotto, S., Faccio, A., and Knox, J. P., Cellulose and pectin localization in roots of mycorrhizal *Allium porrum*: labeling continuity between host cell wall and interfacial material, *Planta,* 180, 537, 1990.

25. Bonfante-Fosolo, P., Tamagnore, L., Perotto, R., Esquerré-Tugayé, M. T., Mazau, D., Mosiniak, M., and Vian, B., Immunocytochemical localization of hydroxyproline rich glycoproteins at the interface between a mycorrhizal fungus and its host plants, *Protoplasma,* 165, 127, 1991.

26. Roland, J. C., Reis, D., Vian, B., and Roy, S., The helicoidal plant cell wall as a performing cellulose-based composite, *Biol. Cell,* 67, 209, 1989.

27. Revel, J. P., A stain for the ultrastructural localization of acid mucopolysaccharides, *J. Microscopie,* 3, 535, 1964.

28. Cabin-Flaman, A., Jauneau, A., and Morvan, C., Use of an ENDOPGH-gold complex: Localisation of pectic acid compounds in the cell walls of flax hypocotyl, in *VI Cell Wall Meeting,* Sassen, M. M. A., Derkesen, J. W. M., Emins, A. M. C., and Wolters, A. M. C., Eds., Neimegen, 1992, 134.

29. Frens, G., Kolloid, *Z.V.Z. Polymere,* 250, 736, 1972.

30. Sargent, C., *In situ* assembly of cuticular wax, *Planta,* 129, 123, 1976.

31. Wattendorf, J. and Holloway, P. J., Studies on the ultrastructure and histochemistry of plant cuticles: the cuticular membrane of *Agave americana* L. *in situ, Ann. Bot.,* 46, 13, 1980.

32. Albersheim, P., A cytoplasmic component stained by hydroxylamine and iron, *Protoplasma*, 60, 131, 1965.
33. Albersheim, P., Mühlethaler, K., and Frey-Wyssling, A., Stained pectin as seen in the electron microscope, *J. Biophys. Biochem. Cytol.*, 8, 501, 1960.
34. Albersheim, P. and Killias, U., Histochemical localization at the electron microscope level, *Am. J. Bot.*, 50, 732, 1963.
35. Catesson, A. -M. and Roland, J. C., Sequential changes associated with cell wall formation in the vascular cambium, *IAWA Bull. n.s.*, 2, 151, 1981.
36. Holloway, P. J. and Watterndorf, J., Cutinized and suberized cell walls, in *Handbook of Plant Cytochemistry*, Vol. II, Vaughn, K. C., Ed., CRC Press, Boca Raton, FL, 1987, 1.
37. Gahan, P. B., *Plant Histochemistry and Cytochemistry*, Academic Press, Florida, 1984, App. 7.
38. Jewell, G. G. and Saxton, C. A., The ultrastructural demonstration of compounds containing 1, 2-glycol groups in plant cell walls, *Histochem. J.*, 2, 17, 1970.
39. Joseleau, J. -P. and Ruel, K., A new cytochemical method for ultrastructural localization of polysaccharides, *Biol. Cell*, 53, 61, 1985.
40. Reis, D., Vian, B., Darzens, D., and Roland, J. -C., Sequential patterns of intramural digestion of galactoxyloglucan in tamarind seedlings, *Planta*, 170, 60, 1987.
41. Štasná, J. and Travnik, Electron microscopic detection of PAS-positive substances with thiosemicarbazide, *Histochemie*, 27, 63, 1971.
42. Thiery, J. P., Mise en évidence des polysaccharides sur coupes fines en microscope électronique, *J. Micrscopie*, 6, 987, 1967.
43. Hopkin, A. A. and Reid, J., Cytological studies of the M-haustorium of *Endocronortium harknessii*: morphology and ontogeny, *Can. J. Bot.*, 66, 974, 1988.
44. Roland, J. -C. and Sandoz, D., Detection cytochimique des sites de formation des polysaccharides pré -membranaires dans les cellules végétales, *J. Microsc.*, 8, 263, 1969.
45. Craig, A. S., Sodium borohydride as an aldehyde blocking reagent for electron microscopic histochemistry, *Histochemistry*, 42, 141, 1974.
46. Silva, M. T. and Macedo, P. M., Improved Thiéry staining for the ultrastructural detection of polysaccharides, *J. Submicrosc. Cytol.*, 19, 677, 1987.
47. Owen, T. P., Jr. and Thomson, W. W., Structure and function of a specialized cell wall in the trichomes of the carnivorous bromeliad *Brocchinia reducta, Can. J. Bot.*, 69, 1700, 1991.
48. Parameswaran, P. and Liese, W., Ultrastructural localization of cell wall components in wood cells, *Holz Roh-Werkstoff*, 40, 139, 1982.
49. Freundlich, A. and Robards, A. W., Cytochemistry of differentiating Plant vascular cell walls with special reference to cellulose, *Cytobiologie*, 8, 355, 1974.
50. Ruel. K., Barnoud, F., and Eriksson, K. -E., Ultrastructural aspects of wood degradation by *Sporotrichum pulverulentum*. Observations on spruce wood impergenated with glucose, *Holzforschung*, 38, 61, 1984.
51. Els Bakker, M. and Gerritsen, A. F., A suberized layer in the cell wall of mucilage cells of *Cinnamomum, Ann. Bot.*, 63, 441, 1989.

52. Sexton, R., Burdon, N., Reid, J. S. G., Durbin, M. L., and Lewis, L. N., Cell wall breakdown and abscission, in *Structure, Function and Biosynthesis of Plant Cell Walls*, Dugger, W. M. and Bartnicki-Garcia, S., Eds., American Society of Plant Physiologists, Rockville, MD, 1984, 195.

53. Joseleau, J. -P., Cartier, N., Chambhat, G., and Ruel, K., Xyloglucan deposition in suspension-cultured Rubus fruticosus cells, in *V Cell Wall Meeting*, Fry, S. C., Brett, C. T., and Reid, J. S. G., Eds., Edinburgh, 1989, 187.

54. Meikle, P. J., Bonig, I., Hoogenrad, N. J., Clarke, A. E., and Stone, B. A., The localization of (1→3)- β-glucans in the walls of pollen tubes of *Nicotiana alata* using a (1→3)- β-glucan specific monoclonal antibody, *Planta*, 185, 1, 1991.

55. Ruel, K., Joseleau, J. -P., and Franz, G., Cytological aspects of xyloglucan deposition in *Naturtium* storage-tissue, in *V Cell Wall Meeting*, Fry, S. C., Bret, C. T., and Reid, J. S. G., Eds., Edinburgh, 1989, 62.

56. Roth, J., The colloidal gold marker system for light and electron microscope cytochemistry, in *Techniques in Immunocytochemistry*, Vol. 2, Bullock, G. R. and Petrutz, P., Eds., Academic Press, London, 1983, 217.

57. Popper, H., Distribution of vitamin A in tissue as visualised by fluorescence microscopy, *Physiol. Rev.*, 24, 205, 1944.

58. Vian, B., Reis, D., Mosiniak, M., and Roland, J. C., The glucoronoglycans and their heliocoidal shift in cellulose microfibrils in linderwood: Cytochemistry in muco and on isolated molecules, *Protoplasma*, 131, 185, 1986.

59. Bonfante-Fasolo, P., Vian, B., and Testa, B., Ultrastructural localization of chitin in the cell wall of fungal spore, *Biol. Cell*, 57, 265, 1986.

60. Chamberland, H., Charest, P. M., Ouellette, G. B., and Pauze, F. J., Chitinase-gold complex used to localize chitin ultrastructurally in tomato root cells infected by *Fusarium oxysporum* f. sp. *radicis-lycopersici*, compared with a chitin specific gold-conjucated lectin, *Histochem. J.*, 17, 313, 1985.

61. Hall, J. L. and Sexton, R., Cytochemical localization of peroxidase activity in root cells, *Planta*, 108, 103, 1972.

62. Mueller, W. C. and Greenwood, A. D., The ultrastructure of phenolic-storing cells fixed with caffeine, *J. Expt. Bot.*, 29, 757, 1978.

63. Al-Azzawi, M. J. and Hall, J. L., Effects of aldehyde fixation on adenosine triphosphatase and peroxidase activities in maize root tips, *Ann. Bot.*, 41, 431, 1977.

64. Sexton, R. and Hall, J. L., Enzyme Cytochemistry, in *Electron Microscopy and Cytochemistry of Plant Cells*, Hall, J. L., Ed., Elsevier, North Holland, Amsterdam, 1978, 63.

65. Frederick, S. E., The cytochemical study of Diaminobenzidine, in *Handbook of Plant Cytochemistry*, Vol. I., Vaughn, K. C., Ed., CRC Press, Boca Raton, FL, 1987, 3.

66. Negrel, J. and Lherminier, J., Peroxidase -mediated integration of Tyramine into xylem cell walls of tobacco leaves, in *V Cell Wall Meeting*, Fry, S. C., Brett, C. T., and Reid, J. S. G., Eds., Edinburgh, 1989, 175.

67. Hanker, J. S., Yates, P. E., Metz, C. B., and Rustioni, A., A new specific, sensitive and non-carcinogenic reagent for the demonstration of horseradish peroxidase, *Histochem. J.*, 9, 789, 1977.

68. Papadimitriou, J. M., Van Duijn, P., Brederoo, P., and Streefkerk, J. G., A new method for cytochemical demonstration of peroxidase for light, fluorescence and electron microscopy, *J. Histochem. Cytochem.*, 24, 82, 1976.
69. Nir, I. and Seligman, A. M., Ultrastructural localization of oxidase activities in corn root tip cells with two new osmophilic reagents compared to diaminobenzidine, *J. Histochem. Cytochem.*, 19, 611, 1971.
70. Graham, R. C. and Karnovsky, M. J., The early stages of absorption of injected horseradish peroxidase in the proximal tubules of mouse kidney. Ultrastructural Cytochemistry by a new technique, *J. Histochem. Cytochem.*, 14, 291, 1966.
71. Henry, E. W., Peroxidase in tobacco abscission zone tissue. III. Ultrastructural localization in thylakoids and membrane-bound bodies of chloroplasts, *J. Ultrastruct. R.*, 52, 289, 1975.
72. Tice, L. W. and Barrnett, R. J., The fine structural localization of glucose-6-phosphatase in rat liver, *J. Histochem. Cytochem.*, 10, 754, 1962.
73. Halperin, W., Ultrastructural localization of acid phosphatase in cultured cells of *Daucus carota, Planta,* 88, 91, 1969.
74. Marty, F. M., Peroxisomes et compartiment lysosomal dans les cellules du meristeme radiculaire d' *Euphorbia characias* L., *Une etude Cytochimique.*, C.R. Acad. Sci. (Paris), 273, 2504, 1971.
75. Berjak, P., Lysosomal compartmentalization, ultrastructural aspects of the origin, development and function of vacuoles in roots of *Lepidium sativum, Ann. Bot.,* 36, 73, 1972.
76. Poux, N., Localisation des activites phosphatasiques acides et peroxydasiques au niveau des ultrastructures végétatales, *J. Microsc.,* 21, 265, 1974.
77. Poux, N., Localization d'activites enzymatiques dans le meristeme radiculaire de *Cucumis sativus* L. III. Activite phosphatasique acide, *J. Microsc.,* 9, 407, 1970.
78. Poux, N., Localisation de la phosphatase et de la phosphatase acide dans les cellules meristematiques de ble (*Triticum vulgare* Vill.) lorsde la germination, *J. Micros.,* 2, 485, 1963.
79. Poux, N., Localisation d'activites enzymatiques dans les cellules du meristeme radiculaire de *Cucumis sativus* L., *J. Microsc.,* 6, 1043, 1967.
80. Bowen, I. D., A high resolution technique for the fine structural localization of acid hydrolases, *J. Microscopy,* 94, 25, 1971.
81. Charvat, I. and Esau, K., An ultrastructural study of acid phosphatase localization in the *Phaseolus vulgaris* xylem by the use of an azo-dye method, *J. Cell. Sci.,* 19, 543, 1975.
82. Smith, R. E. and Fishman, W. H., *p*-(Acetoxymercuric) aniline diazoate: A reagent for the visualizing the napthol A5-BI product of acid hydrolase action at the level of light and electron microscope, *J. Histochem. Cytochem.*, 17, 1, 1969.
83. Barka, T. and Anderson, P. J., Histochemical methods for acid phosphatase using hexazonium pararosanalin as coupler, *J. Histochem. Cytochem.*, 10, 741, 1962.
84. Santos, I. and Salema, R., Cytochemical localization of malic dehydrogenase in chloroplasts of *Sedum telephium,* in *Electron Microscopy*, Vol. 2., Proc. 7th Eur. Congr. Electron Microscopy., Brederoo, P. and de Prister, Eds., European Congress on Electron Microscopy, Leiden Retain, 1980, 246.
85. Wenzel, J. and Behrlisch, D., Elektronenmikroskopischer Nachweis von oxydo-Reduktasen im Herzmuskel der Ratte, *Z. Mikrosk.-Anat. Forsch.*, 84, 372, 1971.

86. Berjak, P., A lysosome-like organelle in the root cap of *Zea mays*, *J. Ultrastruct. Res.*, 23, 233, 1968.

87. Coulomb, P. J., Localisation de l'esterase thiolacétique dans le méristème radiculaire de la courge (*Cucurbita pepo* L. Cucurbitaceae), *C.R. Acad. Sci.*, 268, 656, 1969.

88. Bal, A. K., Cellulase, in *Electron Microscopy of Enzymes*, Vol. 3, Hayat, M. A., Ed., Van Nostrand-Reinhold, Princeton, NJ, 1974, 68.

89. Bal, A. K., Verma, D. P. S., Bryne, H., and MacLachlan, G. A., Subcellular localization of cellulases in auxin-treated pea, *J. Cell Biol.*, 69, 97, 1976.

90. Ekés, M., Electron-microscopic-histochemical demonstration of succinic-dehydrogenase activity in root cells of yellow lupine, *Planta*, 94, 37, 1970.

91. Lieberman, S. J., Vadovinos, J. G., and Jensen, T. E., Ultrastructural localization of cellulases in abscission zones of tobacco flower pedicels, *Bot. Gaz.*, 143, 32, 1981.

92. Nessler, C. L. and Allen, R. D., Pectinase, in *Handbook of Plant Cytochemistry*, Vol. I, Vanghn, K. C., Ed., CRC Press, Boca Raton, FL, 1987, 149.

93. Rambourg, A., Detection des glycoproteines en microscopie electroniques par l'acide phosphotunngstique a bas pH, *Electron Microsc. (Rome)*, 2, 57, 1968.

94. Swift, J.A., *J. Roy. Micr. Soc.*, 88, 449, 1968.

95. Wattendorf, J. and Holloway, P. J., Periclinal penetration of potassium permanganate into mature cuticular membranes of *Agave* and *Clivia* leaves: new implications for plant cuticle development, *Planta*, 161, 1, 1984.

Chapter **5**

Lectin Cytochemistry

Contents

5.1 Introduction

Lectins are proteins or glycoproteins of plant, animal, or bacterial origin that bind to cell surfaces through specific carbohydrate-containing receptor sites. They are of non-immune origin. Because of this unique property, as well as because of the possession of two or occasionally more binding sites in their molecular configuration, lectins have been widely used as probes for cell surface (in animals) and cell wall (in plants) studies in the localization and identification of carbohydrate molecules.

Several lectins are now known but only a few of them have been exploited for marking cell wall carbohydrates. Table 5.1 provides a partial list of some of the more important lectins that have been/are potentially capable of being used in cytochemical localization of cell wall carbohydrates.

Lectins, by themselves, are colorless. Therefore, for cytochemistry they must be tagged onto a probe that can be visibly seen. Fluorochromes serve this purpose in LM. The most commonly used flurochrome is fluorescein isothiocyanate (FITC); other fluorochromes such as tetramethylrhodamine isothiocyanate (TRITC), dichlorotriazinylaminofluorescein (DTAF), or Texas Red can also be used. Since lectins are neither enzymes nor electron-opaque, they must be conjugated to an

TABLE 5.1
Details on Lectins Used in Cytochemistry

Name of lectin	Type	Molecular weight	Recommended haptens (= inhibitors) (for control procedure)	Specific affinity
1. Concanavalin	I	102,000	α-methyl-D-mannoside	(a) O-methyl-α-D mannopyranosyl (b) Non-reducing δ-D-glucopyranosyl (c) (β-D-Fructofuranosyl (d) δ-D-mannopyranosyl
2. Helix pomatia (snail) agglutin	II	79,000		N-Acetylgalactosamine
3. Peanut agglutin	II	106,000	D-lactose or D-galactose	(a) Galactose (b) Lactose
4. Soybean agglutin	II	120,000	α-and β-N-acetyl-D-galactosamine	(a) N-acetyl-α-D-galactosa-saminide (b) α-D-Galactopyranoside (c) β-D-Galactopyranoside
5. Ricinus communis agglutin	II	120,000	D-galactose or D-lactose	(a) α-D-galactose (b) β-D-galactose
6. Griffonia simplicifolia Lectin - I	II	114,000	D-lactose	α-D-galactose
7. Ulex europaeus agglutin	III	46,000	α-L-fucose	α-L-fucose
8. Lotus tetragonolobus lectin	III	120,000	α-L-fucose	α-L-fucose
9. Wheat germ agglutin	IV	36,000	N-acetyl-D-glucosamine or N,N'-diacetyl chitobiose	di-N-acetyl chitobiose

electron-opaque enzyme molecule like horse-radish peroxidase or alcohol dehydrogenase or to a metal particle like ferritin or colloidal gold for use in EM cytochemistry. The same may, however, be used for LM also, with the latter (colloidal gold) needing a silver enhancement treatment (see Page 265 for details).

The use of lectins thus far for probing cell walls of plants has been less than satisfactory. Only a few lectins have been harnessed to yield information and hardly a few systems have been studied. Although lectin probes offer interesting possibilities in cell wall carbohydrate cytochemistry, many problems are to be sorted out. The first is the specificity of the lectin probe to be used. The second pertains to the ease of penetration of lectins into specimens. Although simple sugar binding properties of many lectins are quite well understood, a better understanding of the complex interactions is still needed. Lectin binding is dependent on a number of factors. These factors include, but are not limited to: stearic hindrance (how accessible is the particular sugar residue?); orientation of the sugar (is it linked in an alpha or beta configuration and which carbon atoms are involved?); and location of a specific sugar residue in the carbohydrate chain. The last factor is important because some lectins can recognize only terminal carbohydrate residues. Hence, the lectins have not been of significant service in plant cell wall studies thus far due to the limited selection of lectins relative to the complexities of the various polysaccharides present in the cell wall. Even within the short list of lectins mentioned in Table 5.1, lectins often differ in what particular details of a sugar they recognize.

5.2   Procedures Involving Lectin Probes

5.2.1 Preparation of Lectin-Colloidal Gold Complex

Colloidal gold has a relatively large surface area and it can adsorb irreversibly a wide variety of lectins through a noncovalent process. Lectins adsorbed onto gold particles retain their sugar-binding capacity, although their precise specificity may be modified. Gold particles labeled with lectins can be used in pre- or post-embedding techniques, but post-embedding is superior.

Procedure

1. Prepare 15 nm diameter gold particles as per procedure outlined on Page 64.
2. Adjust the pH of the colloidal gold solution to 7.0 to 7.5 using 0.1 M K2CO3.
3. Add 8 ml of gold sol to 1 ml of lectin solution (20 ug/ml distilled water) in polycarbonate centrifuge tubes.
4. Stir the mixture for 5 min at room temperature.
5. Add 1ml of PEG (0.1% w/v) to this mixture.
6. Stir the mixture for 5 min.
7. Centrifuge at 100,000 g for 30 min at 4°C.
8. Discard the supernatant and resuspend the pellet 1:1 in 0.9 M mannitol and 0.1 M Na phosphate buffered saline, pH 7.4.
9. The gold-lectin complex is ready for use.

5.2.2 Lectin-Colloidal Gold Conjugate Method for Detecting Carbohydrate Monomers in Cell Walls

Tissue Preparation

Fix the tissue in 2% glutaraldehyde in 0.066 M Sorensen's sodium phosphate buffer, pH 6.8 at room temperature for 2 h. Rinse the tissue in the buffer, dehydrate in ethanol, and infiltrate and embed with Spurr's resin. Cut sections of 0.5 to 1 μm thickness and mount in carbon-colloidion coated nickel grids.

Procedure

1. Treat the grids with a 1% H_2O_2 solution for 10 min.
2. Rinse several times in distilled water.
3. Incubate for 45 min at 37°C with colloidal gold conjucated lectin.
4. Wash thoroughly in distilled water.
5. Stain with uranyl acetate and lead citrate.

Result

Regions of cell wall containing the specific sugar monomers are marked by the presence of colloidal gold particles.

Control

Employ incubation medium of gold-lectin complex containing 0.2 M of the concerned hapten mentioned in Table 5.1

Application of This Procedure

L-Fucose, glucose, fructose, mannose, galactomannans, etc. have been localized in different plant systems using the concerned lectin probes. For example, Vithanage and Knox[1] used Con. A probes to study the carbohydrate distribution on the stigma surface pellicle of sunflower, while Hopkin and Reid[2] used colloidal gold–wheat germ agglutin conjugate to study the monokaryotic haustoria of the rust fungus, *Endocronartium harknessii*, infecting the seedlings of *Pinus banksiana*. The exact wall layer of the fungus which is chitinous has been elucidated using this technique.

5.2.3 Fluorescein-Lectin Conjugate Method for Demonstrating Carbohydrate Monomers in Cell Walls

Tissue Preparation

Fresh tissue is preferable.

Procedure

1. Mount the tissue on glass slides containing 10 mM N-2-hydroxy-ethylpiperazine-N'-2-ethanesulphonic acid (HEPES) buffer, pH 6.8.

2. Treat the tissue for 30 min with OsO_4 vapor to quench autofluorescence. Ensure that there is no autofluorescence.

3. Incubate the tissue in a moist dark environment at 37°C for 30 min in 10 mM HEPES buffer, pH 6.8, containing purified FITC or rhodamine-conjugated lectin, both diluted to 200 µg/ml before treating the tissue.

4. Observe under epifluorescence microscope using blue-violet light (FITC-lectin) and green light (rhodamine-lectin).

Result

Regions of the cell wall containing concerned carbohydrate monomer fluoresce to the characteristic color.

Control

Add a 0.2 M solution of the concerned hapten to the dye-lectin conjugate prior to staining (see Table 5.1).

Application of This Procedure

This technique was employed to demonstrate the distribution of fucose containing xyloglucans in the primary cell walls of Azuki bean epicotyls and stems.[3,4] It was shown that xyloglucan was found both on and between cellulose microfibrils (see also Acebes et al.[5]).

Using FITC-Ricin (from the seeds of *Ricinus communis*) conjugate, the galactose residues of fucoidans (the sulphated polysaccharides of brown algae) have been labeled by Quatrano and his associates.[6,7]

References

1. Vithanage, H. I. M. V. and Knox, R. B., Development and cytochemistry of stigma surface and response to self and foreign pollination in *Helianthus annuus, Phytomorphology*, 27, 168, 1977.
2. Hopkin, A. A. and Reid, J., Cytological studies of the M-haustorium of *Endocronortium harknessii*: morphology and ontogeny, *Can. J. Bot.*, 66, 974, 1988.
3. Hayashi, T. and MacLachlan, G., Pea xyloglucan and cellulose. I. Macromolecular organisation, *Plant Physiol.*, 76, 596, 1984.
4. Hayashi, T. and MacLachlan, G., Pea xyloglucan and cellulose. III. Metabolism during lateral expansion of pea epicotyl cells, *Plant Physiol.*, 76, 739, 1984.
5. Acebes, J. L., Lorences, E. P., Revilla, G., and Zarra, I., The cell wall loosening in *Pinus pinaster*. Role of Xyloglucan, in *V Cell Wall Meeting*, Fry, S. C., Brett, C. T., and Reid, J. S. G., Eds., Edinburgh, 1989, 13.
6. Quatrano, R. S., Brawley, S. H., and Hogsett, W. E., The control of the polar deposition of a sulphated polysaccharide in *Fucus* zygote, in *Determinants of Spatial Organiazation*, Subtelny, S. and Konigsberg, I. R., Eds., Academic Press, London, 1979, 77.
7. Quatrano, R. S., Hogsett, W. S., and Roberts, M., Localization of a sulfated polysaccharide in the rhizoid wall of *Fucus distichus* (Phaeophyceae) zygotes, in *Proc. 9th Int. Seaweed Symp.*, Jensen, A. and Stein, J. R., Eds., Princeton University Press, Princeton, NJ, 1979, 113.

Chapter 6

Immunocytochemistry

Contents

6.1   Introduction

It has been known for a long time that the extracellular matrix (= cell wall) of the
plant cells is organized into distinct wall layers. It has also been known that each
layer is constituted of definite macromolecules of the polysaccharide and glycoprotein and often of phenolic and lipoidal categories. However, only recently, through
the use of immunocytochemical studies, has it been possible to establish unambig-
uously where these specific macromolecules reside in these layers thereby providing
crucial information for the development of cell wall models.[1] For example, we know
that the cellulose microfibrils surround the protoplast in multilayered sheath.[2] Xylo-
glucon, the major hemicellulose of dicot cell walls, is hydrogen-bonded to the surface
of cellulose[3] and restricted to the cellulose layer of the wall;[4] neighboring cell walls
share a common middle lamella, where the acidic pectins rhamnogalacturonan I and
homogalacturonans are exclusively localized;[4-6] and interspersed throughout the
cellulose-containing regions are the structural glycoproteins.[7] We have also gained
a lot of insight into the activity and location of many cell wall enzymes mainly
through immunocytochemical studies. In other words, immunocytochemical methods have clarified the location, metabolism, and functions as well as the structure
of the cell wall polymers.[8]

This chapter will attempt to give a general introductory account on the basis of
immunocytochemistry, the theory and protocols for the various procedures employed
and the localization techniques for the various cell wall polymers. It should, however,
be mentioned that the immunolabeling methods outlined in this book are by no
means definitive and those that have worked satisfactorily for the people who
formulated these methods are given. A number of these methods have been tried by
many researchers and have proved to be more or less reliable. However, it should
be mentioned that many of these methods and protocols are constantly evolving.
There is probably no protocol that will work as it is for every plant tissue. There
have to be adjustments in time of treatment, concentration of reagents etc. It should
also be mentioned that not all variables in the protocols outlined are of equal
importance. The most crucial is the reactivity of the anitbody used. It should have
reasonably high titer and, ideally, react with antigens, in many cases, after fixation
with the common fixatives that are used to get the best structural preservation. When
one or both of these criteria are not met, the chances of obtaining highly specific
labeling diminish accordingly.

6.2   Basics of Immunocytochemistry

Immunocytochemistry is a very recent discipline and had its genesis due to the
presence of an immune system in the animals. The immune system is responsible
for the protection of the animal from infectious foreign agents and their toxic
products. During the evolution of animals, a number of powerful mechanisms have
been developed to not only locate precisely the foreign agents finding entry into
them but also to neutralize and, if possible, to eliminate them from their body.[9]

Animal immune systems can be *nonadaptive* or *adaptive*. The first are mediated in a non-specific manner by the animal cells through such processes as *phagocytosis*, secretion of lysozymes, cell lysis, etc. The second type of response is direct and specific against particular foreign molecules and are enhanced by re-exposure to the foreign agents. This is moderated by the lymphocytes of the blood or similar systems.[9] The lymphocytes synthesize and secrete cell-surface receptors (usually proteins) that bind specifically to foreign molecules. The secreted receptors are called *antibodies* and the foreign agents that bind to them are called *antigens*. Any molecule that can be used to induce an adaptive response (i.e., production of antibodies) is called an *immunogen*. Although the words "antigen" and "immunogen" are often used interchangably, they refer to different properties of the same molecule. *Immunogenicity* is not an intrinsic property of any molecule, but it is defined only by its capacity to produce an adaptive response. In the same way, *antigenicity* also is not an intrinsic property of the molecule but is defined by its ability to be bound by an antibody. The word *"immunoglobulin"* is often used interchangeably with "antibody".[10]

It was mentioned earlier that adaptive immunoresponses are mediated   by lymphocytes. The lymphocytes are   distributed throughout the animal body and consequently an immunoresponse can be initiated promptly at any site. During the actual process of immune response, the immunogens accumulate within the lymph nodes or spleen of the animals and these regions serve as the focal points of adaptive immune response.

Many types of lymphocytes with different functions have been identified but three basic types are important: *B cells*, *T cells*, and *helper T cells*.[9] The production of a large number of antibodies that bind specifically to the antigen is the culmination of a series of interactions between all these reacting cells to the presence of the foreign antigen. The first exposure almost always induces a weak response known as *primary response*. During the primary response, the animal, through a series of specialized events, prepares itself to any subsequent exposure to the same antigen. Therefore, when the antigen is reintroduced subsequently, a rapid and intense *secondary response* results. The events that take place during primary and secondary responses, in sequence, are as follows:[9,11]

1.   Once the antigen enters the animal, it is non-specifically engulfed through*phagocytosis* by *macrophages* or other cells of the reticulo-endothelial system (like Langerhans cells in the skin, dendritic cells in the spleen and lymph nodes, and monocytes in blood). As all these cells process antigens in the same way, they are called collectively *antigen processing cells (APCs)*. For the majority of the antigens, this nonspecific uptake is important to initiate effective immune response. The degree of immune response is proportional to the efficiency of phagocytosis. Therefore, sites of active phagocytosis are to be chosen for introducing the antigens into the animals. The best sites have a high number of APCs and low rates of antigen degradation. These include subcutaneous, intradermal, intramuscular, intraperitoneal, and intravenous sites.

2.   Once the antigen is engulfed by an APC, the phagocytic vesicle fuses with a lysosome and the antigen is partially degraded. Fragments of the antigen appear on the cell surface of the APC where they complex with a cell-surface glycoprotein known as *MHC (major histocompatibility complex)* Class II Protein.

3.   In the next stage, helper T cells bind to APCs. This binding induces a series of events leading to helper T cell and T cell proliferation and differentiation. This binding also induces the division of antigen-specific B cells.

4.   B-cells start processing antigens in much the same way as APCs; but unlike APCs, the uptake of antigen by B cells is specific.

5.   In the next step, helper T cells bind to the B cells. This binding is required for a strong antibody response and, therefore, is an important factor regulating antibody production. The binding of helper T cells to B cells stimulates the proliferation of the latter and thus of increased antibody production.

6.   B cells subsequently differentiate into *plasma cells*. The latter are highly specialized to secrete large amounts of antibody of the *IgMs* type; plasma cells are terminally differentiated and live only for 3 to 4 d in the lymphoid organs.

7.   B cells also produce *memory cells* whose life-span is much greater than the plasma cells. These memory cells do not have the capacity to produce antibodies but they remain in circulation. They help in the much faster, more potent, and more persistent immune response and antibody production when the animal is given a second injection with the same antigen.

The cellular and molecular events that take place during secondary response are the same as in primary response. The only major difference is that the antibodies produced during secondary response are predominantly of the *IgG* type.

At any given time an animal can be subjected to the entry of many different foreign agents. Consequently a vast number of antigens challenge the immune system of the animal at the same time. This is taken care of by the animal, through the synthesis of a variety of antibodies, each with a distinct antigen binding site. This is possible because any one lymphocyte can recognize only one antigen.

Detailed chemical structure of five major classes of anibodies (immunoglobulins) is known in considerable detail. They are *IgG, IgM, IgA, IgE*, and *IgD. IgG* is the most important and most abundant antibody found in the serum and its structure is described in the following. This molecule has three protein domains which are joined to form a structure similar to the letter Y. Two identical domains form the arms and a third forms the stem of the Y. The two identical arms each contain a site that can bind to an antigen. Hence, these two arms are called *Fab domains* (i.e., *f*ragment having *a*ntigen *b*inding site). The stem domain is called *FC domain* (i.e., *f*ragment that *c*rystallizes) and this is involved in immune regulation. The region between the Fab and FC domains is called *hinge* and it allows, by its flexibility, lateral as well as rotational movements of the Fabs. The three domains can be separated from one another by cleavage with a protease enzyme such as papain. Two identical heavy chain polypeptides are present in the molecule and they have a molecular weight of approximately 50,000 Da. Two identical light chain polypeptides are also present in the molecule and they have a molecular weight of 20,000 to 25,000 Da. The amino terminal ends with 105 to 115 amino acid residues of both the light and heavy chain polypeptides are variable, while the amino acid sequence of the remaining part of the chains is constant. The variable regions of the one heavy chain and one light chain together form one antigen binding site. The four polypeptide chains are held together by disulphide bridges and non-covalent bonds.[12]

All immunocytochemical techniques are based on the specific interaction of an anitbody with an antigen. This interaction has been understood by the following techniques:[13]

1. Measurement of the affinity of binding between an antibody and a series of related antigens.
2. Use of affinity labeling reagents.
3. Studying site-directed mutagenesis at the antibody-combining sites.
4. Molecular conformational modeling methods.
5. Studying antibody-antigen cocrystals using X-ray diffraction techniques.

All these methods have yielded valuable information about (1) the identity of the antibody region that binds the antigen, (2) the density of the antigen region that binds to the antibody, and (3) the molecular basis for antibody specificity. As already stated, the antigen binding site of the antibody is formed by the heavy- and light-chain variable regions. The amino acids forming the antigen binding site are derived from both the heavy and light chains. These amino acids correspond to the amino acids of the hyper-variable regions within the variable region, as was determined by protein sequencing methods. These hyper-variable regions are known as *comple-mentarity determining regions* (*CDRs*). There are six CDRs, three in each chain, forming discrete loops.[13]

The region of an antigen that interacts with an antibody is known as an *epitope*. The epitope is not an intrinsic property of any antigen but is defined only by reference to the binding site of an antibody. The size of the epitope also varies as it depends on the size of the combining site.

The antibody–antigen interactions can induce either large structural changes in the antibody or the antigen or with no detectable changes. The binding between antibody and antigen is entirely dependent on noncovalent interactions, covalent interactions like hydrogen bonds, van der Waal's forces, coulombic forces and hydrophobic bonds between side chains or the polypeptide backbones, and the complex is in equilibrium with the free components. The combining site milieu can accomodate highly charged as well as hydrophobic molecules. Even small changes in the epitope structure can prevent antigen recognition. *Affinity* is a measure of the strength of the binding of an epitope to an antibody.[13] The degree of affinity is affected by the amount of antibody–antigen complex that will be found at equilibrium. With suitably high affinities, it is possible to bind essentially all of the available antigens by adding excess antibody. For reasonably good immunocytochemical staining, the affinity value for a weak signal is 10^6 mol^{-1} and for a strong signal 10^8 mol^{-1}. *Avidity* is a measure of the overall stability of the antibody–antigen complex and is governed by (1) the intrinsic affinity of the antibody for the epitope in the antigen, (2) the valency of the antigen and antibody, and (3) the geometry of the interacting components. Avidity ultimately decides the success of any immunocytochemical technique.[13]

6.3   Immunizations

It is theoretically possible to produce antibodies with almost any specificity since there is unlimited potential for antigen-combining sites. The correct choice of the immunogen and manipulating techniques, however, is very vital for antibody generation; proteins, carbohydratres, nucleic acids, lipids, and many other naturally occurring or synthetic substances can act as potential immunogens, but to elicit a primary response followed by a strong secondary one, the antigen should have the following: (1) an epitope that can bind to the cell surface antibody of a virgin B cell, (2) ability to promote cell-to-cell communication between B cells and helper T cells, (3) the appropriate minimum size >3000 to 5000 Da, and (4) a degradable nature.[14]

To make an animal respond strongly and appropriately, factors such as purity, dose and form of immunogen, proper use of adjuvants (see below), and potential modifications of the immunogen are essential. The degree of purity of the immunogen to be used depends largely on the intended use of the resulting antibodies. If highly specific antibodies that will only recognize the appropriate antigen or monoclonal antibodes are needed, then highly pure antigen is required. If purification is needed, techniques such as column chromatography, differential extraction, subcellular fractionation, electrophoresis on SDS–polyacrylaminde gels, etc. can be used. Details on these purification methods are beyond the scope of this book.

Healthy animals are very essential to get a good immune response. A wide range of vertebrates can be used. The most used are rabbits, mice, rats, hamsters, guinea pigs, horses, sheep, goats, pigs, and donkeys. The choice of the animal depends on (1) the amount of serum required, (2) antigen source, (3) amount of antigen available, and (4) whether polyclonal or monoclonal antibodies are needed. Larger animals are to be used if greater volumes of sera are needed. Among the smaller animals, rabbits yield a maximum of 500 ml of serum while mice yield only about 2 ml. Monoclonal antibodies are possible in mice and rats but not in rabbits, hamsters, or guinea pigs. The animals reach immune maturation at different periods after birth. While mice and rats take about 6 weeks, rabbits reach immune maturation by 12 weeks after birth. At least two animals should be used for immunization in the case of rabbits, although three to four are preferable; in the case of other animals, three to six are to be used.

During antigen injection, the animals are to be under sedation using anesthetics such as ether and CO_2 (both by inhalation), sodium pentobarbitone, fentanyl/flu-anisone, and fentanyl/droperiodal (the last three by injection).

An *adjuvant*, a non-specific stimulator of the immune response in animals, is often mixed with an antigen for injection into the animal body. An adjuvant has two components: (1) a depositing substance such as mineral oils (used in *Freund's adjuvant*) or alminium hydroxide precipitates to protect the antigens from being rapidly catabolized, and (2) a substance such as heat-killed bacteria, lipopolysaccharides, lipidA (a component of the lipopolysaccharide), SAF-1 and RAS (synthetic adjuvant system), that will non-specifically stimulate the immune response through the stimulation of the activity of the antigen processing cells.[14] A judicious use of adjuvants can induce a strong immune response, especially when the antigen is in a soluble form.

As already stated, the dose at which the antigen is administered is very crucial for proper immune response. An optimal amount of antigen that can produce the strongest response is desirable. Effective dose for producing the best response should be calculated by trial and error, since the injected antigen may be catabolized before reaching the target cells in the animal system. In rabbits, for example, the dose of soluble particulate or insoluble protein antigens to be injected varies between 50 and 1000 µg, while in mice and rats it varies between 50 and 100 µg. The form of antigen in which it is administered is also important. Particulate antigens are often better than soluble ones.

Weak and more soluble antigens can be made more antigenic by (1) making small changes in the structure of an antigen, (2) coupling them to specific couplers such as agarose beads, activated carbon bentonite, red blood cells, etc., (3) converting them into larger compounds by self-polymerization, or (4) binding them to carrier proteins such as BSA. The rationale of the coupling is as follows: For a strong antibody response, a cell-to-cell contact between B cells and helper T cells or between helper T cells and antigen-presenting cells is required. This contact is mediated by a fragment of the antigen that has binding sites for both class II protein binding and T cell receptor binding. The coupling substances add these sites to the antigen molecules so as to make them elicit a stronger response. Some of the procedures for coupling are provided later.

The route of antigen injection is to be decided by the volume of the antigen to be delivered, by the buffers and other components that go with the antigen, and by the speed with which the antigen is to be released into the lymphatics or circulation. Subcutaneous, intramuscular, intradermal, intravenous, lymph nodal, and intraperitonial routes are the ones that are routinely followed.

Disposable syringes are the best for injections. Plastic syringes can be used but these should be avoided when Freund's adjuvant is employed. Syringes with luer tips and luer locks are preferable.

The earliest detection of antibodies against the inoculated antigens is on the seventh day after primary injection and a peak antibody production is noticed around the tenth day. The booster injection is recommended after 2 to 3 weeks. Antibodies in the serum reach a peak by 10 to 14 d after the booster dose; additional booster doses, if desired, can be given.

Serum samples should be collected 7 to 14 d after an injection (normally after booster injection). Usually, small samples are collected until the desired antibodies at the desired concentrations become available. Test bleeds are normally done from the ear vein for rabbits (5 to 10 ml of serum) and from the tail vein for mice or rats (200 to 400 µl serum), since these sites are easily accessible and do not have high numbers of nerve endings. Analysis of test bleeds would indicate the strength of the antibody and also would identify the correct time for collecting large volumes of serum. Subsequently, regular boots and bleeds are performed so as to collect the maximum amount of serum.

6.4   Storing and Purifying Antibodies

A fairly long-term storage of antibodies can be done because they resist denaturation, due to their compact and stable protein domains. The only problem that is encountered during storage is their contamination by microbes. This can be overcome by adding 0.02% sodium azide to the antibody solution. But since sodium azide often interferes with cytochemical localizaton, it should be removed before use of antibodies, by gel filtration or dialysis.

A convenient way of storing the antibodies is to keep them at -20ºC in the serum in which they are collected. Salt concentrations between 0 and 150 mM are suitable for most applications. Repeated freezing and thawing of antibodies, which lead to their aggregation and loss of activity, should be avoided during storage.

Purified antibodies are required for a number of direct and indirect immunolocalization procedures. Several methods are followed for purification and the choice of the method depends on the use for which the antibodies are intended, the species of animal on which it was raised, its class and subclass, if monoclonal, and the source* of the starting material for purification. The commonly employed methods of purification of antibodies are as follows: ammonium sulphate, caprylic acid, DEAE, hydroxyapatite, gel filtration, ammonium sulphate DEAE, ammonium sulphate-caprylic acid, protein A beads, antigen affinity column, and anti-Ig affinity column techniques. Each of the above techniques have both advantages and disadvantages. The majority of workers have found the protein A bead as the most useful technique. During any purification, the purity, the amount, and the antigen-binding activity of the antibody are to be constantly monitored.

6.5   Labeling Antibodies

Immunocytochemical localization studies require the use of labeled antibodies. Labeling enables a researcher to detect the antibodies in a cell. Labeling and, therefore, detection may be *direct* or *indirect*. In the first method, the antibody is purified, labeled, and used to bind directly to the antigen. In the second method, the antigen-specific antibody is unlabeled and need not be purified; its antigen-binding is detected by a secondary reagent, such as labeled anti-immunoglobulin antibodies, labeled protein A or G, etc.[15] A slight variation that uses aspects of both direct and indirect methods is to modify the primary antibody by coupling it to substances such as biotin or dinitrophenol (DNP) and, thus, the modified primary antibody subsequently can be detected by biotin binding proteins such as avidin or streptavidin or by hapten-specific antibodies such as anti-DNP antibodies. Although direct labeling requires fewer steps and is less prone to background problems, it is less sensitive than indirect methods. It also requires a new labeling step for every new antibody to be studied or for every labeling method tried. On the other hand,

---

* The source can be serum (polyclonal), tissue culture supernatants with 10% FBS, or serum free media and ascites (monoclonal in the last three).

indirect methods have the advantage of widely available labeled reagents; also the primary antibody is not modified so as to cause the loss of activity.[15]

The *choice of the label* is the most important exercise, whether it be for direct or indirect methods. The most commonly used labels are enzymes, metals, biotin, flurochromes, etc. Avrameas and Uriel[16] were the first to report labeling of antibodies by covalent coupling to enzymes (see also Avrameas[17] and Farr and Nakane[18]). When enzymes are used as labels, detection of antibodies is done by employing chromogenic substrates of these enzymes that can be visualized by the eye. A large number of enzymes have been tried but the best are peroxidases (from horse radish is often preferred) alkaline phosphatase, β-galactosidase, urease, and glucose oxidase. The advantages of enzyme labeling include long shelf life, high sensitivy, possibility of direct visualization, preparations permanent and working at any level including EM. The disadvantages include the involvement of multiple steps, employment of some hazardous substrates, possible interference by endogeneous enzymes, and the poor resolution in cytochemistry. The biotin label, which involves detection through avidin/streptavidin coupled to various labels, has a long shelf life, high sensitivity, and universal detection while the problems include involvement of multiple steps, employment of hazardous substrates, etc. Fluorochromes as labels have gained tremendous importance because of their long shelf life, capacity to stain living cells and good resolution, but their disadvantages include the possible interference by autofluorescence, quenching, involvement of multiple step protocols, requirement of special equipments for detection, and low sensitivity. Four fluorochromes are in common use: fluorescein, rhodamine, Texas red and phycoerythrin. Metal labels are suitable inert substrates that can go well with antibodies. The advantages include high resolution and permanancy of preparations. Colloidal gold, among the metals, is being increasingly used now-a-days in electron microscopic immunocytochemistry.[19,20] Gold is biologically inert, has very good charge distribution, and is commercially available in discrete and uniform size ranges. Colloidal gold labeling can be quantified and an approximation of the realtive density of antigenic determinants at different sites can be made by actually counting the number of particles in a given area. For light microscopy, its detection can be enhanced by silver deposition methods (see Pages 265). The so far available cell wall cytochemical protocols involve only enzymes, metals, and fluorochromes and, therefore, procedures involving these labels alone are discussed in this book.

6.6   Tissue Preparation

Individual cells and tissues from plants or from their *in vitro* culture preparations as well as peels and sections of plant parts can be used as the materials of study in immunocytochemistry. The materials can be fresh, frozen, freeze-dried, freeze-substituted, or chemically fixed. Paraffin, resin, or plastic embedded tissue can also be used.

6.6.1 Conventional Fixation

All fixation protocols must (1) immobilize the antigen and prevent its leakage and redistrubution during subsequent operation, (2) permeabilize the cell to allow access of the antibody and other reagents needed to capture it, (3) keep the antigen in such a form that it can be recognized efficiently by the antibody, and (4) maintain the cell structure. However, fixation of the antigen comes at a price because many antigenic sites can be destroyed by fixation. As the fixation becomes more stringent and, therefore, superior, the loss of antigenicity becomes severe.[21] The present day immunocytochemist, therefore, often faces the choice of good structural presentation or retention of immunoreactivity within a tissue because an improvement in one of these is usually achieved at the expense of the other.[22]

A wide range of fixatives are in common use and the correct choice of method will depend on the nature of the antigen being examined and on the properties of the antibody preparation. Fixation by organic solvents such as alcohols and acetone removes lipids, dehydrates the cells and precipitates proteins on the cellular architecture, while cross-linking fixatives, like aldehydes, form intermolecular bridges normally through free amino groups, thus creating a network of linked antigens. Both types of fixatives denature protein antigens and, for this reason, antibodies prepared against denatured proteins may be more useful in immunocytochemical localizations. In some instances, antidenatured protein antigens are the only ones that can work.

The most commonly preferred fixatives for immunocytochemistry are formaldehyde and glutaraldehyde. However, fixation in these may mask or even change some epitopes. Of the two, formaldehyde fixation preserves most antigenic sites,[21] but it is reversible.[23] It does not maintain good ultrastructure. Glutaraldehyde fixation is often used as a compromise between antigenic and structural preservation although it maintains only about one half as many antigenic sites as in formaldehyde fixation.[21] Osmium post-fixation often irreversibly destroys antigenic sites.[24]

6.2.2 Cryofixation

This is also called *freeze fixation*. It is the most important alternative to conventional chemical fixation. The main reasons for preferring this are:

1. Fixation is fast and is completed in milliseconds.
2. There is a simultaneous stabilization of cellular components.
3. The images reflect the native structure of living cells.
4. There is an improvement in the Immunocytochemical localization.[25-28]

Under ideal cryofixation conditions, one would like to have cells with their water frozen in the vitreous (non-crystalline) state, but this is not really possible except with very thin (1 to 5 μm) samples.[29] In practice, well-frozen tissues probably contain very little ice crystals (5 nm or less), which do not affect the cell's ultrastructure and with the most readily available cryofixation methods, this is difficult to achieve beyond a depth of about 10 μm without using cryoprotectants.

The following are the most readily available cryofixation methods:

a. **Plunge freezing:** The basic protocols are described in Costello.[30] It is the simplest of methods and the least expensive. The sample is plunged by hand into some type of liquid cryogen at a rapid rate. The cryogens include various halocarbons, propane, liquid helium, and Freon 22. A variety of devices have been built to facilitate rapid immersion of the specimen (see Reference 31). A major limiting factor of this method is that good freezing is generally limited to a few millimeters from the surface. Therefore, only very small specimens can be studied effectively.

b. **Cold metal block freezing:** The basic protocols are described in Boyne.[32] In this method, the specimen is slammed (also called *slamfreezing*) onto a cold metal surface (usually silver or copper) that is cooled by liquid nitrogen or helium (the latter preferred since the sample can be cooled more rapidly). Specimens can be preserved up to a depth of 10 to 15 μm; the area of immediate impact with the metal block may be damaged.

c. **Propane jet freezing:** The basic protocols are described in Gilkey and Staehelin.[25] Here a specimen is sprayed from two sides by jets of liquid propane, which is cooled by liquid nitrogen. The specimen is in a "sandwich" formed by two copper specimen holders. A depth of 20 to 40 μm of good preservation can be obtained. This method has been used by Fernandez and Staehelin[33] and Staehelin and Chapman.[34] A potential problem is that the jets must hit the specimen holder simultaneously. In order to extend the depth of good preservation, many laboratories have treated cells and tissues with cryoprotectants such as glycerol and DMSO before freezing their specimens. However, cryoprotectants cause numerous artefacts and defeat the very purpose of crypreservation. Dextran or sucrose pretreatment before propane jet freezing can increase the diameter of freezing to 100 μm without any apparent artefact.[35]

d. **High pressure freezing:** It is similar to propane jet freezing. Immediately before freezing, the sample is pressurized to 2100 atmospheres to improve the depth of preservation. It is a very costly technique and requires a relatively expensive machine but one can obtain a depth of preservation up to 600 μm.[36-39] In samples with high water content, the depth of preservation is only a few hundred microns (in most plants). The common instrument used is the Balzers HPM 010 high pressure freezing apparatus.[40] In high pressure freezing, specimen handling and preparation before cryofixation takes place is on the order of milliseconds. If the tissue or cell is disrupted by handling, artificial images may result. Therefore, specimen loading must be carefully done.

6.6.3 Freeze Substitution

The major ways to process a specimen after cryofixation include:

1. Direct viewing of frozen specimens
2. Cryoultramicrotomy
3. Freeze fracture
4. Freeze substitution

The first three methods are difficult to combine with immunocytochemical localization and, therefore, freeze substitution is normally followed (see Reference 35).

Freeze substitution is the process of dissolution of ice in a frozen specimen by an organic solvent such as methanol or acetone (with 8% 2,2-dimethoxypropane) at low temperature and usually takes place in the presence of a secondary fixative like OsO_4.[35,41] The temperature at which freeze substitution occurs should be low enough (below -70°C) to avoid secondary ice crystal growth. After freeze substitution is completed, the temperature can be raised without risk of ice crystalization since H_2O is now absent from the specimen.

6.6.3.1   Substitution Media
The regularly used media are acetone and methanol. The latter is advantageous because it can substitute specimens in the presence of a substantial amount of water and that substitution is faster at low temperature in methanol than in acetone. In typical protocols (performed around -80°C), freeze substitution in methanol can take about 18 h, while in acetone it requires 2.53 d. Substitution can also be done in ethanol, diethyl ether, etc.

6.6.3.2   Substitution Equipment
There are a large number of devices both commercial and homemade that can be used as freeze-substitution chambers. Two simple ways to perform freeze-substitution are the use of either a dry ice/acetone bath or a -80°C freezer. The first one gives excellent results as it provides good thermal contact with the cryovials at a constant temperature of -78.5°C. For lower substitution temperatures, a homemade device can be used. This consists of an aluminium block which has holes drilled in it to accomodate standard 1.5-ml cryotubes and which has heater wire around it that is connected to a temperature regulator. This assembly is lowered into a dewar which is filled with liquid nitrogen to a level just below the aluminium block. The temperature regulator is set to -90°C, and the sample is left at this temperature for 2 to 3 d. When the nitrogen runs out, the samples begin to warm slowly; it can be removed at any temperature for further processing. For low temperature embedding, it could be removed at -70 or -35°C; for embedding in conventional resins, it is removed at 0°C or room temperature.

6.6.3.3   Fixatives
Fixative chosen for use during freeze substitution can have profound effects on the amount and specificity of immunogold labeling as well as the quality of ultrastructural preservation.[42] The main variable that determines that fixation is to be used is the sensitivity of the antibody. Freeze substitution in 5% OsO_4 in acetone and embedded in epon works well with many antibodies. If tissues are to be embedded in a low temperature resin by UV polymerization, it is necessary to omit Os or use a very low concentration (1 to 2% in acetone) because Os will block the polymerizing action of UV. Addition of uranyl acetate or tannic acid[43] to OsO_4 in the substitution medium will often enhance contrast. Glutaraldehyde can also be used instead of OsO_4.[26]

If one has to use a protocol in which freeze substitution is done in acetone or methanol alone (in order to maintain antigenicity), it is necessary to check the quality of freezing.

6.6.3.4   *Embedding Resins*

The resins of choice for immunolabeling are LR white and lowicryl resins. LR white is commonly used in immunolocalization for both LM and EM. It is possible to stop alcohol dehydration at 70% or less and go from there directly into LR white. The disadvantage is that it is very viscous below -20°C. LR gold can substitute LR white.

Epon has also been successfully used by Zhang and Staehelin[35] and many others.[27]

6.7   Immunolabeling or "Staining" of Cells and Tissues

Immunolabeling or staining of cells and tissues demonstrates both the presence and subcellular localization of an antigen. Double labeling techniques permit the simultaneous detection of two antigens, allowing comparisons of the relative distribution of different antigens. Many cell staining methods can also be used in conjunction with conventional histological stains to compare the localization of the antigen with other markers. Cell staining is a versatile technique and can be used to determine the approximate concentration of an antigen. Recent improvements in antibody labeling methods, microscopes, cameras, and image analyzers are rapidly extending the sensitivity of cell staining procedures and are making these techniques more quantitative.

Antibody binding and detection form the principal steps in cell and tissue immunostaining. Once the cells or tissues are properly processed and permeabilized, the antibodies are added. As already stated, the antibodies can be labeled *directly* or they can be detected by using a labeled secondary reagent (anti-immunoglobin antibody, Protein A, etc.) that will bind specifically to the primary antibody. Both the direct and indirect detection methods are in common use and the choice of the method will depend on the experimental design. The advantage of indirect detection is that one set of labeled reagents can be used for a number of primary antibodies. Indirect methods will normally give stronger signals but the backgrounds may be worse.

Because the antigen in cell staining will be fixed to a solid phase, the time needed for the antibody to find the antigen will be longer as it is in solution. The incubation time, therefore, has to be adjusted for the material in question, but seldom will time less than 30 min yield efficient binding. The incubation time can be lowered by increasing the concentration of the antibodies, but this will also increase the background. Usually, some compromise needs to be reached between using enough antibody to achieve a good signal and keeping the background to an acceptable level. In all cases, the antibodies should be diluted in buffers containing high concentrations of non-specific proteins. Proteins that are commonly used are bovine serum albumin

(BSA), fetal bovine serum (FBS), non-fat dry milk, or serum from the same species as the labeled antibody.

Specific cell staining reactions should always be compared with control reactions. If indirect detection is followed, the secondary reagent should be tested on its own. When using enzyme-linked detection, the enzyme reaction should be done on the specimen without the addition of any antibodies. This will demonstrate the presence and location of any endogenous enzyme activities.

Three methods are commonly employed in immunolabeling for electron microscopic observations:

1. **Labeling of cryosections or Tokuyasu method**[44]—This method can give excellent results when it works, but it is not an easy method to master. Here the tissue is fixed in a fixative (usually aldehyde) infiltrated with 2.3 M sucrose, frozen-sectioned on a cryomicrotome and then thawed. The sections are immunolabeled, then stabilized with LR white or methyl cellosolve prior to examination in EM.[45] The advantages of this method are that (1) a large tissue sample can be used, (2) this method does not require freeze substitution, (3) we can label antigens that are in the cell interior, and (4) there is a greater retention of antigenicity. The disadvantages include: (1) the requirement of expensive cryomicrotome which is tricky to use, (2) the occurrence of images, especially in plant cells, which are often difficult to interpret because they are in negative contrast (i.e., the cell walls are white against a dark background), (3) the poor structural preservation (probably due to freeze-thawing) observed, (4) the limited resolution, and (5) because of the small size of cell wall pores (5 to 10 nm[46]), the penetration of the antibodies into the thick sections during labeling is limited to a few micrometers only. More information on its application can be found in several review articles.[47-50]

2. **Preembedding labeling**—It is the simplest method for EM preparation. In this method the tissues are chemically fixed, lysed, labeled with primary and secondary antibody antibodies and then treated as a routine for EM preparation. The main disadvantage of this method is that the lysis procedure can cause considerable disruption of the cell's ultrastructure. It also necessitates the diffusion of antibodies into tissue.[51] Diffusion of large molecules such as antibodies or antibody-colloidal gold probes into plant cells is constrained by the small pore size of the cell wall (5 to 10 nm).[46]

3. **Post-embedding labeling**—Here labeling is done in embedded tissues that have been ultrarapidly frozen (i.e., cryofixed), freeze-substituted, and embedded in resins/plastics. This method allows antibodies access to antigens exposed at the surface of the sections. The drawback of this technique is the potential loss of antigenicity of the molecules of interest due to fixation, dehydration, and embedding. Carbohydrate antigens are less sensitive than protein antigens to commonly used fixation protocols but may still be affected by the removal of water and the heat of the resin polymerization. The choice of the resins may also have a bearing on the levels of immunostaining. Often, materials embedded in resins, which show good structural preservation, are no longer able to bind to antibodies. LR white is the resin of choice for those who follow this method of labeling.[4]

6.8   Mounting Immunolabeled Preparations

Mounting media for immunocytochemistry must be compatible with the detection method used. Mounting media can be aqueous or non-aqueous. Suitable aqueous media are made from gelvatol or mowiol. If these are not available, glycerol can be substituted, but permanent mounts are not possible. A suitable non-aqueous medium is DPX, which is available commercially. DPX mounting needs prior air-drying (or drying with graded ethanol) of the specimens.

6.9   Methods in Immunocytochemistry

6.9.1 Methods in Antibody Production and Tissue Preparation

Procedure 1: Freund's Adjuvant[52,53]

Freund's adjuvant is a water-in-oil emulsion [the oil commonly used is paraffin oil and the emulsifiers used are lanolin, lanolin derivatives (like Aquaphor, Falba, Protesin-X, Mannid monoleate, Arlacel A, etc.)] prepared with nonmetabolizable oils. If the mixture contains killed *Mycobacterium tuberculosis* (0.5 mg/ml), it is referred to as *Complete Freund's Adjuvant (CFA)*. Without the bacteria it is known as *Incomplete Freund's Adjuvant (IFA)*.[14]

1. Mix the protein antigens, preferably in saline, with an equal volume of the adjuvant oil and form an emulsion through prolonged and vigorous mixing.
2. Resuspend the killed *M. tuberculossis* bacteria and again vigorously mix. A thick emulsion should develop. It should not also disperse when a drop of it is placed on the surface of a saline solution.
3. Transfer to a glass* syringe. Remove all air. Fix an appropriately sized needle.
4. Now inject the samples into the animal.

Comments

1. Freund's adjuvant (both complete and incomplete) is available commercially, and it is better to use it from a standard supplier rather than to prepare it. However, the worker can mix various proportions of CFA and IFA to have the desired concentration of the bacterium.
2. It is the most commonly used adjuvant for research work, for it stimulates strong and prolonged responses. The main disadvantage of this adjuvant is that it can cause very

---

* Do not use a plastic syringe.

aggressive and persistent granulomas. Therefore, the possible side-effects should be monitored carefully during antibody production in animals. Side-effects can be avoided if the primary injection is given in CFA, and all boosts in IFA.

Procedure 2: Aluminium Hydroxide Adjuvant[54]

1. Prepare 10% potassium alum [aluminium potassium sulphate, $AlK(SO_4)_2 \cdot 12H_2O$] in distilled water.

2. To 10 ml of 10% potassium alum dropwise add, while vortexing, 22.8 ml of 0.25 N NaOH.

3. Incubate at room temperature for 10 min.

4. Centrifuge at 1000 g for 10 min. Discard the supernatant. Add 50 ml of distilled water to the pellet and resuspend the $Al(OH)_3$.

5. Centrifuge at 1000 g for 10 min.

6. Since 1 mg $Al(OH)_3$ will bind approximately 50 to 200 µg of protein antigen, appropriately dilute your antigen in 0.9% saline. If antigen is abundant, set up a titration of different amounts of the $Al(OH)_3$ vs. a constant amount of antigen.

7. Incubate at room temperature for 20 min. Spin at 10,000 g for 10 min. Test the supernatant for the presence of the antigen to be certain that it has bound.[14]

8. Now inject the animal with the samples in any site.

Comment

Here the immunogen is adsorbed onto an aluminium salt, aluminium hydroxide. It avoids the harmful side effects of Freund's adjuvants. The immunogen is either allowed to adsorb to the preformed aluminium salt or can be trapped in the salt during precipitation. When injected, the precipitate provides the depot effect.[9]

Procedure 3: Modifying Antigens by Dinitrophenol (DNP) Coupling[14,55]

1. Dialyze the antigen extensively against 0.5 M sodium carbonate (pH 9.5) to make the final concentration of the antigen approximately 2 mg/ml.

2. Prepare a 4 mg/ml solution of sodium dinitrobenzene sulphonic acid (DNBS) in 0.5 M sodium carbonate (pH 9.5) just before injecting the antigen

3. Mix the antigen and DNBS solution 1:1. Stir overnight at 4°C in the dark.

4. Dialyze this mixture extensively against PBS.

5. Now use the modified antigen for injection.

Comment

Addition of DNP to antigens modifies antigens to make them more immunogenic.

Procedure 4: Modifying Antigens by Arsynyl Coupling[14,55]

1. Dialyze the protein antigen against several changes of 100 mM sodium borate (pH 9.0). Adjust the concentration to approximately 5 mg/ml.

2. Dissolve 1 mM of p-arsanilic acid (0.21 g) in 30 ml of 80 mM HCl containing 8 mM sodium bromide (0.25 g/30 ml).

3. Place the acidic arsanilic acid solution in an ice/salt bath and add 10 ml of freshly prepared and chilled aqueous 0.7% sodium nitrite. This yields diazotized arsanilic acid.

4. Incubate on ice for a further 30 min. Add 60 ml of cold water.

5. Slowly add 0.5 ml of the cold diazotized arsanilic acid per 1 ml of the protein solution. Adjust pH to 9.0.

6. Incubate at 4°C for 4 h. Periodically check the pH and keep it at 9.0 by the addition of 0.2N NaOH if necessary.

7. Dialyze against PBS or saline with several changes overnight.

8. Now inject the samples.

Comment

The diazotized arsanyl group reacts primarily with tyrosine side chains, but also binds less frequently to histidine, free amino groups, and sulphydryl groups, thus making the proteins more immunogenic.

Procedure 5: Modifying Antigens by Denaturation[14]

1. Adjust the concentration of the antigen to 0.5 to 2 mg/ml in any convenient buffer (PBS is good).

2. Heat to 80°C for 10 min.

3. Cool the antigen solution to room temperature. The samples are ready for injection.

Comments

Heat denatures many protein antigens and makes them more immunogenic by exposing new epitopes.

Procedure 6: Coupling* Antigens (Like Xyloglucans) to Protein Carrier (Ovalbumin) by Periodate - Lysine Fixation[56]

1. Prepare the periodate-lysine-paraformaldehyde (PLP) solution as follows: To 0.2 M lysine-HCl in distilled water, add 0.1 M dibasic sodium phosphate until the pH is 7.4.

---

* It is often difficult to generate antibodies against carbohydrates in general as compared to protein antigen. To generate cell wall polysachharide-specific polyclonal antibodies, polysaccharides are coupled to a protein carrier prior to injection into a suitable animal. The coupling depends on the overall electrical charge of the polysaccharides. The commonly employed protein carriers are ovalbumin and BSA.

Dilute the solution to 0.1 M lysine with 0.1 M sodium phosphate buffer, pH 7.4. Dissolve paraformaldehyde in distilled water. Just before use, combine 3 parts of the lysine-phosphate buffer with 1 part paraformaldehyde solution and add solid sodium meta periodate. (The periodate concentration varies from 0 to 0.1 M and the paraformalde-hyde concentration varies from 0 to 4%). Remove the excess fixative by dialysis overnight. The pH should be around 6.2.

2. Incubate the ovalbumin-xyloglucan (or any neutral carbohydrate antigen) mixture in the PLP solution. PLP solution brings about a complexing of ovalbumin with xyloglucan which can then be used after emulsification in Freund's adjuvant for injection to raise antibodies against the the carbohydrate.

Rationale

Xyloglucan (XG) is not negatively charged. So, when it is mixed with ovalbumin by periodate-lysine fixation, periodate apparently oxidises carbohydrate moities on XG and on the oligosaccharide side-chains of ovalbumin to form aldehyde groups. These groups can then be cross-linked via lysine, a divalent amine.

Application of This Procedure

This method has been used by Lynch[57] to couple xyloglucans for immunocytochem-ical localization. In her work, she found that the monosaccharide residues of xylo-glucan and the oligosaccharide side chains of ovalbumin were oxidized by periodate and subsequently cross-linked by lysine, which has two amine groups. The xyloglu-can-ovalbumin complex was then used for injection.

Procedure 7: Coupling Antigens (Like Alginic Acid) to Protein Carrier (Methylated Bovine Serum Albumin) (MBSA)[58]

Alginic acid (of brown algal cell walls) and RG-I (of the cell walls of higher plants) are polysaccharides with negative charge. Therefore, it is easy to couple them to methylated BSA. The negatively charged polysaccharides presumably form ionic complexes with the positively charged protein (i.e., BSA).

1. Mix equal weights of RG-I (or alginic acid) and methylated BSA.
2. Emulsify the mixture in Freund's complete adjuvant.

Procedure 8: Coupling Antigens to Red Blood Cells[14,59]

1. Wash the RBCs three times by centrifugation at 800 g for 5 min and resuspend in PBS. Sheep RBCs are preferred by many as they are available commercially in citrate buffer.
2. Resuspend the RBCs to a final concentration of 5% (v/v).
3. Add an equal volume of freshly prepared 0.005% tannic acid. Mix well and incubate at 37°C for 15 min. Centrifuge at 800 g for 5 min. Remove the supernatant and add the antigen in PBS. Incubate for 15 min at 37°C with occasional mixing or add an

equal volume of 1% glutaraldehyde in PBS to the RBCs. Incubate at room temperature for 1 h with shaking. In this step, an option is given to choose the correct coupling method depending on the antigen. The RBS surface has a complexity of proteins, and several chemical groups are available for coupling. The investigator can choose tannic acid, glutaraldelyde, or even chromic chloride as a coupling agent.

4. Centrifuge the cells at 800 g for 5 min. Wash twice with PBS. The samples are now ready for injection.

Comments

It is one of the good ways of strengthening an immune response. RBCs are large and particulate, making them good targets for phagocytosis. Their size also slows the dispersal. If the source of the cells is different from the animal to be injected, they can provide good targets for Class II-T-cell receptor binding. They are also easy to handle for coupling and storage.[14]

Procedure 9: Subcutaneous Injections for Rabbits[14]

1. Anesthesize the rabbit.
2. Take approximately 400 µl of the antigen to be injected.
3. Pull the skin away from the body near the back of the neck and insert the needle (25-gauge) into the space that has been created. Ensure that the needle is not inserted into the muscle or body wall.
4. Inject the desired amount of antigen. Wait for a few seconds. Withdraw the needle and gently rub the hole to stop any of the inoculum from oozing.
5. Move to the next site and repeat; up to 10 sites per animal can be chosen.

Procedure 10: Intramuscular Injection for Rabbits[14]

1. Anesthesize the rabbit.
2. Take 200 to 400 µl of the antigen to be injected.
3. Place the rabbit on a wire or solid surface that is rough enough.
4. Select the thigh muscle of a rear leg after grasping the leg from the front, and insert the 25-gauge needle of the syringe.
5. Try to withdraw the plunger slowly. If there is resistance, the injection can proceed. If blood appears, withdraw the syringe and move to a nearby site and try again.
6. Withdraw the needle once the injection is over and gently rub the site of injection.

Procedure 11: Intradermal Injections for Rabbits[14,60]

1. Anesthesize the rabbit.
2. Shave an appropriately sized area to expose the skin between the ribs and hip.

3. Hold the skin between the forefinger and thumb.

4. Insert the 25-gauge needle with the bevel side up slightly under the skin and continue pushing the needle between the layers of the skin for at least 0.5 to 0.7 cm depth. Inject about 100 µl of the antigen. The inoculum should form a blister under skin.

5. Withdraw the needle and simultaneoulsy compress the skin gently.

6. Repeat the procedure at the next site. You can inject up to 40 sites in an animal.

Comments

1. If complete Freund's adjuvant is used, small sores will appear at the site of injection. These will clear within a few weeks

2. Under no circumstance should complete Freund's adjuvant be used for more than the first injection.

Procedure 12: Intravenous Injections for Rabbits[14]

1. Place the rabbit in a restraining device.

2. Take 500 to 1000 µl of the antigen.

3. Pull the ear out straight.

4. Make the marginal ear vein* stand out strongly by restricting the return of the blood to the body by pressing lightly on the base of the ear.

5. Insert the 25-gauge needle into the vein. Gently pull back on the plunger and if blood appears in the syringe, proceed with the injection. If no blood is noticed, move to another location on the vein and try again.

6. Inject the antigen slowly and evenly in the vein. Wait for a few seconds and remove the needle. Place a piece of cotton over the injection site as the needle is removed. Hold the site tightly for a few seconds to stop any bleeding.

Comments

Intravenous injection may cause an anaphylactic reation in some animals. This can be prevented by a prior injection of an antihistamine.

Procedure 13: Test Bleed on Rabbits[14]

1. Wrap the rabbit in a towel.

2. Locate the marginal ear vein (see earlier procedure) and shave a patch around the vein about two-thirds of the distance from the head to the top of the ear.

3. Cut the vein with a clean razor at a 45° angle to the vein. Incise just through the top of the vein so that the cut should not be too deep.

---

* It is on the far inside edge of the ear.

4. Collect the blood by allowing it to drip into a clean test tube. If the blood clots on the cut before the desired amount (50 ml) is collected, wipe the ear with warm clean water and continue to collect.

5. Stop the blood flow by gentle pressure to the cut with a sterilized cotton or bandage. Apply pressure 10 to 20 sec and then check to be sure the flow has stopped.

Procedure 14: Serum Preparation[14]

1. Collect the blood in a clean test tube and allow it to clot for 30 to 60 min at 37ºC.

2. Separate the clot from the sides of the test tube using a pasteur pipette.

3. Place the clot at 4ºC overnight to allow it to contract.

4. Remove the serum from the clot. Remove any remaining insoluble material by centrifugation at 10,000 g for 10 min at 4ºC .

5. Store, if need be, for many years at -20ºC or below.

Procedure 15: Storing Sera Containing Antibodies[61]

1. Take the collected serum and add, if necessary, 0.02% sodium azide.*

2. Dispense the antibodies in convenient volumes and store at -20ºC (lasts years) (most antibodies can be stored conveniently at 4ºC for at least 6 months).

Comments

1. Some monoclonal antibodies and some components of the serum are cryoproteins. They are sensitive to low temperature and they get precipitated. If the cryoproteins are a contaminant, they can be removed easily when precipitated.

2. Many antibody solutions will generate an insoluble liquid component with prolonged storage. This can be removed by centrifugation at 10,000 g. In cases where the lipids form a layer over the solution, this layer can be removed.

Procedure 16: Storing Purified Antibodies[61]

1. Adjust the pH of the antibody solution to around 7. Phosphate buffered saline (PBS) or similar isotonic solutions can be used for storing purified antibodies.

2. Store these solutions at relatively high concentrations (i.e., >1 mg/ml). Concentrations up to 10 mg/ml can be commonly used. Antibodies at lower concentrations should be concentrated before storage.

3. If the purifed antibodies are not to be labeled, they can be stored at lower concentrations with the addition of 1% BSA.

4. If necessary, add 0.02% sodium azide.

* Sodium azide is poisonous.

5. Dispense the antibodies in convenient volumes and store at -20°C. Most antibodies are stable for years when stored at -20°C. Working solutions can be stored at 4°C when they are stable for at least 6 months.

Procedure 17: Purification of Antibodies on Protein A Beads (Low Salt)[61,62]

1. Adjust the pH of the crude antibody preparation to 8.0 by adding 1/10 volume of 1.0 M tris (pH 8.0).

2. Pass the antibody solution through a protein A bead column. These columns can bind approximately 10 to 20 mg of antibody per milliliter of wet beads.

3. Wash the beads first with 10 column volumes of 100 mM tris (pH 8.0) and then with 10 column volumes of 10 mM tris (pH 8.0).

4. Elute the column with stepwise addition of 100 mM glycine (pH 3.0) buffer approximately 500 µl per sample. Collect the elute in tubes containing 50 µl of 1 M tris (pH 8.0). Mix each tube gently to bring the pH back to neutral, avoiding bubbling or frothing as these denature the proteins.

5. Identify the immunoglobulin-containing fractions by any suitable method.

Comments

Only antibodies with high-affinity binding sites for protein A such as human, horse, cow, donkey, rabbit, mouse, dog, pig, and guinea pig polyclonal antibodies. Monoclonal antibodies from the mouse IgG_{2a} and IgG_{2b} subclasses can be purified by this method. Antibodies of IgG_3 subclass bind with intermediate affinity, but IgG_1 molecules bind with low affinity. Rat monoclonal antibodies cannot be purified by this method.

Procedure 18: Purification of Antibodies on Protein A Beads (High Salt)[61,62]

1. Add NaCl to a concentration of 3.3 M to crude antibody solution. Add 1/10 volume of 1.0 M sodium borate (pH 8.9). 1.5 M glycine can also be added, if desired, along with NaCl.

2. Pass the antibody solution through a protein A bead column.

3. Wash the beads with 10 column volumes of 3.0 M NaCl, 50 mM sodium borate (pH 8.9) and subsequently with 10 column volumes of 3.0 M NaCl, 10 mM sodium borate (pH 8.9).

4. Elute the column with 100 mM glycine (pH 3.0) by adding this buffer stepwise, approximately 500 µl per sample. Collect the elute in tubes containing 50 ml of 1 M tris (pH 8.0).

5. Mix each tube gently to bring the pH back to neutral, avoiding bubbling or frothing as these denature the proteins.

6. Identify the immunoglobulin-containing fractions by any suitable method.

Comment

This method is good for low affinity antibodies.

Procedure 19: Purification of Antibodies by Ammonium Sulphate Precipitation[61]

1. Determine the volume of antibody solution.

2. Centrifuge at 3000 g for 30 min and transfer the supernatant to an appropriate container. Add a stirring bar and place on a magnetic stirrer.

3. During stirring, slowly add enough saturated ammonium sulphate solution to bring the final concentration to 50% saturation. Add an equal volume of the starting solution. Transfer to 4°C for 6 h or overnight.

4. Centrifuge the precipitate at 3000 g for 30 min.

5. Carefully remove the supernatant. Drain well. For serum, resuspend the pellet in 0.3 to 0.5 volumes of the starting volume in PBS. For monoclonal antibody tissue culture supernatants, resuspend the pellet in 0.1 volume of the starting volume in PBS. Avoid bubbles and frothing.

6. Dialyze the antibody solution vs. three changes of PBS overnight. Be sure to allow enough space for expansion of the antibody solution (twice the resuspended volume) during dialysis.

7. Remove the antibody solution from the dialysis tubing. Centrifuge to remove any remaining debris.

8. Determine the concentration and purity. Store like pure antibodies.

Comments

This method is one of the most commonly used methods. Proteins in solution form hydrogen bonds with H_2O through their exposed polar and ionic groups. When high concentration of small, highly charged ions such as ammonium or sulphate are added, these groups compete with the proteins for binding to water. This removes the water molecules from the protein and decreases its solubility, usually in precipitation. One disadvantage of ammonium sulphate precipitation of antibodies is that the resulting antibodies will not be pure. They will be contaminated with other high-molecular weight proteins. Therefore, this procedure is not suitable for a single-step purification.

Procedure 20: Coupling Antibodies to Horseradish Peroxidase (HRP) by Glutaraldehyde[15,63-65]

1. Dissolve 10 mg of HRP in 0.2 ml of 1.25% glutaraldehyde in 100 mM sodium phosphate (pH 6.8). [The HRP must be pure. This can be determined by measuring the ratio of the HRP absorbance at 403 and 280 nM (RZ = OD 403/OD 280 nM). The ratio should be at least 3.0.]

2. After 18 h at room temperature, remove excess free glutaraldehyde by gel filtration. To make the column easier to load and run, first add 20 µl of glycerol and 20 µl of 1% xylene cylanol. Use a gel matrix with an exclusion limit of 20,000 to 50,000 for globular proteins. Use 100 mm diameter beads. Prepare a column with 5 ml of bead volume according to the manufacturer's instructions. Prerun the column with a minimum of 10 column volumes of 0.15 M NaCl until the buffer level drops just below the top of the bed resin. Stop the flow of the column. Carefully load the column with glutaraldehyde-treated HRP. Release the flow and allow the HRP to run into the column. Just as the level of the HRP solution drops below the top of the column, carefully add 0.15 M NaCl. Run the column with 0.15 M NaCl.

3. Pool the brown fractions which contain the active enzyme.

4. Concentrate the enzyme solution to 10 mg/ml (1 ml final volume) by ultrafiltration or by dialysis against 100 mM sodium carbonate-sodium bicarbonate buffer (pH 9.5) containing 30% sucrose. Change the buffer to 100 mM sodium carbonate-bicarbonate (pH 9.5) either by dialysis or by washing on the ultrafiltration membrane.

5. Add 0.1 ml of antibody (5 mg/ml in 0.15 M NaCl) to the enzyme solution with the pH > 9.0.

6. Incubate for 24 h at 4°C.

7. Add 0.1 ml of 0.2 M ethanolamine (pH 7.0) and further incubate at 4°C for 2 h.

8. At this stage, there will be in the solution the uncoupled HRP, the uncoupled antibody and the HRP-antibody conjugate. For some reactions, no further purification is necessary. In these cases, the uncoupled HRP will not bind to any antigen and will be lost during any washes prior to enzyme detection. The uncoupled antibody potentially may block some antibody binding sites, but it will not score in the enzyme reaction.

Procedure 21: Coupling Antibodies to Horse Radish Peroxidase by Periodate Coupling[15,66,67]

1. Resuspend 5 mg of HRP in 1.2 ml of water and add to this 0.3 ml of freshly prepared 0.1 M sodium periodate in 10 mM soidum phosphate (pH 7.0).

2. Incubate at room temperature for 20 min.

3. Dialyze the HRP solution vs. 1 mM sodium acetate (pH 4.0) at 4°C with several changes overnight.

4. Prepare an antibody solution of 10 mg/ml in 20 mM carbonate (pH 9.5) and add 0.5 ml of this solution to the HRP solution removed from the dialysis tubing.

5. Incubate the mixture at room temperature for 2 h.

6. Reduce the Schiff's bases that have formed by periodate treatment by adding 100 µl of sodium borohydride (4 mg/ml in water).

7. Incubate at 4°C for 2 h.

8. Dialyze vs. several changes of PBS.

9. Now the HRP-antibody conjugates are ready for use.

Comments

Periodate treatment of carbohydrates opens their ring structure and allows these moities to bind to free amino groups.

Procedure 22: Coupling Antibodies to Alkaline Phosphatase[15,63,64]

1. Mix 10 mg of antibody with 5 mg of alkaline phosphatase (usually available as a suspension in 65% saturated ammonium sulphate) in a final volume of 1 ml.

2. Dialyze the mixture against four changes of 0.1 M sodium phosphate buffer (pH 6.8) overnight to remove the free amino groups present in the ammonium sulphate precipitate.

3. Stir the enzyme-antibody mixture in suitable containers in a fume hood on a magnetic stirrer. Add 0.05 ml of a 1% solution of EM grade glutaraldehyde and stir for 5 min.

4. Keep the mixture for 3 h at room temperature and then add 0.1 ml of 1 M ethanolamine (pH 7.0).

5. After a further 2 h incubation at room temperature dialyze overnight at 4°C against three changes of PBS.

6. Centrifuge the mixture at 40,000 g for 20 min.

7. Store the supernatant at 4°C in the presence of 50% glycerol, 1 mM $ZnCl_2$, 1 mM $MgCl_2$, and 0.02% sodium azide.

Comments

1. The relatively high cost of the enzyme and the need to use highly concentrated solutions of enzyme and antibody are the major problems in exploiting this method.

2. The procedure can be scaled down to the 1-mg antibody level by reducing the antibody and enzyme concentration by a factor of 10. Under these circumstances, the time allowed for coupling should be increased to at least 24 h and the yield of conjugate may be reduced.

Procedure 23: Coupling Antibodies to β Galactosidase by Glutaradehyde Coupling[15,63,64]

1. Add 10 mg of antibody to 5 mg of β-galactosidase in a final volume of 2 ml PBS.

2. Dialyze the mixture against four changes of 0.1 M sodium phosphate buffer (pH 6.8) overnight to remove the free amino groups.

3. Stir the enzyme-antibody mixture in a suitable container in a magnetic stirrer in a fume hood. Add 0.1 ml of 1 M solution of EM grade glutaraldehyde. Store for 5 min.

4. Keep the mixture for 3 h at room temperature and then add 0.1 ml of 1 M Methanolamine (pH 7.0).

5. After a further 2 h incubation at room temperature dialyze overnight at 4°C against three changes of PBS.

6. Centrifuge the mixture at 40,000 g for 20 min.

8. Store the supernatant at 4°C in the presence of 50% glycerol and 0.02% sodium azide.

Comment

The procedure can be scaled down to the 1 mg antibody level by reducing the antibody and enzyme concentrations by a factor of 10, but the time allowed for coupling should be increased to at least 24 h.

Procedure 24: Coupling of Antibodies to β-Galactosidase by Maleimido-bensoyl-N- Hydroxysuccinimide (MBS) Ester Coupling[15,68,69]

1. Make a solution of antibodies 1 mg/ml in 50 mM NaCl, 100 mM sodium phosphate (pH 7.0).

2. Add MBS at 20 mg/ml in dioxane.

3. Add 10 μl of MBS solution/ml of antibody and mix well.

4. Incubate for 1 h at 30°C.

5. Dialyze several times for several hours against 10 mM sodium phosphate (pH 7.0) containing 10 mM MgCl$_2$ and 50 mM NaCl.

6. Add β-galactosidase in 10 mM sodium phosphate to a final concentration of 1.5 mg/ml. Then add 1 ml of β-galactosidase per milligram of starting antibody.

7. Incubate for 1 h at 30°C.

8. Add B-mercaptoethanol to a final concentration of 10 mM.

Rationale

MBS is a heterobifunctional reagent and therefore links proteins having cysteine with proteins having free amino groups. Since antibodies normally do not contain free cysteines, the reagent is first bound to antibodies through its free amino groups. The excess cross-linker is removed by dialysis and the antibody-MBS added to β-galactosidase.

Procedure 25: Isothiocyanate Labeling of Antibodies[15,70,71]

1. Prepare an antibody solution of 2 mg/ml in 0.1 M sodium carbonate, pH 9.0.

2. Dissolve fluorescein isothiocyanate (FITC) in dimethyl sulphoxide at 1 mg/ml. Prepare fresh for each labeling reaction.

3. For each milliliter of antibody solution, add very slowly 50 ml of the dye solution with occasional gentle stirring and leave the preparation in the dark for 8 h at 4°C.

4. Add NH$_4$Cl to 50 mM and then incubate at 4°C for 2 hr.

5. Add xylene cylanol to 0.1% and glycerol to 5%.

6. Separate the unbound dye from the conjugate by gel filtration and the conjugated antibody elutes first.

7. Store the conjugate at 4°C in the column buffer in a lightproof container, and if needed, add sodium azide to 0.02%.

Procedure 26: Colloidal Gold Labeling of Antibodies[19,72,73]

1. Adsorb 4 ml of antiserum to 2 ml of CNBr-activated sepharose gel, to which the immunoglobulin has been bound, for 2 h at room temperature.

2. Wash the gel with PBS.

3. Elute the anti-animal* immunoglobulin with 4 ml of 3 M KCNS.

4. Adjust protein concentration to 1 mg/ml in 3 M KCNS.

5. Dialyze the solution against 2 mM borax, pH 9.0 at room temperature.**

6. Carry out coupling by adding a sufficient amount of immunoglobulin to 30 ml of gold solution (12 nm) at pH > 9.0 for the concentration of the former to exceed the stabilization point by 10%.

7. After 1 min, add a sufficient volume of 10% aqueous BSA, to get a final concentration of 0.25%, to stabilize gold particles.

8. Centrifuge this suspension at 50000 g for 45 min. A pellet of a large loose portion and a small compact portion are formed.

9. Remove supernatant without disturbing the pellet.

10. Resuspend the loose pellet in the residual supernatant.

11. Layer this over a 10 to 30% continuous sucrose or glycerol gradient (volume 10.5 ml, length 8 cm) in TBS (0.01 M tris, 0.15 M NaCl, pH 8.2).

12. Centrifuge the gradient at 20000 rev/min for 30 min.

The immunoglobulin-gold complex can be stored for months with 0.02% sodium azide at 4°C. For longer storage it can be frozen in small samples at - 70°C.

Procedure 27: Preparation of Protein A–Gold Complex

Protein A is a 42,000 Da polypeptide that is a normal constituent of the cell wall of the bacterium, *Staphylococcus aureus*. This protein is a highly stable one and is resistant to heat and denaturing reagents. Affinity of protein A for antibodies varies with the class, subclass, and species of the immunoglobulin but sera from humans, donkeys, rabbits, dogs, pigs, and guinea pigs can be used without any problem at all for all tests that rely on protein A. Four characters make protein A particularly valuable: (1) The interaction of antibody with protein A does not change the ability

---

* Can be any animal in which the antibody has been raised.

** Immunogold preparations readily form aggregates, some of which precipitate after centrifugation at 50,000 g for 30 min but the smaller ones remain in suspension. The degree of aggregate formation depends on the immunoglobulin used. Aggregation can be minimized by diluting the immunoglobulin before dialysis or by slightly raising the pH of the borax solution.

of the antibody to combine with the antigen; (2) even a highly denatured protein A molecule is easily renatured; (3) although the affinity for antibody is high, the antibody-antigen bond can be broken effectively by lowering the pH; and (4) protein A is functionally bivalent, i.e., it has two binding sites expressed and so it can be double-labeled with two sizes of colloidal gold particles. Since adsorption of protein A onto the surface of the gold particles is by electrostatic interaction between the negatively charged surface of gold particles and positively charged groups of protein A, and the binding is non-covalent, the conjugate does not interfere with the bioactivity of the protein. The binding of the two is pH-dependent, and stable complexes can be achieved at a pH equal to or slightly higher than the isoelectric point of the protein involved. A pH of 6.0 is ideal.

Determination of Optimal pH for Preparing ProteinA–Gold Complex[74]

1. Prepare protein-A solution at a concentration of 5 to 10 mg/ml distilled water.
2. From this stock solution obtain the desired protein concentration of 50 to 1000 µg/ml of distilled water.
3. Add about 7 µl of the working protein solution to the pH-buffered colloidal gold in the range of those wells showing a red color (showing colloidal gold stability).
4. Mix the two solutions by tapping the tray and by allowing them to remain at room temperature for 10 to 15 min.
5. Determine visually the stability of the protein-gold mixture with pH change. A red color indicates that the mixture is stable and unflocculated, while a blue color indicates that it is flocculated.

Labeling (for EM)

1. Float the nickel grids with mounted thin sections on a drop of 1% chick ovalbumin in PBS (pH 7.4) for 5 min at room temperature to block non-specific attachment of antibodies to residual glutaraldehyde.
2. Transfer grids onto drops of the antibody solution for 1 to 2 h at room temperature or overnight at 4°C.
3. Rinse the grids thoroughly in four changes of PBS for a total period of 5 to 10 min.
4. Dry the grids and then transfer them to drops of protein A–gold solution for 6 h at room temperature.
5. Rinse the grids extensively in several changes of PBS and then in distilled water.
6. Post-stain in 5% aqueous uranyl acetate followed by lead citrate.

Comment

The background staining, if any, is usually the result of high concentration of antisera of gold. This can be minimized by (1) preincubating the tissue with albumin or

Tween solution, (2) using lowicryl instead of epoxy resins, and (3) avoiding immersion of grids on reagents (but using only floating).[75]

Procedure 28: Double Immunolabeling Procedure[76-79]

1. Treat the grids first with 0.1 N HCl for 5 min.
2. Block the grids in 5% low-fat dried milk powder solution in PBST (PBS + 0.5% Tween - 20) for 20 min.
3. Incubate the grids in the first antiserum (usually the serum which recognizes the least abundant antigen) for 1 h.
4. Wash the first antiserum from the grids with PBST .
5. Treat with the smaller protein A–gold probe (7.5 nm).
6. Rinse the grids again thoroughly with PBST
7. Incubate in an excess of protein A (0.2 mg/ml in PBST).
8. Incubate in the second antiserum for 1 h.
9. Wash with PBST.
10. Incubate in the larger protein A–gold probe (17 nm).
11. Wash in PBST.
12. Wash in water.
13. Counterstain in uranyl acetate and lead citrate.
14. Observe in EM.

Comment

Lynch and Staehelin[76] used a blocking step between 7 and 8.

Procedure 29: Binding Antibodies to Cells/Tissues[80]

1. Keep the experimental material ready for antibody binding on a clean slide or grid.
2. Add the first antibody solution. All dilutions must be carried out in 3% BSA-containing PBS. For unlabeled primary antibodies: Monoclonal antibodies are applied as 20 to 50 μg/ml while polyclonals are diluted and tested at various dilutions of 1/10, 1/100, 1/1000, and 1/10,000. For labeled primary antibodies: Primary antibodies labeled with enzymes, fluorochromes, gold, etc. should be assayed at several dilutions in preliminary tests to determine the correct working range.
3. Incubate the slides/grids at room temperature in the humified chamber for a minimum of 30 min and in some cases up to 24 h.
4. Wash in three changes of PBS within 5 min. Supplement the buffer with 1% Triton X-100 or NP-40 to avoid background problems. If the first antibody is labeled, the specimen is now ready for the detection step (See Procedures 30 through 37).

5. Apply the labeled secondary reagent* and incubate for a minimum of 20 min at room temperature in humidified chamber. For gold-labeled reagents, observe periodically under the microscope until a satisfactory signal is obtained.

6. Wash in three changes in PBS (or Tris saline) for 5 min. The specimen is now ready for the detection step.

Comment

If background problems are seen, the non-specific binding can be inhibited by preincubating the specimen with protein (i.e., after Step 1). Commonly used proteins are BSA at 3%, fetal bovine serum at 10%, 10% dry milk, or purified antibodies (used at 1%) from the same species as the detection reagent. The blocking protein can also be added to the antibody preparation (i.e., at Step 2).

Procedure 30: Detection of Horseradish Perioxidase-Labeled Reagents Using DAB[80,81]

A range of substrates for peroxidase are useful, including diminobenzidine (DAB), chloronaphthol, and aminoethyl-carbazole. DAB is the most commonly used substrate and one of the most sensitive for horseradish perioxidase. It yields an intense brown product that is stable in both water and alcohol. The method is good for both light microscopy and TEM.

1. Dissolve 6 mg of DAB in 10 ml of 0.05 M tris buffer (pH 7.6).

2. Add 0.1 ml 3% H_2O_2. H_2O_2 generally is supplied as a 30% aqueous solution and should be stored at 4°C at which it will last for about 1 month. Mix well and if a precipitate appears, filter through Whatman No.1 filter paper.

3. Add the solution to the specimen, incubate for 3 to 15 min at room temperature.

4. Rinse with distilled water.

5. Counterstains, if needed, and dehydrate.

6. Mount in DPX.

Procedure 31: Detection of Horse Radish Perioxidase-Labeled Reagents Using DAB-metal[80]

The DAB substrate for HRP can be made more sensitive by adding metal salts such as cobalt or nickel to the substrate solution. The reaction product is slate gray to black, and the products are stable in both water and alcohol.

1. Dissolve 6 mg of DAB in 9 ml of 0.05 M tris buffer (pH 7.6).

2. Add 1 ml 0.3% w/v aqueous stock solution of nickel or cobalt chloride.

---

* Useful secondary agents are anti-immunoglobin antibodies, protein A, or protein G. The secondary reagent can be labeled with enzymes, flurochromes, or colloidal gold. Carry out all dilutions in solutions such as 3% BSA/PBS or 1% immunoglobulin/PBS.

3. Add 0.1 ml of a 3% aqueous solution of H_2O_2 in H_2O. Mix well and if a precipitate appears, filter through Whatman No.1 filter paper.

4. Apply to specimen, incubate for 3 to 15 min at room temperature.

5. Rinse with distilled water, counterstain if desired, and dehydrate.

6. Mount in DPX.

Procedure 32: Detection of Horseradish Peroxidase-Labeled Reagents Using Chloronaphthol[80,81]

Chloronaphthol gives a blue–black end product. It is less sensitive than DAB, end products tend to diffuse from the site of precipitation, and the products are soluble in alcohol and other organic solvents.

1. Prepare a stock solution of chloronaphthol by dissolving 0.3 g of chloronaphthol in 10 ml of absolute ethanol and store it at -20°C.

2. Add 100 ul of chloronaphthol stock solution with stirring to 10 ml of 0.05 M tris buffer (pH 7.6).

3. Add 0.1 ml 3% aqueous solution of H_2O_2. Mix well and if a white precipitate forms, remove it by filtering through Whatman No.1 filter paper.

4. Incubate the specimen for 10 to 40 min at room temperature.

5. Wash well. Do not use any counterstain prepared in organic solvents and do not dehydrate using the same.

6. Mount in Gelvatol or Mowiol.

Procedure 33: Detection of Horseradish Peroxidase-Labeled Reagents Using Aminoethylcarbazole[80,81]

3-Amino-9-ethyl carbozole (AEC) yields a rose-red product upon oxidation. It is less sensitive than DAB, but can be used if the DAB reaction gives undesirable background. The products are soluble in alcohol, but not in water.

1. Dissolve 4 mg of AEC in 1 ml of n,n-dimethylformamide (DMF).

2. Add 1.0 ml of the AEC solution to 10 to 15 ml of 0.1 M sodium acetate buffer (pH 5.2) with stirring.

All other steps as in previous procedure.

Note: Rose-red color fades if oxidized, therefore, take special care to avoid air bubbles on mounting.

Procedure 34: Detection of Alkaline Phosphatase-Labeled Reagents Using Naphthol-AS-BI- Phosphate/New Fuchsin (NABP/NF) [80,81]

Naphthol-AS-BI-phosphate/new fuchsin (NABP/NF) produces an intense red end product that is stable in alcohols as well as in aqueous solutions.

1. Dissolve 1 mg of new fuchsin in 0.2 ml of 2 N HCl.

2. Dissolve 1 mg of sodium nitrate in 0.25 ml of H_2O.

3. Dissolve 1 mg of naphthol AS-BI phosphate (sodium salt) in 0.2 ml of dimethylformamide.

4. Mix the new fuchsin and sodium nitrate solutions, shake for 1 min, and add the mixture to 40 ml of 0.2 M tris (pH 9.0).

5. Now add the napthol AS-BI solution to the above mixture.

6. Incubate the tissue for 10 to 40 min at room temperature in this mixture and then wash with 20 mM EDTA.

7. Mount in DPX or aqueous mountant.

Procedure 35: Detection of Alkaline Phosphatase-Labeled Reagents Using Bromochloroindolyl/nitro Blue Tetrazolium (BCIP/NBT)[80,81]

BICP/NBT substrate produces an intense black–purple product at the site of enzyme binding. The method is a highly sensitive one.

1. Prepare the following three stock solutions:
    (a) NBT: Dissolve 0.5 g of NBT in 10 ml of 70% dimethylformamide.
    (b) BCIP: Dissolve 0.5 g of BCIP (disodium salt) in 10 ml of 100% dimethylformamide.
    (c) Alkaline phosphatase buffer: 100 mM NaCl, 5 mM $MgCl_2$, 100 mM tris (pH 9.5)
    All stocks are stable at 4°C for at least 1 year.

2. Just before use, add 66 µl of NBT stock to 10 ml of alkaline phosphatase buffer. Mix well and add 33 µl of BCIP stock. Use within 1 h.

3. Add enough substrate solution to cover the tissue at room temperature for about 30 min. Gently agitate.

4. Wash tissue in PBS containing 20 mM EDTA.

5. Counterstain if necessary.

6. Wash in water and mount in any suitable mountant.

Procedure 36: Detection of β-Galactosidase-Labeled Reagents Using Bromochloroindolyl-β-D-Galactopyranoside (BCIG)[80,81]

Only recently β-galactosidase has become popular in immunocytochemistry. The substrate BCIG gives an intense blue product. The product is stable and insoluble in alcohol as well as in water.

1. Dissolve 4.9 mg of BCIG in 0.1 ml of dimethylformamide.*

2. Add the 0.1 ml of the BCIG solution to 10 ml of PBS containing 1 mM $MgCl_2$ and 3 mM potassium ferrocyanide.

---

* Boenisch[81] suggested dissolving 10 mg of BCIG in 0.5 ml of dimethyl formamide; 7 ml PBS containing 1% mM $MgCl_2$ to be added to 0.5 ml of 50 mM potassium ferricyanide and 0.5 ml of 50 mM potassium ferrocyanide. To prepare working solution, mix 0.05 ml of first with 2.276 ml of second. The mixture must be stored at least for 8 weeks at -18°C.

3.   Filter through Whatman No.1 filter paper.

4.   Apply the working solution ready at Step 2 to the specimen, incubate for 10 to 40 min at room temperature, and then wash in water.

5.   Mount in a suitable mountant.

Procedure 37: Detecting Gold-Labeled Reagents by Silver Enhancement[80, 82-84]

1.   Briefly rinse the specimen in distilled water. The samples can be examined directly or can be processed for silver enhancement. For direct observation, mount in Gelvatol or Mowiol.

2.   For silver enhancement, transfer the sample to a dark room equipped with a safe light. To prepare the developer, mix two parts of 0.5 M sodium citrate (pH 3.5) with three parts of 5.6% hydroxyquinone and 12 parts of water. Immediately before use, add three parts of 0.73% silver lactate.

3.   Immerse the slides with the gold-labeled specimens into this developer. Incubate for 2 to 3 min at room temperature.

4.   Rinse briefly in 1% acetic acid and fix in a standard photographic fixative for several minutes.

5.   Counterstain if necessary.

6.   Samples are now ready for mounting in DPX.

Comment

Recently, several companies have introduced silver enhancement kits that do not require using a dark room, permitting development to be moinitored under the microscope.

Procedure 38: Preparation of Anti-Immunoglobulin Antibodies[85]

It must be decided first as to the host in which anti-immunoglobulin antibodies will have to be raised. Generally the animal must be a distant species. Larger animals should be used, if possible, as they yield a greater amount of serum. The suitable animals are rabbits, goats, or sheep. For protein A only rabbits are suitable; they are also commonly used to prepare anti-mouse, -rat and -human immunoglobulin antibodies; goat, sheep, or pigs are used to produce anti-rabbit immunoglobulin antibodies.

Anti-immunogloblin antibodies are available commercially, and can be preferred unless large quantities are needed or special antibody couplings are planned. However, the preparation protocol is given in what follows:

1.   Prepare a large batch of antibodies. Purify them. Mix them at approximately 2 mg/ml in PBS.

2.   At the time of each injection, make an emulsion at 500 µl of the above solution with 500 ml of Freund's adjuvant. The first injection should be done with complete Freund's adjuvant and subsequent ones with incomplete Freund's adjuvant.

3. Inject subcutaneously at several places the emulsion in rabbit; for most purposes, several rabbits should be injected at the same time. Allow the rabbits to rest for 6 weeks after every injection.

4. Take sample test bleeds from these rabbits 14 d after each injection.

5. When the antibody concentration has reached the usable level, each rabbit should be bled to obtain approximately 50 ml of blood. Continue to repeat the injections while collecting the serum 14 d after each injection. Rabbits can continue to yield good anti-imunoglobulin antibodies for about a year. When the last injection has been made, sacrifice the rabbit and collect all the blood.

Procedure 39: Preparation of Fixatives for Immunocytochemical Methods[80]

A. 4% Paraformaldehyde

Paraformaldehyde	8 g
Distilled water	100 ml

Heat to 60°C in a fume cupboard. Add a few drops of 1 N NaOH to help dissolve. Allow the solution to cool to room temperature. Add 100 ml of 2 × PBS. Prepare fresh every time. Incubate material for 10 min at room temperature.

B. 1% Glutaraldehyde

1% (EM grade) glutaraldehyde in PBS. Incubate material for 1 h at room temperature.

Simple fixation with the above two does not allow access of the antibody to the specimen and, therefore, is followed by a permeabilization step using an organic solvent.

Permeabilization is done as follows:

0.2% Triton X-100 in PBS for 2 to 15 min at room temperature depending on the antigens, or

50% Methanol for 2 min at room temperature, or

50% Acetone for 30 sec at room temperature.

Procedure 40: Preparation of Mounting Media

(i) Gelvatol or Mowiol for Mounting Immuno-Stained Preparations[80,86,87]

1. Add 2.4 g of Gelvatol 20-30 (Monsanto chemicals) or Mowiol 4-88 (Hoechst) to 6 g of glycerol; mix and then add 6 ml of water and leave for at least 10 h at room temperature. Add 10 to 12 ml of 0.2 M tris (pH 8.5) and heat to 50°C for 10 to 15 min with occasional stirring. Spin at 5000 g for 15 to 20 min. For fluorescence detection,

add 1,4-diazobicyclo-[2,2,2]-octone (DABCO) to 2 to 5% to reduce fading. Store in air-tight containers at -20°C for 2 to 3 d. The mountant is stable at room temperature for several weeks after cold treatment.

2. Add a small drop of mounting media to the specimen and put on a coverslip. The medium will set within 10 h.

(ii) Polyvinyl Mountants

Polyvinyl alcohol 20 g + glycerol 3 g + water 45 ml + paraphenylene diamine to a final concentration of 0.02 to 0.2% + n-propyl gallate at 0.6%. This combination is good for fluorescence detection.

Procedure 41: Phosphate–Buffered Saline (PBS)–Preparation of Stock Solution

Dissove 8.0 g of NaCl, 0.29 g of KCl, 1.44 g of Na_2HPO_4, and 0.24 g of KH_2PO_4 in 800 ml distilled H_2O. Adjust pH to 7.2. Adjust the volume to 1 L. Dispense in convenient volumes and sterilize by autoclaving. Store at room temperature.

Procedure 42: Tris-Buffered Saline (TBS) (25 mM Tris)

Dissolve 8.0 g of NaCl, 0.29 g of KCl , and 3 g of tris base in 800 ml of distilled water. Adjust the pH to 8.0 with 1 N HCl. Adjust the volume to 1 L. Dispense in convenient volumes and sterilize by autoclaving. Store at room temperature. (In Professor Staehelin's lab [University of Colorado, Boulder] it is stored in a cold room.)

Procedure 43: Determination of the Specificity of Antibodies Raised Against Specific Cell Wall Components

Before using an antibody raised against any cell wall component, ascertain its specificity. Examine the specificity by immunodiffusion assays such as double-diffusion, the quantitative precipitation reaction, the complement fixation assay, the hapten test, immunoelectrophoresis, and immuno blot analysis. For protein antigens use western blot analysis. Details of these procedures are found in standard books.

6.10 Application of Immunocytochemistry to the Study of Cell Wall Components

Antibodies raised against specific cell wall polysaccharides were first used to study cell wall structure in brown algae. Vreeland and co-workers have produced both polyclonal[58,88] and monoclonal[89] antibodies to carbohydrates extracted from the brown alga, *Fucus distichus*. Subsequently polyclonal/monoclonal antibodies have been raised against several cell wall components isolated from different plant sources. Most of these antibodies have been purified, characterized, and used for various immunological purposes. However, not all of them have been utilized for immunocytochemical purposes. In the pages that follow, details are provided on the antibodies thus far raised against specific cell wall components and the labeling procedure, wherever followed in immunocytochemistry.

6.10.1 Immunocytochemical Localization of Cell Wall Polysaccharidis

Localization of Polysaccharides Through Antibodies Against α-L-Arabinose and D-Arabinose[90]

Applications

These antibodies have been used to locate arabionogalactans in dicotyledons and arabinoxylans in grasses. The antibodies have been shown to bind to the primary cell wall and middle lamella of soybean cotyledon cells by an indirect immunofluorescence. Similarly, arabinoxylans have been localized in the primary cell walls of rice endosperm. In the pollen tubes of *Nicotiana alata*, α-L-arabinofuranose-containing polysaccharides were immunolocalized at the outer fibrillar wall layer.

Localization of Polysaccharides by Antibodies Specific to Particular Oligosaccharide Linkages[91,92]

Antibodies Raised

Oligosaccharides such as (β-1→Xy1)3-8 and (β-1→3Gle)3,8 were prepared from polysaccharides and separated by chromatography. The sugars were coupled to BSA and the antisera, raised in rabbits, were purified on an affinity column of BSA-sepharose. The sera were characterized by ELISA and complement fixation tests.

Tissue Preparation

Thin sections suitable for EM observation were prepared as per conventional methods (fixation in 2% glutaraldehyde in 0.5 M cacodylate buffer, pH 7.2). Immunolabeling with primary antibody was followed by a secondary goat-antirabbit antibody conjugated to colloidal gold (15 nm).

Application

Immunocytochemical localization studies employing the above antibodies were carried out on intact *Phaseolus vulgaris* plant cells and on synchronously differentiating

Zinnia elegans mesophyll cells. Based on these studies, it was found that arabinogalactans could be detected in the cell plate of dividing bean cells, although they were absent in the secondary walls of bean cells as well as in the primary walls of suspension cultured *Zinnia* cells. Xylans were not present in the cell plate of dividing root cells. $\beta 1 \rightarrow 4$ xylose polymers were located in small amounts in the primary walls but they occurred massively in secondary thickenings (in differentiating tracheary cells of cultured *Zinnia* mesophyll). $\beta, 1 \rightarrow 3$ oligoglucans (callose) were dispersed in the cell plate and the very young primary walls of bean roots; in growing cell walls, callose was specifically located in the plasmodesmata. During cell plate formation, callose was dispersed over the whole cell plate area.[91,92]

Localization of β-1→3 Glucans (Callose)[93,92-98]

Antibodies Raised

Antibodies against β-1→3 glucans have been prepared variously employing different animals by the researchers mentioned above. Meikle *et al.*[93] have raised monoclonal antibodies in mouse using a laminarin (algal β 1→3 glucans)–haemocyanin conjugate. Laminarins were conjugated to BSA and haemocyanin by reductive amination.[99] A commercially available monoclonal antibody raised against β 1→3 glucans and supplied by Biosupplies, Australia was used by Vennigerholz et al.[98] Horiberger and his school[94,95] used antibodies raised in rabbits against a Laminaribiose–edestin conjugate. Northcote et al.[92] immunized rabbits with a β 1→3 oligoglucoside–BSA conjugate. Kishida et al.[97] prepared and characterized a polyclonal antibody to the branched 1→3 β-D-glucans (mycolaminarin?) of *Volvariella volvacea*, an edible mushroom.

Tissue Preparation and Labeling

Meikle et al.[93] fixed the materials in a mixture of 2% formaldehyde, freshly prepared from paraformaldehyde, and 2% glutaraldehyde in tris buffer 0.09 M, pH 7.2, for 4 h at room temperature. The materials were subsequently washed in buffer and post-fixed in 1% OsO_4 followed by several washes in buffer. Dehydration was done in 10 to 100% ethanol series. Embedding was made in Spurr's resin. Vennigerholz et al.[98] used the same fixatives (prepared in phosphate buffer), but mixed them prior to fixation. Post-fixation was also similar but embedding was carried out in LR White.

Meikle et al.[93] used a rabbit-antimouse-secondary antibody coupled to colloidal gold (15 nm) and also employed uranyl acetate and lead citrate as counter stains. Horisberger and Rouvet-Vauthey[95] and Horisberger et al.[94] used gold labeled protein A.

Omission of first antibody or preincubation of antibody with any β-1→3 glucan (about 100 μg/ml) can be used as control procedures.

Applications

Meikle et al.[93] employed immunocytochemical localization studies on the pollen tubes of *Nicotiana alata* and found that only the inner wall of the pollen tube is callose-positive. A similar conclusion was reached with reference to the compatible and incompatible pollen tubes of *Brugamansia suaveolens* growing in the stylar tissue by Vennigerholz et al.[97] and Geitmann et al.[100] Upon incompatible pollination, the callosic wall becomes much more thickened.[100] The results of the studies of Northcote et al.[92] are already detailed on Page 269. In a very interesting study, Turner et al.[101] obtained clear maize cell wall preparation that contained embedded plasmodesmata and demonstrated the presence of callose in the wall and the plasmodesmata through callose antibodies conjugated to colloidal gold; they also showed that callose was not a constituent of the collar region of the plasmodesmata.

Localization of Pectins[4,98,102-112]

Antibodies Raised

1. JIM 5: It is a monoclonal antibody raised in rats, very specific against acidic unesterified or very low esterified pectin. It does not recognize (i.e., react with) the branched RGI. The myeloma line was IR 983F and the antigen used was apple pectin. The monoclonal antibody has been purified, characterized, and specificity tested.[109]

2. JIM 7: It is a monoclonal antibody specific for methylesterified pectins with esterification ranging from 35 to 90%. It does not react with RGI. The antibody was raised in rats. The myeloma used was IR 983F.

3. Liners and Van Cutsem[106] raised the antibody against homopolygalacturonic acid which was coupled to methylated BSA to confer immunity. Although both rabbits and Balb/C mice were injected with the antigen, monoclonal antibodies (2F4) were obtained only in mice; they were purified, characterized, and specificity tested. The antibodies were specific to supramolecular configuration of homopolygalacturonic acid side chains associated through calcium cations, probably according to the egg-box model. The antibodies recognize oligomers of DP>9 and PGA with a degree of methyl esterification of 30% (random) or 40% (block-wise).

4. Polyclonal antibodies were raised against PGA/RGI which were conjugated to methylated BSA in rabbits. These antibodies recognize the long repeating regions of PGA.[4,76,103]

5. The Complex Carbohydrate Research Center (CCRC) at the University of Georgia, Athens, has developed two types of monoclonal antibodies against RG-I in mice, called CCRC-M2 and CCRC-M7. The first one is specific for RG-I but does not bind to PGA; the second recognizes an arabinosyl-containing epitope on RGI, which suggests that it recognizes a side chain of RGI but the precise structure of the epitope has not yet been determined.

Tissue Preparation and Labeling

Tissue is fixed in 2 to 2.5% glutaraldehude in 0.05 M Na-phosphate buffer, pH 7.0 or 0.1 M sodium cacodylate buffer, pH 7.4, at room temperature for 2 h. Tissues are then washed in buffer, dehydrated in a graded ethanol series, and embedded in LR white at low temperature or at 50°C or in a mixture of epon-araldite epoxy resin. Vian and Roland[113] optionally post-fixed the tissue in 1% OsO_4. For light microscopy, sections of 6 to 10 μm thickness were used. Sections for EM were collected on gold/nickel grids. Bradley et al.[104] suggested a procedure in which the tissues were treated at low temperature. Sections obtained through cryomicrotome have also been used, but essentially for fluorescence microscopy.

Thin sections of embedded resin materials meant for LM were resin de-embedded prior to labeling. Blocking, to prevent nonspecific labeling, was done for a period of about 20 to 30 min in normal serum (of the animal in which the antibody was raised) diluted 1:30 in 0.01 to 0.05 M tris-HCl buffer (and 500 mM NaCl was optional), at pH 7.5, containing 0.2% BSA in TBS. This was followed by incubation in primary antibody diluted up to 1/5. Time of treatment can vary depending on material. Subsequently the materials were washed in TBS, distilled water, or PBST and then incubated in secondary antibody (goat-antirabbit, goat-antirat, goat-antimouse), conjugated to colloidal gold (15 nm) with 0.1% BSA (dilution 1:20 in 0.02 M Tris-HCl, pH 8.2) for 1 h in the dark at room temperature; the secondary antibody in some cases was conjugated to fluorescein isothiocyanate (FITC), Texas red, HRP,[106,107] or alkaline phosphatase (work in our laboratory). Post-staining was followed in uranyl acetate and lead citrate.[109] For colloidal gold complex to be observed under LM, silver enhancement procedure with the Inten SE II silver developer kit supplied by Janssen, UK was used;[113] in these cases the sections were also optionally counterstained with Azure II and methylene blue in sodium metaborate.

Fluorescence labeled preparations were mounted in any non-fluorescent mountant and observed under UV light.

Control Procedures

1. Treatment of materials with antibodies preabsorbed with pectin or polypectate or Na-polygalacturonate (1 mg/ml) 12 h before incubation.[4]
2. Omission of primary antibody or its substitution with an irrelavent antibody MAC 83.[109]
3. Pretreatment of sections with sodium metaperiodate (for anti RGI/PGA work).

Applications

McCann et al.[105] studied pectic polymer orientations in elongating cell walls in cultured carrot cells. JIM 5 labels middle lamella of non-elongating cells in patches but strongly labels throughout in elongate cells. Both round and elongate cells get labeled weakly with JIM 7. In normal parenchyma of carrot, JIM 5 labels the middle lamella and JIM 7 strongly labels throughout the cell wall.

Vennigerholz et al.[98] and Geitmann et al.[100] have used JIM 5 and 7 to study the nature of pectins in incompatible and compatible pollen tubes growing in the style of *Brugmansia suaveolens* L. (Solanaceae). Both the pectins were present in the outer fibrillar layer of the wall. In incompatible crosses, pectinaceous particles seemingly get stuck in the callosic wall of pollen tubes and are comparatively less methyl esterified than the pectins of the compatible tubes.

Vian and Roland[113] used this technique to study pectins in cultured cells of melon and in elongating hypocotyl cells of mung bean to compare pectins in normal and elongating cells.

Knox et al.[110] used this technique to detect the degree of pectin esterification in the different root tissues of carrot, oat, maize, spinach, and sugar beet. They found that there was a spatial variation in the degree of pectin esterification.

Bonfante-Fasolo et al.[102,105] used this method to localize pectic components in the host cell walls of leek roots infected with VAM fungi and to compare the distribution of pectic substances in the host cell wall and in the material occurring at the contact zone between fungi and plant, i.e., the interfacial material laid down around the VAM fungus. This material shared some molecules with the host cell wall, demonstrating that it is of host origin. In the host, the epidermal cells gave only weak labeling and the intercellular spaces of hypodermis and cortex were moderately labeled. Van den Bosch et al.[109] studied *Pisum sativum* root nodules caused by *Rhizobium* sp., soybean nodules by *Brachyrhizobium* sp, and *Phaseolus vulgaris* nodules by *Rhizobium* sp., for the distribution of pectin.

Using this technique, Van Custem et al.[112] did EM immunocytochemical studies for the distribution of pectin on suspension culture cells of carrot and showed that pectin was secreted under a largely methylesterified form. PGA was essentially located on the middle lamella material expanded at the three-way junction between cells or lining intercellular spaces. However, primary walls were not recognized. All layers of the cell walls were labeled after on-grid enzymatic deesterification. The primary walls and middle lamellae of dead cells were heavily labeled indicating that pectin methyl esterases were released by the dead cells. These observations suggest that acidic sequences of pectins do not contribute to the wall stability.

Liners and Van Cutsem[107] reported that pectic polysaccharides underwent a decrease of their methylester content as the growth and maturation of plant cells proceeded. Changes in the degree of methyl esterification of pectins are known to modify the gelling properties of these polymers and, therefore, could be one of the mechanisms by which cells control their wall plasticity during growth. They have studied carrot root apex. In root cap and apical meristem, pectins of low methylester content on the middle lamella expansions at the intercellular spaces and three-way junctions were observed. There was no labeling in walls away from the three-way junctions (i.e., pectin here are largely methyl esterified). Mucilage of both root cap and epidermal cells contained both acidic (only on surface layers) and esterified pectins (throughout). A similar pattern of pectic acid distribution was observed in more differentiated regions—3 mm away from root tip. This is at variance with results obtained by Knox et al.[110]

Staehelin et al.[114] and Moore et al.[103] used RG1 antibody which was found to exclusively label the intercellular layer of sycamore cells, especially at cell corners. Staehelin et al.[114] used this to localize RG1 in the middle lamella of cells (cell corners) of clover root tip and carrot tap root.

Lynch and Staehelin[76] studied the expanding and dividing cells of *Trifolium pratense* root tips using RGI antibodies. RGI is synthesised in golgi only during cell expansion but not during cell division. These epitopes were detected solely in the expanded middle lamella of cortical cell corners, even after pretreatment of sections with pectin methyl esterase to uncover masked epitopes. However, it co-localizes with XG in the epidermal cell walls.

Localization of Alginic Acid[58,88,89,115]

Antibodies Raised

Both polyclonal[58,88] and monoclonal[89,115] have been raised in rabbits and mice, respectively. Alginate was extracted in pure form from the brown alga *Fucus distichus* and was conjugated to methylated BSA. The conjugate was then injected into the test animals. The specificity of the polyclonals was determined by double diffusion tests in agarose buffered with PBS, pH 7.4.

Tissue Preparation and Labeling

Tissues were fixed with acrolein and embedded in a glycol methacrylate-PEG mixture (which does not fluoresce). Later on, Karnovsky's reagent was found to be a better fixative than acrolein. Addition of cetavlon or cetylpuridinium chloride to the fixative prevented mucilage loss. The embedment did not affect antigenicity and allowed labeled antibodies as well as plastic to penetrate the tissue.

Sections of 1 μm thickness were incubated for 30 to 60 min in primary antibody (with normal rabbit serum as controls). The secondary antibody (goat-antirabbit) was conjugated to fluorescein. Alginates showed a green fluorescence.

Alternately, the secondary antibody was conjugated to peroxidase and then the peroxidase was localized through the method of Graham and Karnovsky.[116]

Localization of Xyloglucan [1,4,76,103,114,117-120]

Antibodies Raised

The first antibody to be raised against xyloglucans was the one produced by Moore et al.[103] The purified xyloglucans from cultured sycamore maple cells were conjugated to ovalbumin and injected into rabbits to raise ployclonal antibodies.

Antibodies against xyloglucan oligosaccharides were raised by Sone et al.[119,120] These oligosaccharides were obtained from the xyloglucans of tamarind. Initially antibodies were raised against the β-isoprimeverose, the disaccharide unit of xyloglucan (XG2) in rabbits,[120] but subsequently also against the heptasaccharide unit (XG7)[119] and octoaccharide unit (XG8) (see Hoson[117]) in rabbits.

Puhlmann et al.[118] were able to produce monoclonal antibodies against xyloglucans in mice. These antibodies recognize the terminal fucosyl residue of the trisaccharide chain of XG in the context of an extended XG chain but not in the context of an isolated monosaccharide fragment.

Tissue Preparation and Labeling

Tissues were fixed in 2 to 2.5% glutaraldehyde in 10 mM Na-phosphate buffer, pH 7.2 for 15 h at 4°C or for 2 h at room temperature. In some studies the specimens were fixed in 0.2% paraformalde-hyde plus 0.1% glutaraldehyde in 100 mM sodium phosphate buffer, pH 7.2 (vacuum infiltration for 10 min and then at room temperature for 2 h). Post-fixation was done in 1% OsO_4 in 10 to 20 mM Na or K-phosphate, pH 6.9 to 7.2 for 1 h in ice or at room temperature. Dehydration was done in ethanol series and embedment in LR white. Ultrathin sections for EM were picked on formvar/carbon-coated nickel grids and treated with 0.1 N HCl for 10 min to remove glutaraldehyde from the sections and to expose the antigenic sites. Semi-thin sections were used for LM and FM.

Blocking, to avoid non-specific labeling, was done through incubation of tissues in 5% non- or low-fat milk powder in PBST, in 10 mg/ml BSA solution,[103] or in 10% FCS in low salt PBST (Gibco, Richmond, VA).[76] This is followed by treatment in primary antibody diluted 1:4 to 1:20 in PBST for 15 min to 1 h depending on concentration of antibodies used. Protein A-gold (10 nm), donkey antirabbit antibody coupled to Texas red.[76] Goat antirabbit antibody conjugated to alkaline phosphatase were all used to localize the primary antibody. Fluorescence microscope preparations were mounted in antifade mountant and examined in 590 nm.

Antiserum preabsorbed against its antigen was often used as a control.

Applications

Moore et al.[103] worked on suspension-cultured sycamore cells and found, using labeling experiments, that XG was distributed throughout the primary cell wall. Using the same antibody, Staehelin et al.[121] showed that XG was synthesized and processed in golgi (even during cell division and cell elongation) and was transported to the primary wall; they also showed that XG was associated intimately with cellulose in several plant systems such as root tips of *Trifolium pratense*, *Allium cepa*, and *Daucus carota*, leaf tissues of *Phaseolus vulgaris*, and suspension cultured cells of carrot. Moore and Staehelin[4] localized XG in middle lamella but its occurrence there may be a special event only in cultured cells. Lynch and Staehelin[76] successfully used this labeling technique to show that in clover root tips, the XG epitopes were present in cortical cells at a threefold greater density in the newly formed cross walls than in older longitudinal walls.

Sone et al.,[119,120] using the antibodies raised against xyloglucan oligosaccharides, demonstrated that in Azuki bean epicotyls XG was found almost evenly in the walls of epidermal, parenchymtous, and vascular bundle cells; it was also demonstrated by the same group that xyloglucan was restricted to the primary wall and that it was absent from the middle lamella in soybean cotyledons.

Localization of (1→3), (1→4)-β-D-Glucans[122,123]

Antibodies Raised

Pure (1→3), (1→4)-β-D-glucans did not need a protein coupling for tiggering antibody production.

Application

This hemicellulose is characteristic of grass primary cell walls. Hoson and Nevins,[122,123] using the antibodies, found that this hemicellulose was found in the cell walls of epidermis, parenchyma, and vascular bundles in maize. Distinct antibody binding was observed only along the inner layers of the cell wall.

6.10.2 Immunocytochemical Localization of Cell Wall Proteins and Glycoproteins

6.10.2.1 Enzymes

Localization of Cellulases[124-129]

Antibodies Raised

Byrne et al.[125] were the first to raise antibodies against two cellulases obtained from auxin-treated pea epicotyls. Sexton and his colleagues[124,128,129] have raised antibodies against an iosoenzyme 9.5 (with a PI of 9.5) of cellulase. Daniel and Johansen[127] purified EXO-(CBH-I) glucanase and endo (E-I) glucanase from the fungus *Phaenerochaete chrysosporum* and raised polyclonal antibodies against both of them in rabbits.

Tissue Preparation and Labeling

Tissues were fixed in freshly prepared 3.5% formaldehyde in 30 mM sodium cacodylate buffer, pH 7.2 for 3 h at 20°C.[129] They were then frozen onto cryostat blocks using solid CO^2-ethanol slurry. Sections were taken at a thickness of 30 μm at -18°C and immediately thawed to PBS. A treatment in glycine for 1 h was often followed to saturate free aldehyde groups.

Daniel and Johansen[127] followed a post-embedding immunolabeling procedure using colloidal gold or peroxidase conjugated secondary antibodies (goat-antirabbit). Blocking was done in undiluted goat serum to reduce non-specific binding. Peroxidase was subsequently stained as per the procedure of Graham and Karnovsky.[116] Preimmune serum treatment was the control procedure employed.

Applications

Sexton and his colleagues used the antibodies to study the distribution pattern, location, and timing of cellulases during leaf abscission in the abscission zone of bean leaf; they located the enzyme in the cortical separation layer, which was 2 or 3 cells wide.

Both exo- and endo-glucanases were reported in the cells of wood tissues infected by *Phaenero-chaete chyrysosporum.*[127]

Localization of 1, 3-β-D-Glucanase (Callase)[1,4,130-134]

Antibodies Raised

Vogeli et al.[134] produced antibodies against callase isolated from bean leaf and these were employed in the studies of Staehelin and his colleagues.[1,4,132] Polyclonal antibodies against callase isolated and purified from soybean seedlings were produced by Takeuchi et al.[133] Benhamou et al.[130] developed antibodies in rabbits immunized with the enzyme obtained from tobacco.

Tissue Preparation and Labeling

Tissues were fixed in 2 to 3% glutaraldehyde buffered with 0.1 M sodium cacodylate buffer, pH 7.2 for 2 h at 4°C. Dehydration was in graded ethanol and embedding in Epon[130] or LR white.[132] Semithin (for LM) and ultrathin (for EM) sections were used. Non-specific staining was prevented by subjecting the tissue to BSA-gelatin (in PBS) treatment[132] or in normal goat serum (in PBS-ovalbumin mixture).[130] After treatment in primary antibody, the tissues were subjected to secondary antibody (goat-antirabbit) conjugated to colloidal gold (10 nm)[130] or to protein A–colloidal gold complex. Post-staining was carried out in uranyl acetate and lead citrate. The control treatments included incubation with pre-immune serum and omission of primary antibody treatment.

Applications

Staehelin and his school have proved that β-glucanase (along with chitinase or independently of it) is an enzyme associated with stress in plants. It can be made to appear by subjecting the plants to various stresses like infection, ethylene treatment, drought, etc. In ethylene-treated bean leaves the enzyme appeared in the middle lamella.[132]

Benhamou et al.[130] found that this enzyme could be induced in wilt-fungus (*Fusarium* sp.) infected tomato and eggplants; they observed that this accumulated predominantly in host primary cell walls and secondary thickings of xylem vessel elements, while only a low amount was found in the middle lamella. They also showed the presence of the enzyme on the surface of the fungus.

Localization of Chitinase[131,132,134-137]

Antibodies Raised

Polyclonal antibodies were raised in rabbit against purified chitinase from cultured tobacco cells.[137] Vogeli et al.[134] also produced polyclonal antibodies against purified chitinase in rabbits.

Tissue Preparation and Labeling

Tissues were fixed in 2 to 3% glutaraldehyde, post-fixed in OsO_4, and embedded in LR white. Ultrathin sections were taken and mounted in formvar/carbon coated nickel grids.

Non-specific labeling was prevented by blocking the tissues for 15 min in BSA-gelatin (in PBS). After primary antibody treatment at room temperature, the tissues were incubated in protein A–colloidal gold complex. Post-staining was done in uranyl acetate and lead citrate for 5 sec.

Application

Mauch et al.[132] demonstrated that this enzyme could be induced to appear in plants subjected to stress conditions (such as enthylene treatment). Benhamou et al.[136] showed that chitinase accumulated around fungal hyphae in the tomato plants infected with wilt fungus (*Fusarium* sp.). The enzyme was often associated with altered fungal wall structure.

Localization of (1→3) (1→4)-β-D-Glucanase

Antibodies Raised[123,138-141]

Polyclonal antibodies were raised against the enzyme purified from barley aleurone layer.[140,141] Nevins and co-workers developed the polyclonal antibodies against the enzyme endo- and exo-(1,3) (1,4)-β-glucanases purified from maize coleoptile as well as from *Avena* caryopsis.[138,139] Immunocyto-chemical localization work has not been carried out thus far, in spite of raising antibodies.

Localization of (1→3)-β-Glucan Synthase

Antibodies Raised[142]

Antibodies have been raised against this enzyme isolated from cultured soybean cells.[142] No immunocytochemical localization studies have been carried out thus far.

Localization of Polygalacturonase

Antibodies Raised[143-145]

Antibodies against the enzyme purified from tomato fruits were raised.[143-145] No immunocytochemical localization studies have been carried out thus far.

Localization of β-Fructosidase

Antibodies Raised[146,147]

Iki et al.[146] have isolated β-fructosidase from ripening tomato fruit and have raised antibodies against it. Lauriere et al.[147] produced antibodies against this enzyme (xylose specific) purified from cultured carrot cell walls. No immunocytochemical localization studies have been carried out thus far.

Localization of β-Glucosidase[148,149]

Antibodies Raised

The purified enzyme isolated from the cell walls of spruce seedling hypocotyls was injected into rabbits and polyclonal antibodies were developed.

Tissue Preparation and Labeling

Tissues were sectioned in a cryostat, and the sections were subjected to PBS, pH 7.2 for 10 min. They were then incubated into the primary antibody in PBS for 25 min in a moist chamber at room temperature. They were washed for 10 min in PBS and treated with FITC-conjugated goat anti-rabbit antibody in PBS for 20 min and then washed twice for 5 min each in PBS. The control treatments included incubation with serum from non-immunized rabbits, exposure to only secondary antibody and incubation in the absence of antisera. Observations were made in a fluorescence microscope at 450 to 490 nm.

Applications

Schmid and Grisebach[148] studied hypocotyls and roots in the seedlings of spruce and found the activity of this enzyme in the inner layer of secondary cell walls of epidermal cells and prospective xylem cells; they showed the importance of this enzyme in coniferin synthesis during cell wall lignification. Burmeister and Hosel[149] studied this enzyme in the seedlings of *Cicer arietinum* (see also References 150 and 151).

Localization of α-Galactosidase

Antibodies Raised[152]

Hankins et al.[152] have raised antibodies against a galactosidase obtained from mungbean seeds. No immunocytochemical studies have been carried out thus far.

Localization of Cutinase[153-155]

Antibodies Raised

Cutinase was highly purified and antibody was raised against it in rabbits. The cutinase was purified from *Colletotrichum gloeosporioides*.[153]

Tissue Preparation and Labeling

The antibodies were conjugated to ferritin and observed under EM.

Applications

Initial studies were directed to find out whether *Fusarium solani-pisi* was capable of producing cutinase on pea stems which were infected by it; positive results were obtained. Dickman et al.[153] showed that *Colletotrichum gloeosporioides* was capable of producing the enzyme when it infected papaya fruits.

Localization of Lysozyme[144,156]

Antibodies Raised

Polyclonal antibodies were raised in rabbits against the lysozyme enzyme (W_1A) purified from wheat germ.[156]

Tissue Preparation and Labeling

Materials were fixed in 3% glutaraldehyde in 0.1 M sodium cacodylate buffer, pH 7.2 and incubated for 2 h at 4°C; they were post-fixed in 1% OSO_4 in sodium cacodylate buffer for 1 h at 4°C. The tissues were embedded in Epon 812 and sectioned; sections were mounted on nickel grids.

The tissues were first treated for 5 min in 1% ovalbumin (in PBS pH 7.4) to minimize non-specific labeling, then treated for 60 min with primary antibody diluted 1:40 in PBS-ovalbumin and then incubated with protein A–gold complex for 30 min in a moist chamber. Post-staining was done with uranyl acetate and lead citrate. Several control porcedures were employed including omission of primary antibody treatment, treatment with preabsorbed serum, or with non-labeled protein or colloidal gold alone.

Applications

Audy et al.[156] demonstrated that lysozyme was active in cell walls of embryo and coleoptile in wheat germ.

Localization Nitrate Reductase[157]

Raising of Antibody

Nitrate reductase was purified and injected into rabbits; the antibodies were purified by immunoaffinity chromatography and their specificities tested by double immunodiffusion.

Tissue Preparation and Labeling

Tissue was fixed in 3% formaldehyde–0.5% glutaraldehyde in phosphate buffer, pH 7.4 for 3 to 4 h and then for 1 h in 0.3% formaldehyde in the same buffer. After treatment with borohydride for 30 min, the tissue was soaked overnight in 0.8 M sucrose in the same buffer as above. The tissue was then frozen by immersion in liquid nitrogen and ultra thin sections were cut at -83°C. The ultrathin sections were mounted on grids, washed for 30 min in PBS, then washed for 10 min in PBS, and then incubated in 2% gelatin. They were again washed. After treatment in primary antibody for 10 min at room temperature, the tissues were subjected to washing in PBS and then to ferritin-labeled goat-antirabbit secondary antibody for 10 min at room temperature. Control tissues were treated in pre-immune serum.

Applications

The enzyme was localized in the cell wall–plasmalemma region.

Localization of Prolyl Hydrolase[158-160]

Antibodies Raised

Polyclonal antibodies have been prepared in rabbits against prolyl hydroxylase isolated from bean cells,[160] *Chlamydomonas*, and carrot cells.[159] There appears to be no cross-reactivity between the algal and higher plant antigens as judged by activity inhibition experiments, immunoblotting, or immunodiffusion techniques. Pretreatment of sections with α-mannosidase reduced considerably the labeling of the cell wall, suggesting that significant proportions of the polyclonal antibodies have the N-glycosidically linked oligosaccharide side chains of the prolyl hydroxylase as an antigenic determinant.[158] A similar speculation as to the antigenically important epitopes of prolyl hydroxylase from bean cells has been made by Bolwell and Dixon.[160]

Tissue Preparation and Labeling

Tissue was fixed in glutaraldehyde and post-fixed in OsO_4. LR white was used for embedding. The sections were demasked with periodate prior to indirect immunogold labeling.

Sections were first treated with primary polyclonal antibody, followed by goat-anti-rabbit secondary antibodies coupled to colloidal gold (10 nm).

Application

Since this is a key enzyme involved in the formation of hyp-proteins of cell walls (proline hydro-xylation) and recognizes the secondary structure of the substrate, the polyproline II-helix, attempts were made in the past to study the enzyme and its involvement in hyp-protein formation in the cell walls of *Chlamydomonas*, bean cells, and suspension cultured carrot cells. Considerable activity of the enzyme was noticed in the cell wall as well as in ER and golgi.

Localization of Peroxidase[127,161-170]

Antibodies Raised

Polyclonal antibodies have been raised against peanut and horseradish peroxidase,[161] cultured peanut cell peroxidase,[162,169,170] abscissic acid inducible anionic peroxidase associated with suberization in potato,[163,164] and the purified peroxidase (lignin peroxidase) isolated from the fungus *Phaenerochaete chrysoporum* (in rabbits).[127] Monoclonal antibodies have been raised in mice against maize seedling coleoptile peroxidase.[168] Nearly 10 monoclonal antibodies were raised against cationic peroxidases of cultured peanut cells.[166,167]

Tissue Preparation and Labeling

Tissues were fixed for 4 h at 4°C in 2 to 3% formaldehyde, paraformaldehyde, or glutaradehyde in phosphate buffer, pH 7-7.5. Successive fixation in 2% formaldehyde and in 0.17% glutaraldehyde (both in phosphate buffers) was also followed.[168] Dehydration in ethanol series and embedding in LR white or Lowicryl were then followed. Semithin to ultrathin sections (the latter for EM) were used; the former were collected and dried on gelatin-coated slides while the latter were picked upon parlodion-coated nickel grids.

Blocking in 1 to 5% nonfat dry milk in phosphate buffer (pH 7.6) at 5°C for 2 h or PBS (for 30 min) followed by primary antibody (in 10 mM phosphate buffer, pH 7.6 containing 0.25% BSA*) treatment for 1 h to overnight at 5°C was given. Secondary antibody (goat-antirabbit) conjugated to fluorescein or rhodamine, in case of polyclonals, was followed to label the primary antibody. These slides were examined in 450 nm excitation and 490 barrier filters (in case of fluorescein label) and 530 nm excitation and 560 nm barrier filters (in case of rhodamine). For EM studies, Kim et al.[168] used secondary antibodies (goat-antimouse) conjugated to colloidal gold (15 nm) (diluted 1:20 in 0.05% TBST for 30 min). Daniel and Johansen[127] also used colloidal gold labeling.

Applications

Peroxidase is the most widely studied of enzymes immunocytochemically. Since several isoenzymes of this enzyme are known to be involved in many activities of the cell wall (during its formation, maturation, and functioning), investigations have used the antibodies for various purposes. Espelie et al.[163] found the active involvement of this enzyme near the wound suberin synthesis in cell walls. Daniel and Johansen[127] have studied lignin peroxidase during wood degradation in birch by the fungus *Phaenerochaete chrysosporum* and found it in wood cell wall layers at all sites of degradation. Its activity was often closely associated with cellulase.

Localization of Amine Oxidase[171]

Antibodies Raised

Antibodies have been raised against amine oxidase in rabbits.

Tissue Preparation and Labeling

The secondary antibody used was goat-antirabbit antibody conjugated to FITC.

---

* Kim et al.[168] used 0.05% PBST instead of BSA.

6.10.2.2 *Structural Proteins of Cell Walls*

Localization of Wheat Germ Agglutin (Lectin)[45]

Antibodies Raised

Polyclonal antibodies were raised in rabbits against wheat germ agglutin (lectin) purified from wheat kernals.

Tissue Preparation and Labeling

Tissues (approximately 0.5 mm^3) were fixed 5 h to overnight at 4°C in 4% paraform-aldehyde, 0.3% glutaraldehyde, and 0.75% acrolein all in PBS containing 0.1 M sucrose. Tissues were then transferred to the same solution minus sucrose at 4°C for 12 h. They were then frozen and sectioned in a cryostat at 8 to 12 μm thickness. Primary antibody treatment was given (overnight at 4°C) after treatment in plain goat serum (for 30 min) to prevent non-specific labeling. Secondary antibody (goat–anti-rabbit) conjugated to colloidal gold was then provided to the sections overnight at 4°C or room temperature. For EM, the sections should be thinner and post-staining in uranyl acetate and lead citrate was practiced.

Applications

The distribution of the lectin in the root cell walls of adult wheat plants was studied.[45] The lectin was found distributed in the outer cell layers of root proper and in the root cap.

Localization of Extensin in Tissues[1,114,172-182]

Antibodies Raised

Antibodies were first attempted against the extensin-like proteins of *Chlamydomonas* cell walls and subsequently a number of antibodies have been produced (see References 183 and 184) in this alga.

Polyclonal antibodies against highly glycosylated putative precursors P1 and P2 of extensins and their deglycosylated (by HF treatment) derivatives were first raised in rabbits.[177,178] Almost simultaneously polyclonal antibodies were raised in rabbits against purified extensins of the soybean seed coats.[176] Stafstrom and Staehelin[170] produced polyclonal antibodies in rabbits against extensin 1 isolated from carrot root, and polyclonal antibodies in rabbits against a deglycosylated extensin 1 and against extensin 2 from carrot root. Antibodies against an extensin like HRGP (hydroxyproline rich glycoprotein) which increases due to fungal infection in melon callus tissue were raised by Mazau et al.[180,181] Stiefel et al.[182] produced anti-extensin antibodies against the protein obtained from maize. Kieliszewski et al.[173] developed the antibodies against the threonin-rich extensins of maize. Polyclonal antibodies were raised from extensins of melon callus tissue by Bonfante-Fasolo et al.[174] The antibodies produced against extensin 1 in Staehelin's laboratory targeted periodate sensitive epitopes and probably recognized terminal α-1, 3-arabinoside; the antiextensin 2 antibodies recognized periodate insensitive regions on the carbohydrate side chains. The anti-deglycosylated extensin 1 antibodies recognized the protein backbone of the extensin.

Tissue Preparation and Labeling

1. Fix the tissue in 3% glutaraldehyde or in 4% paraformaldehyde in a suitable buffer. The tissue may be subjected to different protocols depending upon the microscope in which it is to be observed. LR white infiltration and embedding is generally advised.

2. Block sections for 30 min in a 3% (W/V) solution of non-fat dried milk in TBS (8 g NaCl, 0.2 g KCl, 3 g tris base in 800 ml of distilled water, pH 8.0, volume adjusted to 1 L).

3. Drain off the blocking solution carefully.

4. Incubate the tissue in the primary antiserum diluted 1:20 overnight at room temperature. For dilution use TBS.

5. Wash the tissue at least two times for 15 min duration each in TBS.

6. Subject the tissue to treatment in the goat-antirabbit secondary antibody conjugated to any of the standard markers such as peroxidase, alkaline phosphatase, FITC, Texas red or colloidal gold for 1 to 2 h at room temperature.

7. Wash the tissue in the buffer solution twice (15 min each time).

8. Observe under EM, if colloidal gold has been used; under FM if FITC or Texas red has been used; if it is enzyme conjugated, proceed to color the enzyme sites following suitable protocols.

Applications

Extensin 1 antibodies were shown to bind only to cellulose-rich regions of the cell wall,[7] while extension 2 antibodies labeled only the expanded middle lamella at the three-way cell junctions.[1] Cassab and Varner[176] mainly localized the extensin in the seed coat, hilum, and vascular tissues; extensin was not detected in the early stages of development of the seed, but accumulated first in the cell wall of the palisadal epidermal sclereids. In the mature seeds, much extensin was present in the walls of both epidermal and hypodermal sclereids. Bonfante-Fazolo et al.[174] studied the distribution of extensins in pea roots infected with the mycorrhizal fungus, *Glomus versiforme*, and found that the same was concentrated at the contact zone between the host and fungus.

Localization of Extensins in Tissue Prints[176,185-189]

It is a variation of the western blot technique[190] to demonstrate extensins in nitro-cellulose imprints of tissues. The fresh cut surface of a plant organ pressed on the surface of a nitro-cellulose paper can transfer metabolites, enzymes, mRNA, organelles, and wall components to the paper with little lateral diffusion.[189] For localizing peroxidase, as soon as a print is made on nitrocellulose paper, the substrate (H_2O_2) and an insoluble colored product (O-phenylenediamine) are to be added. Then the distribution pattern of peroxidase will be evident on the paper. It uses the cut surface of plant tissue printed onto nitrocullulose paper, which is then probed with anti-HRGP antibodies.

Procedure

1. Cut the nitrocellulose paper to the desired size.
2. Soak it in 0.2 M $CaCl_2$ for 20 to 30 min and dry it.*
3. Wash the freshly cut surface of the tissue (3 sec) in d H_2O, dry on paper towels and blot, giving slight pressure onto the prepared nitrocellulose paper for 15 sec to 1 min.**
4. Dry the tissue print with warm air (2 to 3 min) or air dry (10 min).
5. Process the paper for a conventional western blot. [190]
6. Detect the alkaline phosphatase/peroxidase-conjugated secondary antibody.[191]

Comments

To transfer extensin from cell walls to nitrocellulose paper, presoaking of the paper in 0.2 m $CaCl_2$ is an absolute necessity. However, for transferring other prospective enzymes and metabolites, the above treatment is not needed.

Applications

Since its first employment, tissue printing technique has found wide application. In their work, Cassab and Varner[187] prepared soybean seed prints for extensin detection and observed that its distribution was similar to the pattern observed in the same seed sections prepared by traditional immunocytochemical methods.[176]

Tissue printing has been used to follow the pattern of localization of extensin in ethylene-treated etiolated pea epicotyls.[188] It was found that the localization of extensin changes 72 and 96 h after ethylene treatment. Extensin increases in the epidermal and cortical cells upon ethylene treatment.

---

* It is our experience that the paper should not be fully dry but should be slightly damp for very good results.

** In our trials, we got the best results when the freshly cut surface of the tissue is directly blotted onto the nitrocellulose without drying on paper towels.

Ye et al.[185] used tissue printing hybridization methodology to demonstrate HRGPs (see the next protocol) in tobacco, tomato, and petunia. HRGP mRNAs were most abundant in outer and inner phloem regions in tomato petiole and stem; in tobacco and petunia, HRPG genes were expressed in the abaxial and adaxial phloem regions of leaf mid-vein and in the outer phloem of the stem.

Localization of Glycine-Rich Proteins (GRP) in Tissue Prints[185]

Procedure

1. Select nylon membranes.
2. Bake the membrane for 2 h at 80°C after the prints are made.*
3. Wash in 0.2% SSC, 1% SDS for 4 h at 65°C.
4. Prehybridize for 4 h at 68°C in 2% SSC, 1% SDS 5% Denhardt's reagent, 0.1 mg ml⁻¹ SSD and 10 mM DTT.
5. Clone bean GRP 1.8 genomic DNA into a bluescript plasmid vector and synthesize 35 S-labeled sense and antisense RNA probes using T7 or T3 RNA polymerase.
6. Hybridize the probes to the nylon membrane at 68°C for 16 h.
7. Wash the membrane in 2% SSC and 0.1% SDS at 42°C three times for 20 min each and then in 0.2% SSC and 0.1% SDS at 65°C twice for 20 min each.
8. Expose the membrane to Kodak Tmax 400 film at room temperature.

Application

Ye et al.[185] did tissue level mRNA localization for glycine-rich cell wall proteins in tobacco, tomato, and petunia. They found GRP mRNAs in primary xylem regions and cambial cells that may differentiate into secondary xylem. GRP was localized in tobacco and pentunia in the xylem regions of the mid-vein of the leaves and stem.

Localization of Arabinogalactan Proteins[192-194]

Antibodies Raised

1. Monoclonal antibodies (MAC 207) recognizing arobinogalactan epitopes.[193]
2. JIM 13—a monoclonal antibody that recognizes a carbohydrate epitope on arabinoga-lactan proteins.[111]
3. Anderson et al.[192] and Sedgley and Clarke[194] raised the monoclonal antibody against arabinogalactan protein from hybridomas of mice, immunized against style extracts of *Nicotiana alata*. The specificity of antibodies selected by three clonal lines is primarily directed to β-D-galactopyronose and α-L-arabinofuranose. Antibodies from two cell

---

* See the earlier protocol for the methodology of making prints.

lines preferentially bind to β-D-galactopyronose residues and antibodies from the other cell line preferentially bind α-Larabinofuranose.

4.   JIM 8—a monoclonal antibody.[195,196]

Tissue Preparation

Li et al.[193] fixed the tissue in paraformaldehyde.

Applications

1.   Arabinogalactan glycoproteins have been temporally and spatially localized in the pollen tubes of tobacco grown *in vitro* and semi *in vivo* in the style. Immunolabeling in both cases showed a ring-like pattern along the axis of the pollen tube. Distance between rings was about 6 μm while the diameter of individual rings was about 12 μm. After treatment with pectinase and/or cellulase, the rings remained. Therefore, the ring is probably located in the callose layer of wall.[193]

2.   JIM 13 was employed on non-elongating and elongating cell walls of cultured carrot cells. It very strongly labeled the cell walls of both round and elongate cells and it labeled cell corners in the tissue parenchyma of normal cells.[111]

3.   Sedgley and Clarke[194] have localized arabinogalactan proteins in the cell walls and intercellular matrix of the transmitting tissue of *Nicotiana alata* styles.

Localization of Thionins[197]

Antibody Raised

The 15000 M_r precursor polypeptide of leaf-specific thionins of barley was fused with an *E. coli* β-galactosidase. Antiserum was raised against this fusion product.

Tissue Preparation and Labeling

The antibody was allowed to react with ultrathin sections of lowicryl-embedded leaf material of etiolated barley seedlings. Samples were processed with protein A gold (20 nm).

Control

Pre-immune serum was used as a control.

Applications

Apel et al.[197] found the antigens exclusively in the cell wall. The localization was also analyzed by immunofluorescense (antiserum was coupled to FITC). Thionins were found evenly within the walls of all the leaf cells with the exception of the outer cell wall of the epidermal cell layer which contained a much higher concentration of thionins.

Localization of Self-Incompatibility Determining S-Locus Specific Glycoproteins (SLGS)[198-200]

Self-incompatibility is controlled by a single locus, the S-locus, with several naturally occurring alleles. It is believed that the pollen tube is inhibited when identical S-alleles are expressed in pollen and stigma. Self-incompatibility in *Brassica* has been related to the production of S-locus specific glycoproteins (SLSG) in the stigma. Analysis of various S-allele homozygotes by isoelectric focusing (IEF) and SDS-PAGE have demonstrated that SLSG molecules are highly polymorphic and that they exhibit molecular weight heterogeneity. The activity of these genes becomes manifest 1 d prior to flower anthesis.

Antibodies Raised

SLSG antigen was purified from stigma extracts following fractionation of IEF gels.[201] The gels were stained with coomassie blue 250, destained, and SLSG was electroeluted from the appropriate gel slices.[202] The purified SLSG were then used to immunize Balb/c mice. The polyclonal antiserum from the mouse was collected and stored at -20°C in aliquots. Pre-immune sera from individual immunized mice were also stored at -20°C for control experiments. For the production of monoclonal antibodies, spleen cells from mice were fused with myeloma cell lines and hybridoma supernatants were screened by immunoblot analysis of stigma proteins. Goat anti-mouse IgG secondary antibodies conjugated to alkaline phosphatase were used to identify the positive hybridomas. One monoclonal, MAb/H8, was selected for use. The reactivity of MAb/H8 with bacterial β-galactosidase-SLSG fusion protein demonstrates that the epitope recognized by the antibody resides in the polypetide moiety of SLSG and does not involve a carbohydrate component.

Tissue Preparation and Labeling

Tissues were fixed for 4 h in 95% in ethanol and glacial acetic acid (3:1) for LM work. In addition, to improve antigenicity and better morphological preservation, fixation in 2 to 3% formaldehyde/gultaraldehyde with or without osmification was used. For EM work, fixation for 2 h at 4°C was done either in 2.5% glutaraldehyde or in a mixture of 0.5% glutaraldehyde and 4% paraformaldehyde in phosphate buffer, pH 7.2. The tissues were then washed with four changes of excess phosphate buffer at 15-min intervals, post-fixed for 1 h in 1% buffered OsO_4 at 4°C. High pressure freezing and freeze substitution was also done to significantly improve antigenic preservation of the SLSG.[27] Dehydration was carried out in graded ethanol series and embedding in either LR white or Spurr's epoxy resin. After treatment with primary antibody, labeling was done with colloidal gold (for LM enhancement with Inten SE II silver enhancement kit from Janssen was used for 10 min) conjugated to goat-antimouse antibody. A treatment with aqueous sodium metaperiodate for 15 min followed by incubation in 0.1 N HCl for 10 min was done earlier to primary antibody treatment in order to unmask the antigens.[24,203] For EM, sections taken with diamond knife were mounted on Formvar-coated nickel grids.

Controls

1. Labeling of sections of stigmas of S5 and S15 genotypes of *Brassica oleracea* whose SLSG do not cross-react with the antibody.
2. Omission of primary antibody treatment.
3. Use of pre-immune serum.
4. Use of hybridoma medium instead of MAb/H8 supernatant.

Applications

Using this procedure, Kandasamy et al.[198,199] demonstrated that SLSG was uniformly distributed over the walls of the stigmatic papillar cells. SLSG localization was subsequently done on transgenic tobocco plants (into which *Brassica oleracea* SLG-13 and SLG-22 were introduced) immunocytochemically.[198,199] Immunolabeling was observed in the pistil along the path followed by pollen tubes after pollination. S-antigen accumulated in the intercellular matrix of the transmitting tissue of the style and its continuation in the basal portion of the stigma and outside a few special cells of the placental epidermis of the ovary. This pattern of S-antigen distribution closely resembles that described for the S-associated glycoproteins of self-incompatible *Nicotiana alata* and differs from its distribution in *Brassica oleracea*. The transgenic plants were self-compatible, i.e., *Brassica* SLSG may not be competent to mediate an incompatibility response in tobocco or there is an apparent lack of activity of SLG promoter in another tissue.

Umbach et al.[200] immunolocalized SLSG as well as a glycoprotein, SLR1 (a product of SLR1 gene, which is unlinked to the S-locus). The primary antibody used here was trp E-SLR1 (i.e., SLR1 fused to the aminoterminal portion of the *Escherichia coli* trp gene in the expression plasmid pATH 11). This was detected with 15 nm colloidal gold-conjugated antirabbit secondary antibody. SLR1 protein was found in several crucifer species and did not show any variation in the level between compatible and incompatible cases. Both SLSG and SLR1 occurred as glycoforms and were immunolocalized in the stigma papillae cell walls.

Localization of Self-Incompatibility Glycoprotein (S2-Ribonuclease)[204-206]

Antibodies Raised

Anderson et al.[204,205] raised an anti-S_2-peptide antibody in rabbits to synthetic peptide corresponding to the stretch of the major hypervariable region B, which is hydrophilic and, thus, predicted to lie on the surface of the glycoprotein. This antibody binds specifically to the S_2-glycoprotein in protein gel blots under conditions where there was no detectable binding to the other buffer soluble style proteins or to other S-allele glycoproteins. Styles were fixed. For immunofluorescence, sections were incubated with anti-S_2-peptide antibody (30 µg/ml) in SC buffer for 1 h at 20°C. Sections were washed in SC and then incubated with rabbit-antisheep IgG conjugated to fluorescein isothiocyanate (1:50 dilution) for 1 h at 20°C. For EM, ultrathin sections were incubated in the same primary antibody, washed in buffer and then labeled with rabbit-antisheep IgG antibody conjugated to 15 nm colloidal gold.

Tissue Preparation and Labeling

Immunogold (15 nm) as well as fluorescein isothiocyanate markers were used. Both were conjugated to sheep anti-rabbit IgG. For LM, immunogold labeling was enhanced by silver (Janssen silver enhancement kit).

Applications

Anderson et al.[205] showed that stigma cell walls including surface papillae were labeled, while cortical cells and epidermis of style were devoid of labels. Placental epidermis and outer layer of ovules were marked.

References

1. Swords, K. M. M. and Staehelin, L. A., Analysis of extensin structure in plant cell walls, in *Modern Methods of Plant Analysis*. N.S. Vol. 10. *Plant Fibers,* Linskens, H. F. and Jackson, J. F., Eds., Springer Verlag, Berlin, 1989, 219.
2. Roland, J. -C. and Vian, B., The cell wall of the growing plant cell: its three dimensional organisation, *Int. Rev. Cytol.*, 61, 129, 1979.
3. Valent, B. and Albersheim, P., Structure of plant cell walls. V. On the binding of xyloglucan to cellulose fibres, *Plant Physiol.*, 54, 105, 1974.
4. Moore, P. J. and Staehelin, L. A., Immunogold localization of the cell-wall-matrix polysaccharides rhamnogalacturonan I and xyloglucan during cell expansion and cytokinesis in *Trifolium pratense* L.: implications for secretory pathways, *Planta*, 174, 433, 1988.
5. Albersheim, P., Mühlethaler, K., and Frey-Wyssling, A., Stained pectin as seen in the electron microscope, *J. Biophys. Biochem. Cytol.*, 8, 501, 1960.
6. Selvendran, R. R., Development in the chemistry and biochemistry of pectic and hemicellulosic polymers, *J. Cell Sci.*, Suppl. 2, 51, 1985.
7. Stafstrom, J. P. and Staehelin, L. A., Antibody localization of extensin in cell walls of carrot storage root, *Planta*, 174, 321, 1988
8. Hoson, T., Structure and function of plant cell walls; Immunological approaches, *Int. Rev. Cytol.*, 130, 233, 1991.
9. Austyn, J. M., *Antigen-Presenting Cells,* IRL Press, Oxford, 1989, Chap. 1.

10. Harlow, E. and Lane, D. P., *Antibodies. A Laboratory Manual*, Cold Spring Harbor Laboratory, New York, 1988, Chap.1.

11. Harlow, E. and Lane, D. P., *Antibodies. A Laboratory Manual*, Cold Spring Harbor Laboratory, New York, 1988, Chap. 4.

12. Harlow, E. and Lane, D. P., *Antibodies. A Laboratory Manual*, Cold Spring Harbor Laboratory, New York, 1988, Chap. 2.

13. Harlow, E. and Lane, D. P., *Antibodies. A Laboratory Manual*, Cold Spring Harbor Laboratory, New York, 1988, Chap. 3.

14. Harlow, E. and Lane, D. P., *Antibodies. A Laboratory Manual*, Cold Spring Harbor Laboratory, New York, 1988, Chap. 5.

15. Harlow, E. and Lane, D. P., *Antibodies. A Laboratory Manual*, Cold Spring Harbor Laboratory, New York, 1988, Chap. 9.

16. Avrameas, S. and Uriel, J., Methode de marquage d' antigen et danticorps avec des enzymes et son application en immunodiffusion, *C.R. Acad. Sci.*, D.262, 2543, 1966.

17. Avrameas, S., Enzyme markers: Their linkage with proteins and use in Immuno-histochemistry, *Histochem. J.*, 4, 321, 1972.

18. Farr, A. G. and Nakane, P. K., Immunohistochemistry with enzyme labeled antibodies: A brief review, *J. Immunol. Methods*, 47, 129, 1981.

19. Faulk, W. P. and Taylor, G. M., An immunocolloid method for the electron microscope, *Immunochemistry*, 8, 1081, 1971.

20. Horisberger, M. and Rosset, J., Colloidal gold, a useful marker for transmission and scanning electron microscopy, *J. Histochem. Cytochem.*, 25, 295, 1977.

21. Craig , S. and Goodchild , D. J., Post- embedding immunolabeling. Some effects of tissue preparation on the antigenicity of plant proteins, *Eur. J. Biol.*, 28, 251, 1982.

22. Hobot, J. A. and Newman, G. R., Strategies for improving the cytochemical sensitivity of ultrastructurally well-preserved resin embedded biological tissue for light and elec-tron microscopy, *Scanning Microscopy*, Suppl. 5, S27, 1991.

23. Tokuyasu, K. T. and Singer, S. J., Improved procedures for immunoferritin labeling of ultrathin frozen sections, *J. Cell Biol.*, 71, 894, 1976.

24. Bendayan, M. and Zollinger, M., Ultrastructural localization of antigenic sites on osmium fixed tissues applying the proteinA-gold technique, *J. Histochem. Cytochem.*, 31, 101, 1983.

25. Gilkey, J. C. and Staehelin, L. A., Advances in ultrarapid freezing for the preservation of cellular ultrastructure, *J. Electron Microsc. Tech.*, 3, 177, 1986.

26. Ichikawa, M., Sasaki, K., and Ichikawa, A., Optimal preparatory procedures of cryo-fixation for immunocytochemistry, *J. Electron Microsc. Tech.*, 12, 88, 1989.

27. Kandasamy, M. K., Parthasarathy, M. V., and Nasarallah, M. E., High pressure freezing and freeze substitution improve immunolabeling of S-locus specific glycoproteins in the stigma papillae of *Brassica*, *Protoplasma*, 162, 187, 1991.

28. Nicolas, G., Advantages of fast-freeze fixation followed by freeze substitution for the preservation of cell integrity, *J. Elect. Micros. Tech.*, 18, 395, 1991.

29. Dubochet, J., Adrian. M., Chang, J. -J., Lepault, J., and Mc Dowall, A. W., Cryoelectron microscopy of vitrified specimens, in *Cryotechniques in Biological Electron Micros-copy*, Steinbrect, R. A. and Zierold, K., Eds., Springler-Verlag, Berlin, 1987, 114.

30. Costello, M. J., Ultra-rapid freezing of thin biological samples, *Scan. Electron. Microsc.*, 2, 361, 1980.

31. Ridge, R. W., A simple apparatus and technique of the rapid freeze and freeze-substi-
    tution of the single-cell algae, *J. Electron Microsc.*, 39, 120, 1990.
32. Boyne, A. F., A gentle, bounce-free assembly for quick-freezing tissues for electron
    microscopy: application to isolate Torpedine ray electrocyte stacks, *J. Neurosci. Meth.*,
    1, 353, 1979.
33. Fernandes, D. E. and Staehelin, L. A., Structural organization of ultrarapidly frozen
    barley aleurone cells actively involved in protein secretion, *Planta*, 165, 455, 1985.
34. Staehelin, L. A. and Chapman, R. L., Secretion and membrane recycling in plants cells:
    novel structures visualised in ultra rapidly frozen sycamore and carrot suspension -
    cultures cells, *Planta*, 171, 43, 1987.
35. Zhang, G. F. and Staehelin, L. A., Functional compartmentalization of the golgi appa-
    ratus of plant cells. Immunocytochemical analysis of high-pressure frozen-and freeze-
    substituted *sycamore* maple suspension culture cells, *Plant Physiol.*, 99, 1070, 1992.
36. Moor, H., Theory and practice of high pressure freezing, in *Cryotechniques in Biolog-
    ical Electron Microscopy*, Steinbrecht, R. A. and Zierold, K., Eds., Springer Verlag,
    Berlin, 1987, 175.
37. Dahl, R. and Staehelin, L. A., High pressure freezing for the preservation of biological
    structure: Theory and Practice, *J. Electr. Microsc. Tech.*, 13, 165, 1989.
38. Kiss., J. Z., Giddings, T. H., Staehelin, L. A., and Sack, F. D., Comparison of the
    ultrastructure of conventionally fixed and high pressure frozen/freeze substituted root
    tips of *Nicotiana* and *Arabidopsis*, *Protoplasma*, 157, 64, 1990.
39. Michel, M., Hillmann, T., and Müller, M., Cryosectioning of plant material frozen at
    high pressure, *J. Microsc. (Oxford)*, 163, 3, 1991.
40. Craig, S. and Staehelin, L. A., High pressure freezing of intact plant tissue: evaluation
    and characterization of novel features of the endoplasmic reticulum and associated
    membrane systems, *Eur. J. Cell Biol.*, 46, 80, 1988.
41. Steinbrecht, R. A. and Müller, M., Freeze-substitution and freeze-drying, in *Cryotech-
    niques in Biological Electron Microscopy*, Steinbrechtt, R. A. and Zierold, K., Eds.,
    Springler Verlag, Berlin, 1987, 149.
42. Newman, G. R. and Hobot, J. A., Role of tissue processing in colloidal gold methods,
    in *Colloidal Gold: Principles, Methods and Applications*, Vol. 2, Hayat, M. A., Ed.,
    Academic Press, San Diego, CA, 1989, 33.
43. Tiwari, S. C. and Gunning, B. E. S., Development and cell surface of a non-syncytial
    invasive tapetum in *Canna*: Ultrastructural, freeze - substitution, cytochemical and
    immunfluorescence study, *Protoplasma*, 134, 1, 1986.
44. Tokuyasu, K. T., Immunocytochemistry on ultra-thin frozen sections, *Histochemistry*,
    12, 381, 1980.
45. Raikhel, N. V., Mishkind, M., and Palevitz, B. A., Immunocytochemistry in plants with
    colloidal gold conjugates, *Protoplasma*, 121, 25, 1984.
46. Baron-Epel, O., Gharyal, P. K., and Schindler, M., Pectins as mediators of wall porosity
    in soybean cells, *Planta*, 175, 389, 1988.
47. Tokuyasu, K. T., Application of cryo-ultramicrotomy to immunocytochemistry, *J.
    Microsc. (Oxford)*, 143, 139, 1986.

48. Sitte, H., Neumann, K., and Edelmann, L., Cryosectioning according to Tokuyasu Vs rapid-freezing, freeze-substitution and resin embedding, *Immuno-Gold Labeling in Cell Biology*, Verkleij, A. J. and Leunissen, J. L. M., Eds., CRC Press, Boca Raton, FL, 1989, 63.

49. Leunissen, J. L. M. and Verkleij, A. J., Cryoultramicrotomy and immunogold labeling, in *Immuno-Gold Labeling in Cell Biology*, Verkleij, A. J. and Leunissen, J. L. M., Eds., CRC Press, Boca Raton, FL, 1989, 95.

50. Van Bergen en Henegouwen, P. M. P., Immunogold labeling of ultra thin sections, in *Colloidal Gold:Principles, Methods and Applications*, Vol. I, Hayat, M. A., Ed., Academic Press, San Diego, CA, 1990, 191.

51. Craig, S., Moore, P. J., and Dunahay, T.G., Immunogold localization of intra - and extra - cellular proteins and polysaccharides in plant cells, *Scan. Electron Microsc.*, 1, 1431, 1987.

52. Freund, J., The mode of action of immunologic adjuvants, *Adv. Tuberc. Res.*, 7, 130, 1956.

53. Freund, J. and McDermott, K., Sensitization to horse serum by means of adjuvants, *Proc. Soc. Exp. Biol. Med.*, 49, 548, 1942.

54. Chase, M. W., Production of antiserum, *Methods Immunol. Immunochem.*, 1, 197, 1967.

55. Amkraut, A. A., Garvey, J. S., and Campbell, D. H., Competition on haptens, *J.Expt. Med.*, 124, 293, 1966.

56. McLean, I. W. and Nakane, P. K., Periodate-lysine-paraformaldehyde fixative. A new fixative for immunoelectron microscopy, *J. Histochem. Cytochem.*, 22, 1077, 1974.

57. Lynch, M. A., Synthesis, secretion and cell localization of secreted complex polysaccharides in the developing root. Ph.D Thesis, University of Colorado at Boulder, Boulder, CO, 1992.

58. Vreeland, V., Localization of a cell wall polysaccharide in a brown alga with labeled antibody, *J. Histochem. Cytochem.*, 18, 371, 1970.

59. Adler, F. L. and Adler, L. T., Passive hemagglutination and hemolysis for estimation of antigens and antibodies, *Methods Enzymol.*, 70, 455, 1980.

60. Vaitukaitis, J. L., Production of antisera with small doses of immunogen: Multiple intradermal injections, *Methods Enzymol.*, 73, 46, 1981.

61. Harlow, E. and Lane, D. P., *Antibodies. A Laboratory Manual*, Cold Spring Harbor Laboratory, New York, 1988, Chap. 8.

62. Ey, P. L., Prowse, S. J., and Jenkin, C. R., Isolation of pure IgG, IgG_2a and IgG_2b immunoglobulins from mouse serum using Protein A-Sepharose, *Biochemistry*, 15, 429, 1978.

63. Avrameas, S., Coupling of enzymes to proteins with glutaraldehyde. Use of the conjugates for the detection of antigens and antibodies, *Immunochemistry*, 6, 43, 1969.

64. Avrameas, S. and Ternynck, T., The cross-linking of proteins with glutaraldehyde and its use for the preparation of immunoadsorbants, *Immunochemistry*, 6, 53, 1969.

65. Avrameas, S. and Ternynck, T., Peroxidase labeled antibody and Fab conjugates with enclosed intracellular penetration, *Immunochemistry*, 8, 1175, 1971.

66. Nakane, P. K. and Kawaoi., A., Peroxidase-labeled antibody: A new method of conjugation, *J. Histochem. Cytochem.*, 22, 1084, 1974.

67. Tijssen, P. and Kurstak, E., High efficient and simple method for the preparations of peroxidase and active peroxidase antibody conjugates for enzyme immunoassays, *Anal. Biochem.*, 136, 451, 1984.

68. Kitagawa, T. and Aikawa, T., Enzyme coupled immunoassay of insulin using a novel coupling reagent, *J. Biochem.*, 79, 233, 1976.

69. O'Sullivan, M. J., Gnemmi, E., Morris, D., Chieregatti, G., Simmonds, A. D., Simmons, M., Bridges, J. W., and Marks, V., Comparison of two methods of preparing enzyme-antibody conjugates: Application of these conjugates for enzyme immunoassay, *Anal. Biochem.*, 100, 100, 1979.

70. Goding, J. W., Conjugation of antibodies with fluorochromes: Modifications to the standard methods, *J. Immunol. Methods*, 13, 215, 1976.

71. The, T. H. and Feltkamp, T. E. W., Conjugation of fluorescein isothiocyanate to antibodies. II. A reproducible method, *Immunology*, 18, 875, 1970.

72. DeMey, J., Moeremans, M., Geuens,G., Nuydens, R., and de Brabander, M., High resolution light and electron microscopy localization of tubulin with the IGS (immunogold staining) method, *Cell Biol. Int. Rep.*, 5, 889, 1981.

73. Slot, J. W. and Geuze, H. T., Gold markers for single and double immunolabeling of ultrathin cryosections, in *Immunolabeling for Electron Microscopy*, Polak, J. M. and Varndell, I. M., Eds., Elseviere, Amsterdam, 1984, 129.

74. Hodges, G. M., Smolira, M. A., and Livingston, D. C., Scanning electron microscope immunocytochemistry in practice, in *Immunolabeling for Electron Microscopy*, Polak, J. M. and Varndell, I. M., Eds., Elsevier, Amsterdam, 1984, 189.

75. Bendayan, M., Facts and artefacts in colloidal gold postembedding cytochemistry, *Proc. 44th Ann. Meet. EMSA*, San Francisco Press, San Francisco, CA,1986, 44.

76. Lynch, M. A. and Staehelin, L. A., Domain-specific and cell type-specific localization of two types of cell wall matrix polysaccharides in the clover root tip, *J. Cell Biol.*, 118, 467, 1992.

77. Moore, P. J., Immunogold localization of specific components of plant cell walls, in *Modern Methods of Plant Analysis*, N.S. Vol. 10. *Plant Fibres*, Linskens, H. F. and Jackson, J. F., Eds., Springer Verlag, Berlin, 1989, 70.

78. Moore, P. J., Swords, K. M. M. Lynch, M. A., and Staehelin, L. A., Spatial reorganization of the assembly pathways of glycoproteins and complex polysaccharides in golgi apparatus of plants, *J. Cell Biol.*, 112, 589, 1991.

79. Titus, D. E. and Becker, W. M., Investigation of the glyoxisome- peroxisome transition in germinating cucumber cotyledons using double-label immunoelectron microscopy, *J. Cell Biol.*, 101, 1288, 1985.

80. Harlow, E. and Lane, D. P., *Antibodies. A Laboratory Manual*, Cold Spring Harbor Laboratory, New York, 1988, Chap. 10.

81. Boenish, T., Basic enzymology, in *Immunochemical Staining Methods*, Naish, S. J., Ed., DAKO Corporation, 1989, Chap.3.

82. Danscher, G. and Nörgaard, J. O. R., Light microscopic visualization of colloidal gold on resin embedded tissue, *J. Histochem. Cytochem.*, 31, 1394, 1983.

83. Danscher, G., Localization of gold in biological tissue. A photochemical method for light and electron microscopy, *Histochemistry*, 71, 81, 1981.

84. Holgate, C. S., Jackson, P., Cownen, P. N., and Bird, C. C., Immunogold- Silver staining: New method of immunostaining with enhanced sensitivity, *J. Histochem. Cytochem.*, 31, 938, 1983.

85. Harlow, E. and Lane, D. P., *Antibodies. A Laboratory Manual*, Cold Spring Harbor Laboratory, New York, 1988, Chap. 15.

86. Heimer, G. V. and Taylor, C. E. D., Improved mountant for immunofluorescence preparations, *J. Clin. Pathol.*, 27, 254, 1974.

87. Osborn, M. and Weber, K., Immunofluorescence and immunocytochemical procedures with affinity purified antibodies: Tubulin-containing structures, *Methods Cell Biol.*, 24, 97, 1982.

88. Vreeland, V., Immunocytochemical localization of the extracellular polysaccharide alginic acid in the brown sea weed *Fucus disticus*, *J. Histochem. Cytochem.*, 20, 358, 1972.

89. Vreeland, V., Slonich, M., and Laetsch, W. M., Monoclonal antibodies as molecular probes for cell wall antigens of brown alga *Fucus*, *Planta*, 162, 506, 1984.

90. Kaku, H., Shibata, S., Satsuma, Y., Sone, Y., and Misaki, A., Interaction of α-L-arabinofuranose-specific antibody with plant polysaccharides and its histochemical application, *Phytochemistry*, 25, 358, 1986.

91. Northcote, D. H., Localization of polysaccharides by antibodies specific to particular sugars and oligosaccharides linkages, in *V Cell Wall Meeting*, Fry, S. C., Brett, C. T., and Reid, J. S. G., Eds., Edinburgh, 1989, 58.

92. Northcote, D. H., Davey, R., and Lay, J., Use of antisera to localize callose, xylan and arabinogalactan in the cell-plate, primary and secondary walls of plant cells, *Planta*, 178, 353, 1989.

93. Meikle, P. J., Bonig, I., Hoogenrad, N. J., Clarke, A. E., and Stone, B. A., The localization of $(1{\to}3)$- β-glucans in the walls of pollen tubes of *Nicotiana alata* using a $(1{\to}3)$- β-glucan specific monoclonal antibody, *Planta*, 185, 1, 1991.

94. Horisberger, M., Rouvet-Vauthey, M., Richli, U., and Farr, D. A., Cell wall architecture of the halophilic yeast *Saccharomyces rouxii*. An immunocytochemical study, *Eur. J. Cell Biol.*, 37, 70, 1985.

95. Horisberger, M. and Rouvet-Vauthey, M., Cell wall architecture of the fission yeast *Schizosaccharomyces pombe*, *Experientia*, 41, 784, 1985.

96. Latge, J. P., Cole, G. T., Horisberger, M., and Prevost, M. G., Ultrastructural and chemical composition of the ballistospore wall of *Conidiobolus obscurus*, *Expt. Mycol.*, 10, 99, 1986.

97. Kishida, E., Sone, Y., Shibata, S., and Misaki, A., Preparation and immunochemical characterization of antibody to branched β-$(1{\to}3)$-D-glucan of *Volvariella volvacea*, and its use in studies of antitumor actions, *Agric. Biol. Chem.*, 53, 1849, 1989.

98. Vennigerholz, F., Hudak, J., and Walles, V., Immunocytochemical localization of pectin and callose in the pollen tube wall of *Brugmansia suaveolens* L., *VI Cell Wall Meeting*, Sassen, M. M. A., Derksen, J. W. M., Emons, A. M. C., and Wolters, A. M. C., Eds., Neijmegen, 1992, 47.

99. Roy, R., Katzenellenbogen, E., and Jennings, H. J., Improved procedures for the conjugation of oligosaccharides to protein by reductive amination, *Can. J. Biochem. Cell Biol.*, 62, 270, 1984.

100. Geitmann, A., Hadak, J., Vennigerholz, F., and Walles, B., Immunogold localization of pectin and callose in pollen grains and pollen tubes of *Brugmansia suaveolens*- Implications for the self-incompatability reaction, *J. Plant Physiol.*, 147, 225, 1995.

101. Turner, A., Wells, B., and Roberts, K., Plasmodesmata of maize root tips: Structure and Composition, *J. Cell. Sci.*, 107, 3351, 1994.

102. Bonfante-Fasolo, P., Vian, B., Perotto, S., Faccio, A., and Knox, J. P., Cellulose and pectin localization in roots of mycorrhizal *Allium porrum*: labeling continuity between host cell wall and interfacial material, *Planta*, 180, 537, 1990.

103. Moore, P. J., Darvill, A. G., Albersheim, P., and Staehelin, L. A., Immunogold localization of xyloglucan and rhamnogalacturonan I in the cell walls of suspension-cultured sycamore cells, *Plant Physiol.*, 82, 787, 1986.

104. Bradley, D. J., Wood, E. A., Larkins, A. P., Galfie, G., Butcher, G. W., and Brewin, N. J., Isolation of monoclonal antibodies reacting with peribacterioid membranes and other components of pea root nodules containing *Rhizobium leguminosarum*, *Planta*, 173, 149, 1988.

105. Bonfante-Fasolo, P., Vian, B., Perotto, S., Faccio, A., Reis, D., and Knox, J. P., Ultrastructural localization of host cell wall molecules in leek mycorrhizal roots, *V. Cell Wall Meeting*, Fry, S. C., Brett, C. T., and Reid, J. S. G., Eds., Edinburgh, 1989, 61.

106. Liners, F. and van Cutsem, P., A conformation of pectin recoganised by monoclonal antibodies, in *V Cell Wall Meeting*, Fry, S. C., Brett, C. T., and Reid, J. S. G., Eds., Edinburgh, 1989, 60.

107. Liners, F. and van Cutsem, P., Immunogold localization of acidic and esterified pectins in the carrot root apex, in *VI Cell Wall Meeting*, Sassen, M. M. A., Derksen, J. W. M., Emons, A. M. C., and Wolters, A. M. C., Eds., Neijmegen, 1992, 33.

108. Liners, F., Latesson, J. -J., Didembourgh, C., and Van Cutsem, P., Monoclonal antibodies against pectin. Recognition of a conformation induced by Calcium, *Plant Physiol.*, 91, 1419, 1989.

109. Van den Bosch, K. A., Bradley, D. J., Knox, J. P., Perotto, S., Butcher, G. W., and Brewin, N., Common components of the infection thread matrix and the intracellular space identified by immunocytochemical analysis of pea nodules and uninfected roots, *EMBO J.*, 8, 335, 1989.

110. Knox, J. P., Linstead, P. J., King, J., Cooper, C., and Roberts, K., Pectin esterification is spatially regulated both within cell walls and between developing tissues of root apices, *Planta*, 181, 512, 1990.

111. McCann, M. C., Stacey, N. J., Wilson, R., and Roberts, K., Polymer orientations in elongating cell walls, in *VI Cell Wall Meeting*, Sassen, M. M. A., Derksen, J. W. N., Emons, A. M. C., and Wolters. A. M. C., Eds., Neijmegen, 1992, 54.

112. Van Cutsem, P., Messiaean, J., and Liners, F., Monoclonal antibodies to pectin and the study of the cell wall, in *VI Cell Wall Meeting*, Sassen, M. M. A., Deksen, J. W. M., Emons, A. M. C., and Wolters, A. M. C., Eds., Neijmegen, 1992, 11.

113. Vian, B. and Roland, J. -C., Affinodetection of the sites of formation and of the future distribution of polygalactrouranans and native cellulose in growing plant cells, *Biol. Cell.*, 71, 43, 1991.

114. Staehelin, L. A., Giddings, T. H., Levy, S., Lynch, M. A., Moore, P. J., and Swords, K. M. M., Organization of the secretory pathway of cell wall glycoproteins and complex polysaccharides in plant cells, in *Endocytosis, Exocytosis and Vesicle Traffic in Plants*, Hawes, C. R., Coleman, J. O. D., and Evans, D. E., Eds., Cambridge University Press, Cambridge, 1991, 183.

115. Vreeland, V. and Laetsch, W. M., Identification of associating carbohydrate sequences with labeled oligosaccharides: Localization of alginate gelling subunits in cell walls of a brown alga, *Planta*, 177, 423, 1989.

116. Graham, R. C. and Karnovsky, M. J., The early stages of absorption of injected horseradish peroxidase in the proximal tubules of mouse kidney. Ultrastructural cytochemistry by a new technique, *J. Histochem. Cytochem.*, 14, 291, 1966.

117. Hoson, T., Auxin-induced cell wall extension, in *Proc. VII Internat. Symp. Cellular Basis Growth Devt. Plants*, Shibaoka, H., Ed., Osaka University, Osaka, Japan, 1992, 27.

118. Puhlmann, J., Dunning, N., Albersheim, P., Darvill, A. G., and Hahn, M. G., A monoclonal antibody that binds to sycamore maple xyloglucan recognises a fucose-containing epitope, in *3rd Intl. Congress, Intl. Soc. Plant and Mol. Biologists*, Tucson, AZ, 1991, 1028.

119. Sone, Y., Misaki, S., and Shibata, S., Preparation and characterization of antibodies against 6-0-α-D-xylopyranosyl-β-D-glucopyranose (β-isoprimeverose), the disaccharide unit of xyloglucan in plant cell walls, *Carbohyde. Res.*, 191, 79, 1989.

120. Sone, Y., Kuramae, J., Shibata, S., and Misaki, A., Immunocytochemical specificities of antibody to the heptasaccharide unit of plant xyloglucan, *Agric. Biol. Chem.*, 53, 2821, 1989.

121. Benhamou, N., Mazau, D., Grenier, J., and Esquerré- Tugayé, M., Time-course study of the accumulation of hydroxyproline - rich glycoproteins in root cells of susceptible and resistant tomato plants infected by *Fusarium oxysporum* f. sp. *radicis - lycopersici*, *Planta*, 184, 196, 1991.

122. Hoson, T. and Nevins, D. J., ß-D-Glucan antibodies inhibit auxin-induced cell elongation and changes in the cell wall of *Zea* coleoptile segments, *Plant Physiol.*, 90, 1353, 1989.

123. Hoson, T. and Nevins, D. J., Effect of anti-wall protein antibodies on auxin-induced elongation, cell wall loosening and β-D-glucan degradation in maize coleoptile segments, *Physiol. Plant.*, 77, 208, 1989.

124. Sexton, R, Burdon, N., Reid, J. S. G., Durbin, M. L., and Lewis, L. N., Cell wall breakdown and abscission, in *Structure, Function and Biosynthesis of Plant Cell Walls*, Dugger, W. M. and Bartnicki-Garcia, S., Eds., American Society of Plant Physiologists, Rockville, MD, 1984, 195.

125. Byrne, H., Christou, N. V., Verma, D. P. S., and Maclachlan, G. A., Purification and characterization of two cellulases from auxin-treated pea epicotyls, *J. Biol. Chem.*, 250, 1012, 1975.

126. Durbin, M. L., Sexton, R., and Lewis, L. N., The use of immunological methods to study the activity of cellulase isozymes (1:4 glucan 4 glucan hydrolase) in bean leaf abscission, *Plant Cell Environ.*, 4, 67, 1981.

127. Daniel, G. and Johansen, G., In-situ detection of cellulases and lignin degrading enzymes produced by *Phaenerochaete chrysosporum* during degradation of lignocellulose, in *V Cell Wall Meeting*, Fry, S. C., Brett, C. T., and Reid, J. S. G., Eds., Edinburgh, 1989, 70.

128. Sexton, R., Durbin, M. L., Lewis, L. N., and Thomson, W. W., Use of cellulase antibodies to study leaf abscission, *Nature*, 283, 873, 1980.

129. Sexton, R., Durbin, M. L., Lewis, L. N., and Thomson, W. W., The immunocytochemical localization of 9.5 cellulase in the abscission zones of bean *Phaseolus vulgaris* cv. Red Kidney, *Protoplasma*, 109, 335, 1981.

130. Benhamou, N., Grenier, J., Asselin, A., and Legrand, M., Immunogold localization of β-1, 3-glucanases in two plants infected by vascular wilt fungi, *Plant Cell*, 1, 1209, 1989.

131. Keefe, D., Hinz, U., and Meins, F. Jr., The effect of ethylene on the cell-type specific and intracellular localization of β-1,3-glucanase and Chitinase in tobacco leaves, *Planta*, 182, 43, 1990.

132. Mauch. F., Meehl, J. B., and Staehelin, L. A., Ethylene induced chitinase and β-1,3-glucanase accumulate specifically in the lower epidermis and along vascular strands of bean leaves, *Planta*, 186, 367, 1992.

133. Takeuchi, Y., Yoshikawa, M, Takeba, G., Tanaka, K., Shibata, D., and Horino, O., Molecular cloning and ethylene induction of mRNA encoding a phytoalexin elicitor-releasing factor, β-1,3-endoglucanase, in soyabean, *Plant Physiol.*, 93, 673, 1990.

134. Vögeli, U., Meins, F. Jr., and Boller , T., Co-ordinated regulation of chitinase and β-1,3 glucanase in bean leaves, *Planta*, 174, 364, 1988.

135. Mauch, F. and Staehelin, L. A., Functional implications of the subcellular localization of ethylene-induced chitinase and β-1,3-glucanase in bean leaves, *Plant Cell*, 1, 447, 1989.

136. Benhamou, N., Joosten, M. H. A. J., and De Wit, P. J. G. M., Subcellular localization of chitinase and of its potential substrate in tomato root tissues infected by *Fusarium oxysporum* f. sp. *Radicis-lycopersici*, *Plant Physiol.*, 92, 1108, 1990.

137. Shinshi, H., Mohnen, D., and Meins, F, Jr., Regulation of a plant pathogenesis- related enzyme: inhibition of chitinase and chitins mRNA accumulation in cultured tobacco tissues by auxin and cytokinin, *Proc. Natl. Acad. Sci. USA*, 84, 89, 1987.

138. Inouhe, M. and Nevins, D. J., Inhibition of auxin-induced cell elongation of maize coleoptiles by antibodies specific for cell wall glucanases, *Plant Physiol.*, 96, 426, 1991.

139. Labrador, E. and Nevins, D. J., Selected cell wall proteins from *Zea mays*: assessment of their role in wall hydrolysis, *Physiol. Plant.*, 77, 487, 1989.

140. Stuart, I. M., Loi, L., and Fincher, G. B., Development of (1→3, 1→4), β-D-glucan endohydrolase isoenzymes in isolated scutella and aleurone layers of barley (*Hordeum vulgare*), *Plant Physiol.*, 80, 310, 1986.

141. Woodward, J. R. and Fincher, G. B., Substrate specificity and kinetic properties of two (1→3), (1→4) - β-D-glucan endo-hydrolases from germinating barley (*Hordeum vulgare*) endosperm, *Carbohydr. Res.*, 106, 111, 1982.

142. Fink, J., Jeblick, W., and Kauss., H., Partial purification and immunological characterization of 1,3- β- glucan synthase from suspension cells of *Glycine max*, *Planta*, 181, 343, 1990.

143. Ali, Z. M. and Brady, C. J., Purification and characterization of the polygalacturonases of tomato fruits, *Austr. J. Plant Physiol.*, 9, 171, 1982.

144. DellaPenna, D., Alexander, D. G., and Bonnett, A. B., Molecular cloning of tomato fruit polygalacturonase: analysis of polygalactronase mRNA levels during ripening, *Proc. Natl. Acad. Sci. USA*, 83, 6420, 1986..

145. Tucker, M. L., Robertson, N. G., and Grierson, D., Changes in polygalacturonase isoenzymes during the ripening of normal and mutant tomato fruit, *Eur. J. Biochem.*, 112, 119, 1980.

146. Iki, K., Sekiguchi, K., Kurata, K., Tada, T., Nakagawa, H., Ogura, N., and Takehana, H., Immunological properties of β-fructofuranosidase from ripening tomato fruit, *Phytochemistry*, 17, 311, 1978.

147. Laurière, M., Laurière, M. C., Chrispeels, M. J., Johnson,K. D., and Sturm, A., Characterization of a xylose -specific antiserum that reacts with the complex asparagine-linked glycans of extracellular and vacuolar glycoproteins, *Plant Physiol.*, 90, 1182, 1989.

148. Schmidt, G. and Grisebach, H., Immunofluorecent labeling of enzymes, in *Modern Methods of Plant Analysis*, N.S. Vol. 4, *Immunology in Plant Sciences*, Linskens, H. F. and Jackson, J. F., Eds., Springer Verlag, Berlin, 1986, 156.

149. Burmeister, G. and Hösel, W., Immunohistochemical localization of β-glucosidases in lignin and isoflavane metabolism in *Cicer arietinum* L. seedlings, *Planta*, 152, 578, 1981.

150. Marcinowski, S. and Grisebach, H., Enzymology of lignification. Cell-wall bound β-glucosidase for coniferin from spruce (*Picea abies*) seedlings, *Eur. J. Biochem.*, 87, 37, 1978.

151. Marcinowski, S., Falk, H., Hammer, D. K., Hoyer, B., and Grisebach, H., Appearance and localization of a β-glucosidase hydrolysing coniferin in spruce (*Picea abies*) seedlings, *Planta*, 144, 161, 1979.

152. Hankins, C. N., Kindinger, J. I., and Shanon, L. M., Legume lectins: I . Immunological cross reactions between the enzymic lectin from mung beans and other well characterized legume lectins, *Plant Physiol.*, 64, 104, 1979.

153. Dickman, M. B., Patil, S. S., and Kolattukudy, P. E., Purification and characterization of an extracellular cutinolytic enzyme from *Colletotrichum gloeosporidioides* on *Carica papaya*, *Physiol. Plant Pathol.*, 20, 333, 1982.

154. Kolattukudy, P. E., Fungal penetration of defensive barriers of plants, in *Structure, Function and Biosynthesis of Plant Cell Walls*, Dugger, W. M. and Bartnicki-Garcia, S., Eds., American Society of Plant Physiologists, Rockville, MD, 1984, 302.

155. Shaykh, M., Soliday, C., and Kolattukudy, P. E., Proof for the production of cutinase by *Fusarium solani* f. *pisi* during penetration into its host, *Pisum sativum*, *Plant Physiol.*, 60, 170, 1977.

156. Audy, P., Benhamou, N., Trudel, J., and Asselin, A., Immunocytochemical localization of a wheat germ lysozyme in wheat embryo and coleoptile cells and cytochemical study of its interaction with the cell wall, *Plant Physiol.*, 88, 1317, 1988.

157. Roldàn, J. M., Verbelen, J. P., Butler, W. L., and Tokuyasu, K., Intracellular localization of nitrate reductase in *Neurospora crassa*, *Plant Physiol.*, 70, 872, 1982.

158. Robinson, D. G., Andreae, M., and Blankestein, P., Plant prolylhydroxylase, in *Organisation and Assembly of Plant and Animal Extracellular Matrix*, Adair, W. S. and Mecham, R. P., Eds., Academic Press, San Diego, CA, 1990, 283.

159. Andreae, M., Blankestein, P., Zhang, Y. -H., and Robinson, D. G., Towards the subcellular localization of plant propyl hydroxylase, *Eur. J. Cell Biol.*, 47, 181, 1988.

160. Bolwell, G. P. and Dixon, R. A., Membrane-bound hydroxylases in elicitor-treated bean cells. Rapid induction of the synthesis of propyl hydroxylase and a putative cytochrome P-450, *Eur. J. Biochem.*, 159, 163, 1986.

161. Cairns, E., Van Huystee, R. B., and Cairns, W. L., Peanut and horseradish peroxidase isoenzymes. Intraspecies and interspecies immunological relatedness, *Physiol. Plant.*, 49, 78, 1980.

162. Chibbar, R. N. and van Huystee, R. B., Immunocytochemical localization of peroxidase in cultured peanut cells, *J. Plant Physiol.*, 123, 477, 1986.

163. Espelie, K. E. and Kolattukudy, P. E., Purification and characterisation of an abscissic acid-inducible anionic peroxidase associated with suberization in potato, *Arch. Biochem. Biophys.*, 240, 539, 1985.

164. Espelie, K. E., Franceschi, V. R., and Kolattukudy, P. E., Immunocytochemical localization and time course of appearance of an anionic peroxidase associated with suberization in wound-healing potato tubers, *Plant Physiol.*, 81, 487, 1986.

165. Griffing, L. R. and Fowke, L. C., Cytochemical localization of peroxidase in soyabean suspension culture cells and protoplasts: Intracellular vacuole differentiation and presence of peroxidase in coated vesicles and multivesicular bodies, *Protoplasma*, 128, 22, 1985.

166. Hu, C., Carbonera, D., and van Huystee, R. B., Production and preliminary characterization of monoclonal antibodies against cationic peanut peroxidase, *Plant Physiol.*, 85, 299, 1987.

167. Hu, C. R., Smith, R., and van Huystee, R., Biosynthesis and localization of peanut peroxidases, *J. Plant Physiol.*, 135, 391, 1989.

168. Kim, S. -H., Terry, M. E., Hoops, P., Dauwalder, M., and Roux, S. J., Production and characterization of monoclonal antibodies to wall-localised peroxidases from corn seedlings, *Plant Physiol.*, 88, 1446, 1988.

169. Stephan., D. and van Huystee, R. B., Peroxidase biosynthesis as part of protein synthesis of cultured peanut cells, *Can. J. Biochem.*, 58, 715, 1980.

170. Van Huystee, R. B. and Labarzewski, J., An immunological study of peroxidase release by cultured peanut cells, *Plant Sci. Lett.*, 27, 59, 1982.

171. Angelini, R. and Federico, R., Evidence for H_2O_2 production by polyamine oxidation in the cell wall: Spatial and functional correlation between amine oxidase and peroxidase, in *V Cell Wall Meeting*, Fry, S. C., Brett, C. T., and Reid, J. S. G., Eds., Edinburgh, 1989, 178.

172. Stafstrom, J. P. and Staehelin, L. A., A second extension-like hydroxyproline-rich glycoprotein from carrot cell walls, *Plant Physiol.*, 84, 820, 1987.

173. Kieliszewski, M. J., Leykam, J. F., and Lamport, D. T. A., Structure of the threonine-rich extension from *Zea mays*, *Plant Physiol.*, 92, 316, 1990.

174. Bonfante-Fosolo, P., Tamagnore, L., Perotto, R., Esquerré-Tugayé, M. T., Mazau, D., Mosiniak, M., and Vian, B., Immunocytochemical localization of hydroxyproline rich glycoproteins at the interface between a mycorrhizal fungus and its host plants, *Protoplasma*, 165, 127, 1991.

175. Moore, P. J. and Staehelin, L. A., Immunogold localization of the cell-wall-matrix polysaccharides rhamnogalacturonan I and xyloglucan during cell expansion and cytokinesis in *Trifolium pratense* L.: implications for secretory pathways, *Planta*, 174, 433, 1988.

176. Cassab, G. I. and Varner, J. E., Immunocytolocalization of extensin in developing seed coats by immunogold- silver staining and by tissue printing on nitrocellulose paper, *J. Cell Biol.*, 105, 2581, 1987.

177. Conrad, T. A., Lamport, D. T. A., and Hammarschmidt, R., Detection of glycosylated and deglycosylated extensin precursors by indirect competitive ELISA, *Plant Physiol.*, 83, 1, 1987.

178. Kieliszewski, M. and Lamport, D. T. A., Cross-reactivities of polyclonal antibodies against extensin precursors determined via ELISA techniques, *Phytochemistry*, 25, 673, 1986.

179. Ludevid, M. D., Ruiz- Avila, L., Vallés, M. P., Stiefel, V., Torrent, M., Torné, J. M., and Puigdomenech, P., Expression of a cell wall protein gene in dividing and wounded tissues of *Zea mays*, *Planta*, 180, 524, 1990.

180. Mazau, D., Rumeau, D., and Esquerré- Tugaye, M. T., Two different families of hydroxyproline-rich glycoproteins in melon callus, *Plant Physiol.*, 86, 540, 1988.

181. Mazau, D., Rumeau, D., and Esquerré - Tugaye, M. T., Biochemical characterization and immunocytochemical localization of hydroxyproline-rich glycoproteins in infected plants, *Plant Physiol.*, Suppl. 83, 658, 1987.

182. Stiefel, V., Ruiz-Avila, L., Raz, R., Pilar, M. P., Gomez, J., Pages, M., Martinez-Izquierdo, J. A., Ludevid, M. D., Longdala, J. A., Nelson, T., and Puigdomenech, P., Expression of a maize cell wall hydroxyproline-rich glycoprotein in early leaf and root vascular differentiation, *Plant Cell*, 2, 785, 1990.

183. Roberts, K., Gref, C., Hills, G. J., and Shaw, P. J., Cell wall glycoproteins: Structure and functions, *J. Cell Sci.*, Suppl. 2, 105, 1985.

184. Woessner, J. P. and Goodenough, U. W., Molecular characterization of a zygote wall protein: an extensin-like molecule in *Chlamydomonas reinhardtii, Plant Cell*, 1, 901, 1989.

185. Ye, Z. -H., Song, Y. -R., Marcus, A., and Varner, J. E., Comparative localization of three classes of cell wall proteins, *Plant J.*, 1, 175, 1991.

186. Ye, Z. -H. and Varner, J. E., Tissue specific expression of cell wall proteins in developing soybean tissues, *Plant Cell*, 3, 23, 1991.

187. Cassab, G. I. and Varner, J. E., Tissue printing on Nitrocellulose paper: A new method for immunolocalization of proteins, localization of enzyme activities and anatomical analysis, *Cell Biol. Internat. Rep.*, 13, 147, 1989.

188. Cassab, G. I., Lin, J. -J., Lin, L. -S., and Varner, J. E., Ethylene effect on extensin and peroxidase distribution in the subapical region of pea epicotyls, *Plant Physiol.*, 88, 522, 1988.

189. Varner, J. E., Taylor, R., Cassab, G. I., Lin, J. - J., Yuen, H., and Pont-Lezica, R., Cell wall assembly and architecture, *Curr. Topics Plant Biochem. Physiol.*, 7, 134, 1988.

190. Towbin, H., Staehelin, T., and Gordon, J., Electrophoretic transfer of proteins from polyacrylamide gels to nitrocellulose sheets: Procedure and some applications, *Proc. Natl. Acad. Sci. USA*, 76, 4350, 1979.

191. Blake, M. S., Johnston, K. H., Russell-Jones, G. J., and Gotschlich, E. C., A rapid, sensitive method for detection of alkaline phosphatase- conjugated anti-antibody on Western blots, *Anal. Biochem.*, 136, 175, 1984.

192. Anderson, M. A., Sandrin, M. S., and Clarke, A. E., A high proportion of hybridoma raised to a plant extract secrete antibody to arabinose or galactose, *Plant Physiol.*, 75, 1013, 1984.

193. Li, Y., Pierson, E. S., Bruun, L., Roberts , K., and Cresti, M., Periodic ring-like structure in the pollen tube wall of *Nicotiana tabacum*, in *VI Cell Wall Meeting*, Sassen, M. M. A., Derksen, J. W. M., Emons, A. M. C., and Wolters, A. M. C., Eds., Neijmegen, 1992, 62.

194. Sedgley, M. and Clarke, A. E., Immunogold localization of arabinogalacton protein in the developing style of *Nicotiana alata*, *Nord. J.Bot.*, 6, 591, 1986.

195. Pennell, R. I., Janniche, L., Schofield, G. N., Booij, H., de Vries, S. C., and Roberts, K., Identification of a transitional cell state in the developmental pathway to carrot somatic embryogenesis, *J. Cell Biol.*, 119, 1371, 1992.

196. Pennell, R. I., Janniche, L., Kjellbom, P., Schofield, G. N., Peart, J. M., and Roberts, K., Developmental regulation of a plasma membrane arabinogalacton protein epitope in oil seed rape flowers, *Plant Cell*, 3, 1317, 1991.

197. Apel, K., Reimann-Philipp, U., Bohlmann, H., Behnke, S., and Schrader, G., Leaf-specific thionins of barley—A novel class of cell wall proteins toxic to plant-pathogenic fungi: Their possible function and subcellular localization in barley leaves, *Current Topics in Plant Biochem. Physiol.*, 7, 140, 1988.

198. Kandasamy, M. K., Paolillo, D. J., Faraday, C. D., Nasrallah, J. B., and Nasarallah, M. E., The S- locus specific glycoproteins of *Brassica* accumulate in the cell wall of developing stigmatic papillae, *Dev. Biol.*, 134, 462, 1990.

199. Kandasamy, M. K., Dwyer, K. G., Paolillo, D. J., Doney, R. C., Nasrallah, J. B., and Nasarallah, M. E., *Brassica* S-proteins accumulate in the interface matrix along the path of pollen tubes in transgenic tobacco pistils, *Plant Cell*, 2, 39, 1990.

200. Umbach, A. L., Lalonde, B. A., Kandasamy, M. K., Nasrallah, J. B., and Nasrallah, M. E., Immunodetection of protein glycoforms encoded by two independent genes of the self-incompatibility multigene family of *Brassica, Plant Physiol.*, 93, 739, 1990.

201. Nasrallah, J. B., Doney, R. C., and Nasarallah, M. E., Biosynthesis of glycoproteins involved in the pollen-stigma interaction of incompatability in developing flowers of *Brassica oleracea* L., *Planta*, 165, 100, 1985.

202. Hunkapillar, M. W., Lujan, E., Ostrander, F., and Hood, L. E., Isolation of microgram quantities of proteins from polyacramide gels for amino-acid sequence analysis, *Methods Enzymol.*, 91, 227, 1983.

203. Craig, S. and Goodchild, D. J., Periodate- acid treatment of sections permits on -grid immunogold localisation of pea seed vicilin in ER and Golgi, *Protplasma*, 122, 35, 1984.

204. Anderson, M. A., Cornish, E. C., Mau, S. -L., Williams, E. G., Hoggart, R., Atkinson, A., Bönig. I.,Grego, B., Simpson, R., Roche, P. J., Haley, J. D., Penschow, J. D., Niall, H. D., Tregear, G. W., Coghlan, J. P., Crawford, R. J., and Clarke, A. E., Cloning of cDNA for a stylar glycoprotein associated with expression of self incompatibility in *Nicotiana alata, Nature*, 321, 38, 1986.

205. Anderson, M. A., McFadden, G. I., Bernatzky, R., Atkinson, A., Orpin, T., Dedman, H., Tregear, G., Fernley, R., and Clarke, A. E., Sequence variability of three alleles of the self-incompatibility gene of *Nicotiana alata, Plant Cell*, 1, 483, 1989.

206. Murfett, J., Cornish, E. C., Ebert, P. E., Bönig, I., McClure, B. A., and Clarke, A. E., Expression of a self-incompatability glycoprotein (S2-Ribonuclease) from *Nicotiana alata* in transgenic *Nicotiana tabacum, Plant Cell*, 4, 1063, 1992.

Index

A

Abopon, 45
Absorption filter, 152, 153
Acanthaceae, 19, 56
Acetaldehyde, as prefixative, 179
Acetic acid, 35, 37, 38, 48
2-Acetoamido-2-deoxy-D-glucopyranose, 5
Acetolysis, 123, 168
Acetone, 35, 37, 50–51
    for dehydration, 180, 185
    in freeze substitution, 244
Acetylation control procedure, 55, 154, 157
N-Acetyl-D-galactosamines, control procedure, 228
Acid fuchsin, 41
Acidic stains, 41
Acidic waxes, fluorescence microscope procedure, 172
Acid phosphatase
    light microscope procedures, 103–111
    TEM procedures, 208–213
Acriflavine Method, 163–164
Acrolein, 35, 37, 48, 54, 179
Acrylamide-gelatin-jung resin, 183, 187
Adaptive immune system, 235
Adcrusting substances, 3
Additive fixatives, 34
Adjuvant, 237
AEC (aminoethylcarbazole), 263
Affinity, 237
Agar, 8, 32, 39
Agarose, 8
Agglutins (lectins), 9–10, 228
AGP (arabinogalactan proteins), 286–287
Alcian blue, 70–71, 74–76
Alcian Blue - Alcian Yellow Method, 74–75
Alcian yellow, 40, 74–75
Alcohol dehydrogenase, 228
Aldehydes, as prefixative, 179
Algae, 2
    alginic acid in, 8
    cellulose in, 60, 61
    chitin in, 5

    diatoms, 18, 73, 158, 163
    extensins in, 10
    mucilaginous substances of, 3
    newly-formed cells of, 158
    sulfated polysaccharides in, 8–9
Alginic acid, 8
    coupling to protein carrier, 250
    fluorescence microscope procedures, 157, 164
    immunocytochemistry procedures, 273
    light microscope procedures, 70–84
    stains for, 40
Alkaline Hydroxylamine Hydrochloride Method, 81–82
Alkaline phosphatase
    coupling of antibodies to, 257
    detection of, 263–264
Allium cepa, 274
Aluminium: Chrome Azural-S Method, 128–129
Aluminum, 19
Aluminum Hydroxide Adjuvant, 248
Aluminum sulphate, in Toluidine Blue O-Aluminium
    Sulphate Method, 77–78
Amido Black 10 B Method, 87–88
Amine oxidase, immunocytochemistry procedures, 282
Amino acids, 9-10 *See also* Proteins
Aminoethylcarbazole (AEC), 263
N,N' bis (4-Aminophenyl)-1,3, xylene diamine (BAXD),
    89, 92–93
N,N' bis (4-Aminophenyl)-N,N'-dimethyl ethylene
    diamine (BED), 89, 92–93
Ammonia, in Autofluorescence Method Coupled with
    Ammonia Treatment, 167
Ammonium oxalate extraction, 80
Ammonium salts, 41
Aniline Blue-Black Method, 88–89
Aniline Blue Fluorescence Method, 160–161
Aniline Blue Method, 69
Aniline dyes, 42, 46
Aniline oil, as clearing agent, 38
Aniline stains, 45
Aniline Sulphate Test, 119–120
Anionic stains, 41
Anisotropic properties of cellulose, 5
Anthers, 5

Hydrocharitaceae, 123
Hydrochloric acid, in Pholoroglucinol - HCl Method, 121–122
Hydrogen bonds, in staining, 42, 44–45
Hydrogen peroxide
    light microscope procedures, 99–100
    in Silver - Hydrogen Peroxide Method, 132
Hydroquinone, as impurity, 188
Hydroxylamine - Ferric Chloride Method, 81–82
Hydroxyproline-rich glycoprotein (HRGP), 283, 286
Hydroxyproline-rich proteins (extensins), 9, 10–11
Hypericum calycinum, 85, 157
Hyphal balls, 69
Hypochromatic metachromasy, 42
Hyp-protein (HRGP), 9, 10–11

Iodine, 19
    in Potassium Hydroxide-Iodine-Potassium Iodide Procedure, 65–66
    in Potassium Iodide - Iodine - Sulphuric Acid Method, 12–13, 60–61
Iron, in Colloidal Iron Method, 83–84, 195–196
Iron Diamine-Alcian Blue Method, 75–76
Isobutyl methacrylate, 154
Isoditityrosine, 10
Isodityrosyl residues, 12
Isoelectric focusing (IEF), 288
Isopentane, 32
Isopropanol, as clearing agent, 38
Isopropyl alcohol, dehydration procedure, 50–51
Isothiocyanate, in labeling of antibodies, 258–259

I

IgG antibodies, 236
Immersion oil, 154
Immune systems, 234–235
Immunizations, 238–239
Immuno blot analysis, 267
Immunocytochemistry, 234–237
    immunolabeling, 245–248, 258–261
    labeling antibodies, 240–241
    localization of polysaccharides, 268–275
    localization of proteins and glycoproteins, 275–290
    methods in antibody production, 247–267
    storage and purification of antibodies, 238, 240, 253–255
    tissue preparation, 241–245
Immunoelectrophoresis, 267
Immunogenicity, 235
Immunoglobulins, 235, 236, 254
Immunogold preparations, 259–260, 290
Immunolabeling, 245–248, 258–261
Incandescence, 151
Incomplete Freund's Adjuvant, 247
Incrusting substances, 3
Indirect coloring, 41
Indoxyl Acetate Method, 113–114
Induced fluorescence, 152
Infection. *See* Pathogens
Infiltration, 39–40, 185
Inherent fluorescence, 152
Injection methods for rabbits, 251–252
Insects, attractants for, 16
Insoluble polysaccharides
    fluorescence microscope procedures, 154–156
    light microscope procedures, 53–57
    TEM procedures, 192–193
Intercellular transport, 2
    pectic substances and, 7
    phosphatases and, 13
Intussusception, 2
Iodide, in Chlor-Zinc-Iodide Method, 66–67

J

JIM 5, 270, 271, 272
JIM 7, 270, 272
JIM 8, 287
JIM 13, 286

K

Krajcinovic Amine Test, 78–79

L

Labeling
    of antibodies, 240–241, 246, 258–261
    of cells and tissues, 245–246
    immunolabeling, 245–248, 258–261
Lacmoid Blue Method, 67–69, 161, 162
D-Lactose, in control procedures, 228
Lamiaceae, 9
Laminaria, 73, 75
Laminarin, 5, 55, 162
Lanolin, 247
Larix decidua, 170
Lead phosphate, 41
Lead Salt Method, 104–106, 208–209
Leaf abscission zone research, 56
Lectin - Colloidal Gold Complex Preparation, 229
Lectin - Colloidal Gold Conjugate Method, 230
Lectins, 9–10
    cytochemistry, 227–231
    in Fluorescein - Lectin Conjugate Method, 231
    immunocytochemistry procedures, 283–290
Legumes, root-rhizobium recognition, 9
Leguminosae, 9
Lewitsky's Fluid, 50, 128

Y

Z

Milton Keynes UK
Ingram Content Group UK Ltd.
UKHW021625071024
449327UK00020BA/1197